Charles A. Francis, PhD
Raymond P. Poincelot, PhD
George W. Bird, PhD
Editors

Thanks for all you do for MSU!

Best Regards.

Developing and Extending
Sustainable Agriculture
A New Social Contract

George Bird

5/31/12

"**T**his compilation of works from some of the best thinkers and practitioners of sustainable agriculture sets the stage with the rigor required to advance the concept of sustainable agriculture as an essential part of developing a more sustainable world. All of the writers bring perspectives that are necessary in understanding the distance yet to be covered. The editors are to be commended for bringing us this rich portfolio for thought and action. This book contains some of the best thinking from those who were there from the beginning and those that have taken significant professional risks in becoming associated with the advancement of sustainable agriculture as a legitimate science. The book is ecology-based but quickly moves to address the issues of profitability, and ultimately the social dimensions and interconnectiveness of all life. The editors and contributors have brought fresh thinking, free of jargon, to help all of us better understand how we can extend sustainable agriculture into a more equitable contract with society and indeed all life in the universe. This book belongs in the hands of decision makers at every level. I don't know of a better collection of work that meets the goal of scholarship, new thinking, and an implicit call to action. The book is easy to read, comprehensive in scope, and is a reminder of how far we have come."

Jim Horne, PhD
President and CEO,
Kerr Center for Sustainable
Agriculture, Inc.

"**D***eveloping and Extending Sustainable Agriculture* is a must-read for environmental and agriculture educators, students, and anyone involved in sustainable farming that works with farmers and consumers. The A-list of contributing writers in this book have all been deeply involved with helping to develop and pioneer sustainable agriculture practices, research, education, and outreach in the United States during the past twenty years. Their experiences and lessons learned, along with new ideas, will give the reader much to think about regarding where we need to take sustainable agriculture next in America. We must keep looking at the big picture, thinking long-term, working with the whole farm, and working collectively with all of the people involved in the food chain, including the public! Sustainable agriculture goes where large-scale corporate conventional agriculture doesn't. The thread of sustainable agriculture is clearly woven through all of our environments, communities, and economies. We are at a critical moment in time where a book such as this one is a must-read to participate in the ongoing public debate and important work over the future of American agriculture and the health of our communities."

Anthony Rodale
Chairman Emeritus,
The Rodale Institute

Developing and Extending Sustainable Agriculture

A New Social Contract

HAWORTH FOOD & AGRICULTURAL PRODUCTS PRESS™
Sustainable Food, Fiber, and Forestry Systems
Raymond P. Poincelot, PhD
Senior Editor

Biodiversity and Pest Management in Agroecosystems by Miguel A. Altieri

Developing and Extending Sustainable Agriculture: A New Social Contract edited by Charles A. Francis, Raymond P. Poincelot, and George W. Bird

Soil and Water Conservation Handbook: Policies, Practices, Conditions, and Terms by Paul W. Unger

Developing and Extending Sustainable Agriculture
A New Social Contract

Charles A. Francis, PhD
Raymond P. Poincelot, PhD
George W. Bird, PhD
Editors

Haworth Food & Agricultural Products Press™
An Imprint of The Haworth Press, Inc.
New York • London • Oxford

For more information on this book or to order, visit
http://www.haworthpress.com/store/product.asp?sku=5709

or call 1-800-HAWORTH (800-429-6784) in the United States and Canada
or (607) 722-5857 outside the United States and Canada

or contact orders@HaworthPress.com

Published by

Haworth Food & Agricultural Products Press™, an imprint of The Haworth Press, Inc., 10 Alice
Street, Binghamton, NY 13904-1580.

PUBLISHER'S NOTE
The development, preparation, and publication of this work has been undertaken with great care.
However, the publisher, employees, editors, and agents of The Haworth Press are not responsible
for any errors contained herein or for consequences that may ensue from use of materials or
information contained in this work. The Haworth Press is committed to the dissemination of ideas
and information according to the highest standards of intellectual freedom and the free exchange of
ideas. Statements made and opinions expressed in this publication do not necessarily reflect the
views of the Publisher, Directors, management, or staff of The Haworth Press, Inc., or an
endorsement by them.

Cover design by Kerry E. Mack.

Library of Congress Cataloging-in-Publication Data

Developing and extending sustainable agriculture : a new social contract / Charles A. Francis,
Raymond P. Poincelot, George W. Bird, editors.
  p. cm.
  Includes bibliographical references and index.
  ISBN-13: 978-1-56022-331-3 (hc. : alk. paper)
  ISBN-10: 1-56022-331-6 (hc. : alk. paper)
  ISBN-13: 978-1-56022-332-0 (pbk. : alk. paper)
  ISBN-10: 1-56022-332-4 (pbk. : alk. paper)
  1. Sustainable agriculture. 2. Sustainable agriculture—United States. I. Francis, Charles A.
II. Poincelot, Raymond P., 1944- III. Bird, George W.

  S494.5.S86D48 2006
  631.5'8—dc22

                                                                    2005037389

CONTENTS

ABOUT THE EDITORS

Charles A. Francis, PhD, is Professor of Agronomy and Horticulture at the Institute of Agriculture and Natural Resources at the University of Nebraska, Lincoln. He has been Senior Scientist at the International Center for Tropical Agriculture (CIAT, Colombia), Director of the International Program (Rodale Institute, Pennsylvania), and Director of the Center for Sustainable Agricultural Systems (University of Nebraska). He was co-editor of *Sustainable Agriculture in Temperate Zones, Exploring the Role of Diversity in Sustainable Agriculture,* and *Agroecosystems Analysis.* His research is in crop rotations, breeding for sustainable systems, and design of rural landscapes and their communities. He teaches Agroecology, Science-Based Organic Farming, and Urbanization of Rural Landscapes at UNL, and Agroecology and Farming Systems at the Norwegian University of Life Sciences (UMB).

Raymond P. Poincelot, PhD, is Professor of Biology at Fairfield University and is currently Associate Dean for the College of Arts and Sciences. He has been Chair of the Biology Department at Fairfield University and Assistant/Associate Biochemist at the Connecticut Agricultural Experiment Station in New Haven. He served as the founding and continuing Editor in Chief of the *Journal of Sustainable Agriculture.* In addition, Dr. Poincelot is Senior Editor for the Sustainable Food, Fiber, and Forestry Systems book series bearing the Haworth Food & Agricultural Products Press imprint. His books include *Sustainable Horticulture: Today and Tomorrow.* His major area of research is the use of compost, seaweed, and humic acid products in sustainable agriculture.

George W. Bird, PhD, is Professor of Nematology and Associate Chairperson of the Department of Entomology at Michigan State University. He played significant roles in the development of both Integrated Pest Management (IPM) and sustainable agriculture concepts and programs in the United States. He is a former national Director of SARE (Sustainable Agriculture Research and Education, USDA/CSREES). He is currently involved in research, education, and extension-outreach related to the nature of soil quality, with special reference to the role of soil fauna in making nutrients available to plants.

Developing and Extending Sustainable Agriculture
© 2006 by The Haworth Press, Inc. All rights reserved.
doi:10.1300/5709_a

Contributors

John E. Barbuto Jr. is Associate Professor of leadership, University of Nebraska-Lincoln, with interests in antecedents of leadership, work motivation, influence processes, servant leadership, transformational leadership, and pedagogical innovations in leadership education.

Gary Bentrup is a landscape planner with the U.S. Department of Agriculture (USDA) National Agroforestry Center in Lincoln, Nebraska, with interests in conservation corridors and multiple resource planning. His current focus is on developing better ways to design landscapes that simultaneously achieve landowner and societal objectives.

Michael J. Brewer is coordinator of the IPM program and Associate Professor of entomology at Michigan State University. His interests are in the development and implementation of economical and environmentally responsible pest management strategies. He also studies how land-use strategies may contribute to sustainable approaches to pest management.

Lorna Michael Butler is Henry A. Wallace endowed chair for sustainable agriculture, and Professor in the Departments of Sociology and Anthropology, Iowa State University, with interests in the social and cultural dimensions of sustainable agriculture, sustainable rural livelihoods and food insecurity, the rural-urban interface; and family farm entrepreneurship and value-chain alliances.

Heidi Carter is Page County extension education director, Iowa State University, in charge of agriculture and natural resources, 4-H youth development, communities, and family programs; other responsibilities are administering the county budget and staff, and building partnerships.

Jerry DeWitt is Professor of entomology, extension state sustainable agriculture coordinator, and interim director of the Leopold Center at Iowa State University. He works part time for USDA-SARE in Washington, DC. He is an avid photographer of rural America and has published his work in books and magazines across the nation.

Developing and Extending Sustainable Agriculture
© 2006 by The Haworth Press, Inc. All rights reserved.
doi:10.1300/5709_b

xiv DEVELOPING AND EXTENDING SUSTAINABLE AGRICULTURE

Elbert C. Dickey is dean and director of University of Nebraska Cooperative Extension at the University of Nebraska, with administrative responsibility for all aspects of Cooperative Extension, administrative leadership for three major extension areas, and responsibility for program and budget development.

Rick Exner is coordinator of the farming systems program that is a cooperative effort of Iowa State University Extension and Practical Farmers of Iowa. His professional interests include information- and management-intensive solutions to issues in cropping and livestock systems.

Cornelia Butler Flora is Charles F. Curtiss Distinguished Professor of agriculture and sociology, Iowa State University, with interest in building environmental, cultural, human, social, political, built and financial capital to increase economic vitality, ecosystem health, and social well-being.

Terry Gompert is extension educator with University of Nebraska–Lincoln, with interest in grass-based agriculture, intensive grazing management, sustainable agriculture, relationships in direct marketing, and holistic management.

Oran Hesterman is program director of Food Systems and Rural Development at W.K. Kellogg Foundation, with leadership responsibilities for the foundation's programs in sustainable agriculture and community-based food systems. He is also a founder and cochair of the Sustainable Agriculture and Food Systems Funders, an affinity group of foundation professionals active in funding NGO and university programs.

John Ikerd is Professor Emeritus of agricultural economics, University of Missouri–Columbia, with interests in sustainable agriculture, sustainable communities, and sustainable living, and emphasis on the economics of sustainability.

Rhonda R. Janke is Associate Professor and extension specialist in sustainable cropping systems, Department of Horticulture, Forestry, and Recreation Resources at Kansas State University. Research/extension projects include whole-farm planning, soil quality, farmer-friendly water quality test kits, and herbs as an alternative high-value crop. She raises sheep, poultry, herbs, and vegetables on a small farm and markets them locally.

Frederick L. Kirschenmann is the former director of the Leopold Center for Sustainable Agriculture and Professor of religion and philosophy at Iowa State University, with interests in ecological systems, policies, and markets that preserve midsize farms and conserve natural resources. He also manages his family's 3,500-acre organic farm in North Dakota.

Kent McVay is Assistant Professor of agronomy, Kansas State University–Manhattan, with interests in no-tillage cropping systems, soil conservation, and water quality as affected by agriculture.

Richard Olson is the director of the Sustainability and Environmental Studies Program at Berea College in Berea, Kentucky. He teaches in the areas of ecological design, sustainable communities, and Christianity and sustainability.

Michele Schoeneberger is the research program leader and soil scientist at the USDA Forest Service National Agroforestry Center located in Lincoln, Nebraska. Her research is aimed at developing agroforestry and other tree-based buffer technologies to mitigate nonpoint source pollution, sequester carbon, restore other essential ecosystem functions, and provide economic opportunities at the farm/landscape levels.

Richard Straight is the lead agroforester with the USDA Forest Service/Cooperative Forestry at the National Agroforestry Center in Lincoln, Nebraska. He is primarily involved in providing resources and training for natural resource professionals to increase the use and understanding of agroforestry technologies such as windbreaks, alley cropping, and riparian forest buffers on agricultural lands.

Richard Thompson farms in Boone County, Iowa, along with his wife Sharon and son Rex and family. He is a farmer, researcher, and educator who works in solving problems and sharing results with others to improve profitability and balance in farming, protect the environment, and strengthen rural communities.

Shirley K. Trout is a freelance technical writer and researcher who works with her husband on her family's farm near Waverly, Nebraska. She looks at her husband's gradual transition from conventional to more sustainable farming as her action research in producer motivation and diffusion of innovation.

Steven S. Waller is dean of the College of Agricultural Sciences and Natural Resources, University of Nebraska–Lincoln, and Professor of agronomy. He has taught and conducted research in range and forage sciences. He is a former regional coordinator for the North Central Region Sustainable Agriculture Research and Education program.

Foreword

We have witnessed many changes during our lifetimes. While some have improved the quality of life for people around the world, others have not. The global human population has doubled; poverty has increased; ecosystems have grown fragile. Agriculture has changed, too, moving from a preponderance of family farms to large, corporate entities. In recent decades, some agricultural sectors have begun to move toward sustainability.

The 1985 Farm Bill began the process of introducing sustainability into our Land Grant agricultural research, education, and Extension programs. In 1987, the United Nations Commission on Environment and Development provided a formal definition of sustainable development. This was followed by the 1990 Farm Bill authorizing the Sustainable Agriculture Research and Education Program (SARE). That legislation mandated that SARE policy and accountability be established through Regional Administrative Councils composed of farmers and ranchers, non-governmental organizations, private business, government, and academia. In addition to addressing agricultural production and natural resource challenges, this unique structure resulted in projects designed to deal with quality of life issues for farmers, farm families, and rural communities. Many were highly integrated research, education, and Extension projects linking local adoption with production, social, and natural resource sciences. During the past 15 years, SARE has served as a national and international catalyst for evaluating research, education, and Extension in the context of sustainable development.

Developing and Extending Sustainable Agriculture: A New Social Contract goes well beyond SARE. It brings together many key leaders in sustainable agriculture in ways that both document the process of this grassroots innovation and provide a vision for the future without presenting a *specific roadmap*. The authors of the sixteen chapters are recognized authorities in areas ranging from sustainable soil, crop, and livestock manage-

Developing and Extending Sustainable Agriculture
© 2006 by The Haworth Press, Inc. All rights reserved.
doi:10.1300/5709_c

ment systems to questions of ethics and productivity. They are current university deans, foundation administrators, life-long professional farmers, and employees of innovative farmer organizations. This breadth of experience is rarely assimilated into a single volume.

Both generalists and specialists will find *Developing and Extending Sustainable Agriculture: A New Social Contract* valuable. The information it presents is supported in a comprehensive manner through references from both the classical scientific literature and electronic Web sites. In the introductory chapter, the authors make a strong case for the need for education in sustainable development from the earliest years of school through life-long learning. Farmer to farmer education has become an integral part of the sustainable agriculture learning process and needs to be evaluated for its potential in other sectors of society. Likewise, the on-farm research component associated with sustainable agriculture represents a new paradigm that is vastly different from that of previous models for applied research. At our institution, for example, Michigan growers were involved in the design, review, and revision process for an Extension publication series on *Production Ecology.* This process and its widely accepted and highly acclaimed products would not have been possible without the proceeding 10 years of development in sustainable agriculture.

The inclusion of whole-farm planning and implementation strategies as well as economic tools for impact assessment make this book useful for all components of the agricultural community. Similarly, the information it provides on the social capital needed for sustainable development and procedures for creating vibrant linkages between farms and communities is imperative for understanding the current ties between rural, suburban, and urban environments and their relation to present and future generations. First-hand information is provided by Land Grant institution administrators on building and managing regional, state-wide and local research, education, and Extension programs designed to foster the sustainability of U.S. agriculture. While the examples are weighted to the North Central Region, they are fundamental in nature and should be of global value.

In a global community of significantly more than 6 billion people, the issues facing all aspects of our food system and its associated natural resources are major. It often appears that the attributes of modern science, globalization, and an ecological worldview are at odds. The procedures and philosophy of sustainable agriculture as described in *Developing and Extending Sustainable Agriculture: A New Social Contract* provide an excellent model for charting the course for both local and global food systems for the near and distant future. One of the most important aspects of the philosophy of sustainable agriculture is meaningful dialogue: among farmers and ranchers, non-governmental organizations, private business, government and academia. It is through such dialogue that we believe an appropriate, sustainable and meaningful future will evolve.

James Ian Gray
Vice President, Research and Graduate Studies
Michigan State University, East Lansing 48824
(former Director, Michigan Agricultural Experiment Station)

Fred L. Poston
Vice President and Treasurer, Financial Operations
Michigan State University, East Lansing, 48824
(former Dean, College of Agriculture and Natural Resources)

Preface

Developing and Extending Sustainable Agriculture: A New Social Contract is not a formula or road map toward the future, but rather a conceptual and practical collection of chapters on the current state of farming and ranching and how it may be improved. The book relies heavily on ecological principles of management, and many of the contributors describe how farmers and ranchers are looking to natural ecosystems to help inform the design of better agroecosystems. We find the principle of eco-literacy to be essential in this search for improved and more sustainable systems. In spite of the current dominant paradigm of industrial agriculture, we have an abiding faith in the positive potential of the human mind to grasp the impacts of current systems and to seek better ones for the future. We believe in what was stated by Nobel laureate René Dubos, "Trend is not destiny."

Sustainable agriculture is essential for the future of a secure, resilient, and diverse food supply for this country. The progress toward reaching these goals has been slow due to a large investment by most farmers and ranchers in the current industrial agriculture paradigm, and due to a research and Extension establishment that is also heavily invested in the status quo. We feel that it is time for a major change.

The Sustainable Agriculture Research and Education (SARE) Program has brought substantial federal funding to support new directions in agriculture. There has been a renewed focus on such sustainable practices as crop rotations, integrated pest management, alternative sources of nutrients to maintain productivity, and rotational grazing systems, among other moves toward greater efficiency of production and substitution of inputs. Much of the effort to date closely resembles the research and outreach approaches that have been the stock of the land grant universities for the last several decades, and some results have reached the farms and ranches of the Midwest Region. Yet these programs have so far not lived up to their full potential to make an impact on small- and medium-sized farms that will provide a more secure income and better quality of life for these farm families, nor for the rural communities where they live.

There is a critical need for a broader approach that embraces the production, the economics, the environment, and the social challenges facing the

Developing and Extending Sustainable Agriculture
© 2006 by The Haworth Press, Inc. All rights reserved.
doi:10.1300/5709_d

rural United States. *Developing and Extending Sustainable Agriculture: A New Social Contract* has been written by some of the individuals most involved in sustainable agriculture research and Extension in the region. They describe the successes to date, and the impacts of programs on farm economics and the environment. The authors also go beyond what has been done so far to project the needs for the coming decades. They are moving outside the proverbial box to envision how things could be in the future. Although there is a continuing move toward a more industrial agriculture and a global food system, these authors have described the negative impacts and long-term challenges that such a system will bring. In a hopeful vision of the future, they provide alternatives that can solve these problems and help us toward a more secure and equitable food system—one that can be achieved without compromising the natural resource base on which we all depend and the environment where we must live. We owe no less to our children and the generations that follow, and to the natural ecosystem on which we depend.

Chapter 1

Overview of the Educational Social Contract: Building a Foundation for Sustainable Agriculture

Raymond P. Poincelot
Charles A. Francis
George W. Bird

INTRODUCTION

Cooperative Extension was organized and financed by the federal government to bring technology to the rural United States. Recognizing that the Industrial Revolution was contributing to the growth and modernization of larger cities, leaders with foresight and courage launched an initiative with the Smith-Lever Act of 1914 to add a public service mandate to the role of the land grant universities (Hildreth and Armbruster, 1981). This act added an explicit Extension and practical outreach dimension to the already prominent and growing state university system.

Through the past century, Extension has been nearly synonymous with rural development, demonstrating and moving technology to rural areas. It has also been central to improving rural culture and community through promotion of education activities. Among the most widespread activities of Extension are 4-H clubs and county fairs. In financial terms, the investment in coordinated research, development, and Extension appears to return between 20 and 40 percent per year in benefits to farmers, ranchers, and society (McDowell, 2001). During much of the time since its founding, Extension has been seen as one of the most trusted sources of information and applied practices for rural people (Alston and Pardey, 1996).

The evolving role of Extension over the past century and the impact of Sustainable Agriculture Research and Education (SARE) programs on the

Developing and Extending Sustainable Agriculture
© 2006 by The Haworth Press, Inc. All rights reserved.
doi:10.1300/5709_01

directions and projects in Extension are explored in this introductory chapter. With the emergence of a new paradigm in agriculture that emphasizes environmental and social impacts in tandem with production and economics, Extension's role needs to change to adjust to changing demands. Toward that end, we explore several topics in this chapter.

- We describe the growth of industrial farming, and its impact on the traditional U.S. family farm.
- We examine the research foundation necessary to affirm the new recommendations central to development and application of alternative technologies and systems, a growing role of the land grant universities and especially Extension.
- We review the numerous bulletins, journal article sources, and book series initiated to report on sustainable agriculture over the past 15 years, many of which have received SARE support.
- We look at the major changes occurring in mainstream agriculture, owing in no small part to the research in sustainable systems.
- Attention is directed toward marketing and consumer education that has been important in promoting the demand for organic food as well as other food and fiber products that are produced in a sustainable way.
- We conclude this introduction with a short preview of the following chapters, and provide the groundwork for the final chapter on future directions and multifunctional landscapes and communities.

Extension has played a key role in recent change, and the SARE program financing has been instrumental in supporting the researchers, Extension educators, farmers, ranchers, and students who have been active in the process. In this book we answer the intriguing question posed by George McDowell (2001): Are we in the process of "renegotiating or abandoning the social contract?"

EVOLVING ROLE OF EXTENSION

The county Extension agent, today often called an educator, has been a fixture in rural communities. These specialists in agriculture and home sciences have been central and respected figures in helping rural families and residents in smaller communities to interface with the growing world of technology. County Extension offices have been an important window on the outside technology landscape, a resource that was indispensable to often-isolated rural families and communities. The county agent, stereotypically depicted in the famous Norman Rockwell painting showing the agent as authority figure evaluating the young man's calf, was a highly respected leader in the community. Often he or she was one of the few people with a college education, and along with the school teacher represented the intelligentsia and leadership of the rural community. Extension enjoyed a valuable role in the rural United States, and had a recognized and accepted contract with the farming population and small communities.

But "the times they are a-changin'," with credit to Bob Dylan. Today, the county agent or advisor is far from the only information or education provider in town. The Industrial Revolution and now the information technology revolution have brought the seed corn dealers, fertilizer and chemical company representatives, and local cooperative or elevator sales staff prominently into this arena. In addition, crop producers often employ consultants to provide recommendations for hybrids and varieties, fertilizer and herbicide rates and formulations, and insecticide and irrigation application schedules. Precision agriculture offers the promise of unique treatments in every niche in each field, theoretically saving the farmer input costs and helping to protect the environment. With all these resource people available, at a cost to be sure, and information resources coming from the commercial sector including those from the Internet, the entire advising and information domain has changed.

In this new environment, Cooperative Extension is searching for a viable role in which their institution can continue to serve society. County fairs and 4-H represent important activities that meet the expectations of many clients. We supply educational bulletins and fact sheets from county offices, and electronic access to technical educational materials, to answer questions on plant protection, fertilization, irrigation, machinery use, record keeping, and economics in farming. One popular bulletin series in each state reports the results of uniform testing of hybrids and varieties—an unbiased source of information that clients can use to check on similar results from the commercial sector. Pesticide training leads to a required certificate for farmers and commercial applicators who apply restricted-use, toxic chemicals. We deal with structural issues such as family corporations and innovative marketing strategies. Extension is presently expanding its reach to urban audiences with education on care of homes and gardens, and is taking on new topics such as conflict resolution and raising children in challenging times.

In spite of these additional directions, the demand for services, recognition, and funding for Cooperative Extension continues to be on the decline. Some people argue that in this age of specialization, industrialization, and focus on economies of scale on large farms and high levels of production, there is not only a diminishing need for traditional Extension activities but also a decline in capacity and orientation, thus calling into question the tenets of the original contract with the rural United States. As farms consolidate, families and communities disappear from the rural landscape, and consequently the client base declines in number. Often we are challenged by critics who charge that Cooperative Extension has no viable role in an industrial farming's future. Some county educators feel that some of their major roles today are to organize space for other people's meetings and make the coffee for gatherings.

In this milieu the U.S. Congress allocated money to initiate a sustainable agriculture program for research and education, and has continued to maintain or increase funding each year for the past 15 years. The SARE budgets in FY2003 were $12 million for the United States, including $3 million in the North Central Region for research and education, farmer research, educator training, and thesis research grants. Total spending has been $165 million in the United States and $40 million in the North Central Region since the start of the program in FY1987. Through the following chapters, we present many of the contributions to sustainable agriculture that have been made during this period, especially those financed by SARE programs. These stories are told by key farmers, educators, and nonprofit organization representatives in the region.

One of the cornerstones of the North Central SARE program has been a recognition of farmers and ranchers as true agricultural professionals. They have been incorporated as leaders and key members of the decision-making process in the choice of research and education grant proposals, as well as assuming the role of on-farm and on-ranch researchers in their own grant program. Various authors refer to farmers, ranchers, or producers, but in this book we interpret all these terms as synonyms for professional agriculturists who are not only program clients but also full members of the regional team.

INDUSTRIAL AGRICULTURE
AND FAMILY FARMING

A bimodal farming model is emerging in the North Central Region, which consists of a very small number of large farms producing commodities, and a large number of small and medium-sized farms producing a diversity of fruits, vegetables, nut crops, meat and eggs, and value-added specialty crops and products. From farmers in this latter group, complaints arise that not enough research and Extension is available for people on small acreages and not enough attention is given to mid-sized family farms. Moreover, some potential or current students point out that there are few or no courses available in organic gardening or farming, or even in sustainable agriculture, on most of our university campuses.

The demands from the industrial agriculture sector are quite different. There, immense value is placed on education, but generally in economics and business management—including issues such as personnel, budgeting, finance, and macroeconomics. In the real world of industrial commodity production, most farm managers and owners depend little on traditionally delivered Extension programs. Their information providers are consultants or company representatives from the firms that sell inputs. If they do want specific information from Extension, they personally call the right specialist and do not waste time waiting for a meeting or seeking out a publication. Thus the logical target audiences for future programs are the owners and operators of small and medium-sized farms, who badly need attention and relevant information to survive the impacts of consolidation and globalization in agriculture. They are specifically identified as clients for the SARE program. We also recognize that a majority of farm acres are operated by managers with larger holdings, and expect that success by smaller-sized family farm operations will lead to a "trickle-up" effect that will result in rapid adoption by large farmers.

BUILDING AN APPROPRIATE
INFORMATION FOUNDATION

Science-based recommendations for farming practices are based on research in universities, nonprofit organizations, and industry. Replicated experiments are conducted under controlled conditions and responses to isolated systems components are measured and incorporated into farmer bulletins or Extension education programs. In this way, answers are found for appropriate hybrid/variety choices, optimal levels of applied fertilizers, least-cost chemical treatments for weeds and insects, and irrigation scheduling frequency.

Extension has been viewed as an impartial source of information on these practices, not connected in a business relationship with the manufacturers or vendors of any commercial product. Although some questions exist about research grants provided to the university for hybrid testing or screening pesticides, it is generally accepted that the resulting research is objective. The broader question is whether universities have spent disproportionate time on conventional agricultural options, with a large opportunity cost of not looking at nonchemical alternatives. For example, Sooby (2003) reports on the limited attention that land grant universities have given to research on organic agriculture.

Recommendations for practices in sustainable systems need to be developed through a similar rigorous process in order for these to be accepted by Extension specialists and by farmers. The SARE program was funded to fill a perceived gap in research-based recommendations for small and medium-sized family farmers. There has been an impression in academia that sustainable agriculture and organic farming research is more difficult to publish than conventional research results. In our view, there is no evidence of discrimination by reviewers or editors against publishing results in these areas, as long as the research is well-designed and the results analyzed according to the conventional norms of mainstream journals. What appeared to be needed in this arena was a wider range of potential journals, Extension bulletins, and books that were more focused on sustainable systems. Several high-quality international journals and publication series were founded to fill these needs.

Journal Publications

It is important to establish relevant technical journal publishing opportunities for research work in sustainable agriculture, as well as to encourage publication in mainstream journals. Technical articles in refereed, internationally known journals are essential to professional recognition in universities. This venue is especially critical for junior faculty who are not yet tenured or fully promoted in the university system. Involvement by new faculty in sustainable agriculture research and education is essential for building not only a knowledge base, but also the human capital within universities to ensure a strong foundation for the future. It is very fortunate that two new journals initiated at about the same time as the SARE program have continued publication and appear to be viable for the future. A new journal has just begun publication in the United Kingdom, and adds stronger international dimensions to journal opportunities. It is instructive to examine how important these journals have been for extension articles as well as research reports.

American Journal of Alternative Agriculture

Initially published in 1986 by the Institute for Alternative Agriculture, this journal's editor was Dr. Garth Youngberg and its technical editor was Dr. William Lockeretz. The initial mission was to publish a journal "covering the full range of topics falling under the alternative agriculture umbrella" (Youngberg, 1986). The goal was to include articles ranging from production agronomy to economics and policy, with emphasis on "systems-oriented, interdisciplinary approaches." The journal has successfully met these goals with timely technical research reports, commentaries, and book reviews. The editorship was passed to Dr. Robert Papendick (ARS/USDA) at Washington State University in 1997, and the journal was published by the Wallace Institute for Alternative Agriculture. In 2003, the journal was transferred to C.A.B. International in the United Kingdom and the name was changed to *Renewable Agriculture and Food Systems,* under the capable editorship of Dr. John Doran (ARS/USDA) at University of Nebraska, Lincoln. A broader scope of the journal includes a focus on whole-farm and landscape-level systems, as well as articles on more steps in the food chain, environmental impacts of agriculture, and social implications of alternative approaches to a sustainable food system (Doran and Peck, 2003).

Journal of Sustainable Agriculture

This journal was initiated in 1990 and has provided a consistent and high-quality publishing outlet for national and international research papers and reviews in sustainable agriculture. The editorial board includes 42 people from 8 countries. Editor Dr. Ray Poincelot from Fairfield University was founder of the journal and has given it a strong continuity from the beginning. The original mission of the journal was to examine ongoing and future agricultural systems and their relationships to the resource base and the environment (Poincelot, 1990). The focus is to promote worldwide study and application of sustainable agriculture in search of solutions to the problems of resource depletion and environmental damage (Poincelot, 1997). The Haworth Press includes this publication among their many technical journals and provides strong logistical support and advertising as well as an affiliated book series, *Sustainable Food, Fiber, and Forestry Systems.*

International Journal of Agricultural Sustainability

This publication was launched in 2003 as a cross-disciplinary, peer-reviewed journal that helps in furthering our understanding of sustainability in agriculture and the broader food systems. The goal was to include theoretical developments as well as evaluation of current and past agricultural systems, plus ideas on transitions toward sustainability of agriculture and food systems. The strategy is to publish information on what is not sustainable about past and current systems as well as research toward the transitions needed to make future food systems more productive, sustainable, and secure (Pretty et al., 2003). The editor is Prof. Jules Pretty of University of Essex, U.K., with four associate editors and an editorial board of 34 people from 20 countries.

Other Journals

A wide range of additional journals have published systems-oriented articles, as well as reports on sustainable agriculture practices in temperate and tropical systems. For sustainable agriculture practices, the *Journal of Soil & Water Conservation* has published a substantial number of articles, reviews, and commentaries, especially those related to soil and water quality and conservation issues. *Agricultural Systems* has an obvious focus on broader issues rather than individual practices, and has included some articles on sustainable systems. The *Journal of Production Agriculture,* published from 1988 to 1999 by the American Society of Agronomy, included numerous articles on sustainable systems. Although it has ceased publication, the concepts continue in a special section of *Agronomy Journal.* The *African Crop Science Journal* was initiated in 1993, and often includes articles on tropical cropping systems, and some on sustainability. Other journals, including the *Journal of Extension,* include articles on sustainability issues, but the journals highlighted are the ones that most frequently publish articles on sustainable agriculture.

Extension Bulletins and Fact Sheets

To quantify the number of bulletins, fact sheets, and other Extension publications available on sustainable agriculture is difficult, because of the wide range of definitions that are used and the boundaries we could choose. The following bulletins and brochures represent examples of series that are directly related to sustainable practices and systems, and provide models of what can be done in individual states and made available to others.

Michigan State University

A series of Extension bulletins starting with *Michigan Field Crop Ecology* (Cavigelli et al., 1998) has been highly successful with farmer clients and useful as a practical classroom reference. The goals of the team preparing these bulletins were to provide an ecological foundation for environmentally sound and economically viable farming practices, and to show how ecological principles could be used to design systems. The assumption was that readers had a basic prior understanding of biology or practical experience in farming, sufficient to delve into the details of nutrient cycling, plant protection, and efficient water use by crops. Also in this series is *Michigan Field Crop Pest Ecology and Management* (Cavigelli et al., 2000).

Fact Sheets from Other Universities

A number of states have established centers for sustainable agriculture or for the study of agricultural systems. These states have generated fact sheets, bulletins, newsletters, annual reports, and other information and recommendations that have entered the Extension publications series or reached farmers in other ways. Examples are the fact sheets provided from the University of Wisconsin, the University of Minnesota, and Iowa State University, and there are many other items worthy of mention. Specially funded centers have published these recommendations under their banners in order to show the results of that center's work, to justify their budgets or for other reasons, and these have in most cases been embraced by mainstream Cooperative Extension in that state as well as applied elsewhere.

1. Center for Integrated Agricultural Systems (University of Wisconsin, Madison): comprehensive list of research reports and research briefs (more than 65).
2. Minnesota Institute for Sustainable Agriculture (University of Minnesota): series of 16 comprehensive research bulletins over topics of wide interest.
3. Leopold Center for Sustainable Agriculture (Iowa State University): comprehensive series of research fact sheets, research reports, and speeches and papers from staff.

Other University Contributions to Sustainable Agriculture Extension

Many other Extension items have appeared in the last two decades that relate to reduced or more carefully targeted fertilizer and pesticide use, efficient water management, specialty crops, and value-added products—all topics that many associate with sustainable agriculture. Often these topics are imbedded in the traditional lists of Extension bulletins, circulars, and fact sheets in the series that are a central part of the outreach programs of land grant universities. Although there is some loss of identity when materials are not identified with a sustainable agriculture program or center, the benefits may be far greater. When this information enters what most consider the "mainstream" of Extension publications, we have achieved to some degree what the SARE program was designed to accomplish. This change has also been used to justify the discontinuation of some centers or other focused activities in sustainable agriculture in land grant universities in this region, under the assumption that "all of our programs are now engaged in sustainable agriculture." Only time will tell if this type of decision is justified. An alternative view is that such changes are meant to return the focus to industrial agriculture and to reduce any challenges to the current mainstream.

Book Series from Universities

Other important contributions to the scientific foundation for sustainable farming practices and agricultural systems include the textbooks, monographs, symposia proceedings, and other books that summarize and interpret research results. Although the library and its traditional resources are now challenged by the lure of the Internet and its massive and immediate search potentials, a substantial database must exist for a successful search. With some books and many journals now being made available online, there may be new methods of accessing information. But the need for providing research- and experience-based recommendations is no less important. Several book series have emerged in the arena of sustainable agriculture.

Our Sustainable Future

This series from the University of Nebraska is a collection of books on topics ranging from farming practices on organic farms (Bender, 1994) to economics and farm policy (Grant, 2002) to potentials for national green plans (Johnson, 1995). The series was designed to present a wide range of topics on research, development, and field practice in the arena of agricultural and food system sustainability. There are currently 16 books in the series and 5 associated titles (Exhibit 1.1).

Agroecology Series

CRC Press in Florida has launched an impressive series of books in agroecology under the editorship of Clive Edwards of Ohio State University. To date there have been several titles that include agroforestry (Buck et al., 1999), soil tillage (El Titi, 2003), biodiversity (Collins and Qualset, 1998), soil organic matter (Magdoff and Weil, 2004) and agroecosystem interactions with rural communities (Flora, 2001).

Haworth Press Books and Series

The Haworth Press, although not exclusively devoted to sustainable agriculture, publishes a substantial number of books and journals on the topic.

EXHIBIT 1.1. Our Sustainable Future series, available from the University of Nebraska Press

- *Agricultural Research Alternatives,* William Lockeretz and Molly Anderson, 1993
- *Building Soils for Better Crops,* Fred Magdoff, 1993
- *Changing the Way America Farms,* Neva Husainen, 1999
- *A Conspiracy of Optimism: Management of the National Forests Since World War Two,* Paul Hirt, 1994
- *Crop Improvement for Sustainable Agriculture,* Brent Callaway and Charles Francis, editors, 1993
- *Ecology and Economics of the Great Plains,* Daniel Licht, 1997
- *Economic Thresholds for Integrated Pest Management,* Leon Higley and Larry Pedigo, editors, 1997
- *Future Harvest: Pesticide-Free Farming,* Jim Bender, 1994
- *Green Plans: Greenprint for Sustainability,* Huey Johnson, 1995
- *Making Nature, Shaping Culture: Plant Biodiversity,* Lawrence Busch, William Lacy, Jeffrey Burkhardt, Douglas Hemken, Jubel Moraga-Rojel, Timothy Koponen, and Jose de Souza Silva, 1995
- *Ogallala: Water for a Dry Land,* John Opie, 2000
- *Uphill Against Water: The Great Dakota Water Wars,* Peter Carrels, 1999
- *Willard Cochrane and the American Family Farm,* Richard Levins, 2000
- *Family Farming: A New Economic Vision,* Marty Strange, 1988
- *Chernobyl: The Forbidden Truth,* Alla Yaroshinskaya, 1995
- *The Last Harvest: The Genetic Gamble that Threatens to Destroy American Agriculture,* Paul Raeburn, 1995
- *New Roots for Agriculture,* Wes Jackson, 1985
- *The Struggle for the Land,* Paul Olson, editor, 1990
- *Raising a Stink: The Struggle over Factory Hog Farms in Nebraska,* Carolyn Johnsen, 2003
- *Down and Out on the Family Farm: Rural Rehabilitation in the Great Plains, 1929-1945,* Michael Grant, 2002
- *The Curse of American Agricultural Abundance: A Sustainable Solution,* Willard Cochran, 2003

This list contains books in the series edited by Charles Francis, Cornelia Flora, and Paul Olson, plus other books associated with the series. Lincoln: University of Nebraska Press; www.nebraskapress.unl.edu.

Besides the *Journal of Sustainable Agriculture,* they also publish the *Journal of Sustainable Forestry.* Most of the books dealing with food and agriculture are found under their imprint line called Haworth Food & Agricultural Products Press. In this subsidiary can be found two series, *Sustainable Food, Fiber and Forestry Systems* and *Home and Consumer Horticulture.* Dr. Ray Poincelot has served as editor for some of these books. Several relevant titles are shown in Exhibit 1.2.

Sustainable Agriculture Education and Extension Materials

The Center for Sustainable Agricultural Systems at the University of Nebraska, Lincoln used SARE funding to establish an informal publications series, also called the Nebraska Green Books, to bring together appropriate classroom and Extension materials related to sustainable agriculture. Included are resource materials from classroom and Extension work from across the United States, and featured are four volumes that report the results of the SARE-funded workshops in the Professional Development Program for the North Central Region (see Carter et al., Chapter 8).

National SARE Publications

The national SARE program has provided an impressive series of titles that are directed at practical topics and approaches that are accessible to farmers and other practitioners in agriculture. These range from specific field practices (*Managing Cover Crops Profitably,* SARE, 2001; *Steel in the Field,* Bowman, 1997; *Building Soils for Better Crops,* Magdoff and van Es, 2000) to practical guides for direct marketing (*Direct Marketing Resource Guide,* Chaney et al., 2004). The SARE program has also published a series of bulletins on diversification, conducting on-farm research, and other how-to topics that are valuable to farmers practicing sustainable agriculture. Most of these publications are available online for no charge, or on a CD for the cost of duplication. These are found at: http://www.sare.org/publications/.

Other Publishers

Many other books contribute to the field of sustainable agriculture. Examples of these organizations include Chelsea Green Publishing and Island Press (Coleman, 1995; Edwards et al., 1990; Little, 1987; Soule and Piper, 1992) and publishers in the United Kingdom. (Lampkin, 2003). Even major publishing houses issue an occasional book that deals with sustainable food production (Francis et al., 1990; Poincelot, 2003).

EXHIBIT 1.2. Books dealing with sustainable agriculture, available from The Haworth Press, Inc.

- *Allelopathy in Agroecosystems*, Ravinder K. Kohli, Harminder Pal Singh, and Daizy R. Batish, 2001
- *Bacterial Disease Resistance in Plants: Molecular Biology and Biotechnological Applications*, P. Vidhyasekaran, 2002
- *Bacillus thuringiensis: A Cornerstone of Modern Agriculture*, Matthew Metz, 2003
- *Biodiversity and Pest Management in Agroecosystems*, Miguel A. Altieri, 1994
- *Concise Encyclopedia of Bioresource Technology*, Ashok Pandey, 2004
- *Conservation Tillage in U.S. Agriculture: Environmental, Economic, and Policy Issues*, Noel D. Uri, 1999
- *The Contribution of Managed Grasslands to Sustainable Agriculture in the Great Lakes Basin*, E. Ann Clark and Raymond·P. Poincelot, 1996
- *Cropping Systems: Trends and Advances*, Anil Shrestha, 2004
- *Expanding the Context of Weed Management*, Douglas D. Buhler, 1999
- *Fungal Disease Resistance in Plants: Biochemistry, Molecular Biology, and Genetic Engineering*, Zamir K. Punja, 2004
- *Integrating Sustainable Agriculture, Ecology, and Environmental Policy*, Richard K. Olson and Raymond P. Poincelot, 1992
- *Landscape Agroecology*, Paul A. Wojtkowski, 2004
- *Nature Farming and Microbial Applications*, Hui-lian Xu, James F. Parr and Hiroshi Umemura, 2000
- *New Dimensions in Agroecology*, David R. Clements and Anil Shrestha, 2004
- *The Next Green Revolution: Essential Steps to a Healthy, Sustainable Agriculture*, James E. Horne and Maura McDermott, 2001
- *Plant-Derived Antimycotics: Current Trends and Future Prospects*, M. K. Rai and Donatella Mares, 2003
- *Sustainable Aquaculture: Global Perspectives*, B. B. Jana and Carl D. Webster, 2003
- *Sustainable Soils: The Place of Organic Matter in Sustaining Soils and Their Productivity*, Benjamin Wolf and George H. Snyder, 2003
- *Tillage for Sustainable Cropping*, P. R. Gajri, V. K. Arora and S. S. Prihar, 2002
- *Water Use in Crop Production*, M. B. Kirkham, 2000

Binghamton, NY: The Haworth Press; www.haworthpress.com.

MAINSTREAM ACTIVITIES
IN SUSTAINABLE AGRICULTURE

One of the difficult questions to explore is how much influence the activities in sustainable agriculture—research, classroom teaching, Extension programs—have had on mainstream agriculture and on the land grant universities. There is no question about the higher level of interest and concern about environmental impacts, and at times the social impacts, of our agricultural technologies. Unfortunately, these influences are difficult to quantify.

A survey conducted in 1992 assessed the impacts of sustainable agriculture programs on research, teaching, and Extension in the land grant universities across the United States (Francis et al., 1995). There were 146 respondents, a 49 percent return rate of the 300 surveys sent out. Administrators, faculty identified as interested in sustainable agriculture, and random faculty members from agronomy departments in all 50 states were asked to rank the impact of sustainable agriculture programs (from 1 = low to 9 = high) on university activities. We reported the results separately from the four USDA regions. In 1992, there was significantly lower impact on teaching than on the research and Extension programs. Administrators reported a higher impact than either random faculty or interested faculty, and there was no difference between the two faculty respondent groups. Impacts were reported as highest in the Northeast Region, intermediate in the West, and lowest in the North Central and Southern Regions. These results are consistent with the conventional wisdom that change occurs more rapidly on either coast compared to the U.S. heartland. Average rating was 5.2 on the 9-point scale.

When the same survey was repeated in 2000, the same groups (but not necessarily the same people) were queried with the same questions, and 138 of 300 people responded for a 46 percent return rate (Francis, 2000). In spite of eight years of focused programs in sustainable agriculture, the overall rating was 5.1 on the 9-point scale, nearly identical to the previous survey. Teaching again rated significantly lower than other activities, and administrators perceived a higher impact on programs than that reported by either faculty respondent group. The Northeast Region still ranked sustainable agriculture programs as having greater impact on land grant activities compared to the other three regions, which did not differ among themselves.

We can speculate that the increased mainstreaming of sustainable agriculture research and education caused people to consider many of the activities formerly seen as new and unique to now be a part of the routine research and learning landscape, and thus not identified with sustainable agriculture as before. What is clear from the results is that people are familiar with sustainable agriculture, a concept that was not necessarily on their radar screens two decades ago. SARE grants have certainly contributed to this awareness and change.

Teaching of sustainable agriculture principles in mainstream Extension settings has appeared through numerous topics such as reduced and no tillage, integrated pest management, and irrigation scheduling. Some focused programs have made a substantial impact, such as growth of reduced tillage after the introduction of Buffalo no-till planters and cultivators from Fleischer Manufacturing Co. some four decades ago in Nebraska, and the noted reduction in fertilizer nitrogen application in Iowa as a result of excellent research and effective Extension programs. There is no doubt that sustainable agriculture in Extension has focused primarily on the details of production practices, to a limited extent on creative crops and marketing alternatives, and to a much smaller degree on landscape-level issues such as ecosystem services.

One of the major spin-offs or products in the evolution of education in sustainable agriculture is the emergence of agroecology as a popular topic for systems education. With roots in Chile and Mexico, and more recently in California (Altieri, 1983; Gliessman, 1998), university agroecology courses have grown rapidly in acceptance over the past decade. Many of the instructors who were teaching sustainable agriculture courses now label them agroecology. With the broadening of the definition to include all steps in the food system from production to the consumer, agroecology in the Nordic Region and the North Central Region of the United States is now often called "the ecology of food systems" (Francis et al., 2003). Many of the topics addressed by SARE-funded research—field practices to reduce pesticides and chemical fertilizers, diversity in enterprises and products, on-farm processing, direct marketing—could be considered under the current umbrella of agroecology and represent one contemporary impact of sustainable agriculture on the learning landscape.

CLOSING THE LOOP:
EDUCATING THE CONSUMER

Sustainable agriculture outreach must involve all parties, from professional researchers, educators, and Extension workers to commercial producers, processors, and marketers to consumers, if sustainable agriculture is to become the dominant agricultural paradigm. Sustainable agriculture professionals have done a commendable job with research, education, and Extension outreach, especially given the constraints imposed by limited staff and funds. Extension outreach is the critical connection, where we must assure successful communication among researchers, educators, and farmers.

To date, consumers have received little attention, other than from the organic farming component of sustainable agriculture.

The next challenge for acceptance must be to capture the hearts and minds of consumers, many of whom have no idea where or how their food is produced. This campaign will require considerable time, patience, education, and outreach. Closing the loop by including consumers is absolutely essential, as consumers control the economics of success for sustainable agriculture. In particular, we need to get the attention of both urban and suburban consumers, groups that will require some differences in approach. In general, these two consumer groups place different weights on motivators when it comes to food purchases. Different income levels and types of education also further complicate the picture. For example, the ranks of environmental, health, and cost factors are influenced by one's disposable income, level of education, and neighborhood location (e.g., see this website: http://www.organicagcentre.ca/ Research Database/res_food_consumer.html). To persuade the consumer to buy sustainably produced products, "one size fits all" is a strategy that is doomed to fail.

Success in changing consumer habits will require educational outreach on two fronts: in school and postschool. The term *school* is used broadly to include education through college years, and postschool in the lifelong learning we work to encourage. Extension can help guide sustainable agriculture outreach through 4-H groups, curriculum units in the sciences for K through 12 levels, and through classes and activities at land grant universities.

In K-12 the potential for expanded curricula is substantial, given the numerous resources available today. The decade-old Wisconsin curriculum for grades 9 through 12, *Toward a Sustainable Agriculture,* is the oldest sustainable agriculture course (CIAS, 1991). Numerous more recent print and Web resources for K-12 teachers also exist (Gold, 2002; Seyler, 1995; ATTRA, http://attra.ncat.org/attra-pub/PDF/k12.pdf; AFSIC, http://www.nal.usda.gov/afsic/AFSIC_ pubs/k-12.htm). An example of a curricular module at the high school level is *The Living Soil: Exploring Soil Science and Sustainable Agriculture with Your Guide, the Earthworm,* which is produced by the Agricultural Education Materials Services (AEMS) at Iowa State University. Considerable opportunities exist for the enhancement of sustainable agriculture in 4-H, as their current programs are primarily dominated by conventional agricultural underpinnings. The same is true for national "Ag in the Classroom" courses offered for enrichment to younger students in most states.

Numerous undergraduate programs offering courses and degrees in sustainable agriculture now exist at land grant universities. Lists of educational and training opportunities can be found at the Alternative Farming Systems Information Center Web site and in print (http:// www.nal.usda.gov/ afsic/ AFSIC_pubs/edtr.html; AFSIC, 2003). Examples in the North Central Region include programs in the Universities of Illinois, Iowa State, Michigan State, Minnesota, and Wisconsin. Outside the region, other well-known sustainable agriculture programs can be found at the Universities of California at Davis and at Santa Cruz, Maine, Texas A&M, and Vermont. A few private universities and colleges, such as Hampshire College in Amherst, Massachusetts, and Sterling College in Craftsbury, Vermont, have initiated undergraduate degrees in sustainable agriculture. Graduate degrees are now emerging such as the MS and PhD graduate programs in sustainable agriculture launched at Iowa State University in 2000. An excellent teaching resource for the organic farming part of sustainable agriculture exists: *Teaching Organic Farming and Gardening Resources for Instructors* (http://zzyx .ucsc.edu/casfs/ training/manual/index.html), available from the program at U.C. Santa Cruz in California.

Teaching faculty at undergraduate institutions observe that students are frequently underinformed about the food system, especially from an environmental, energy, and health perspective. One way to reach students at non–land grant colleges is through courses in horticulture, which are typically offered in the sciences. Such a course is also a good means of conveying sustainable agricultural information at land grant universities for students in nonagricultural degree programs, a majority of the undergraduate population. A good resource for this direction is a new book, *Sustainable Horticulture: Today and Tomorrow* (Poincelot, 2003). Other courses in which sustainable agricultural information can be utilized include ecology, environmental science, plant science, and those that deal with contemporary science and technology issues that impact society. As students graduate and move into mainstream society, knowledge about sustainable agriculture will heighten their awareness and likely increase future demands for food produced in a more sustainable way.

More consumers need to understand the connections among agriculture, their food, their health, and their environment. One approach is to wait it out, since modern societies are organized around fossil fuels as the driving force. As impacts of current systems begin to threaten the life support base, consumer interest will increase concerning natural and contemporary energy resources. In turn, this questioning will lead to an examination of other natural resources such as soil and water. This trend will increase public interest in rural areas and agricultural landscapes, and make the sustainable agriculture message more palatable to urban audiences.

Another approach is to reach out to consumers now. We believe in the need to raise current sustainable food system awareness. To reach out to present-day consumers, we need to emphasize the personal health and environmental benefits for them and their families that are offered through a sustainable food system. The general public does understand and respond to health issues such as benefits of fiber, antioxidants, and health-promoting phytochemicals from vegetables and fruits, as well as the relationship between pesticides and cancer. There is a rapidly growing awareness of obesity as a national epidemic in the United States. Once informed, consumers are receptive to information about local food systems including backyard gardens, community gardens, farm stands, compost, organic products, and community-supported agriculture. They also can understand more complex ecosystem functions and services such as wetlands for clean water, windbreaks for clean air, intact soils that do not erode, oxygen from green plants, and the reduction of global warming by carbon dioxide capture by plants and soils (Daily, 1997). Consumers will want to learn more about their food supply and the rural landscape, and ultimately will want information and choices from the food systems that are connected to sustainable agriculture practices.

Organic food producers and processors have already been successful in getting information to the public about the certified labels that identify their products in the market, as well as making direct connections for sale from the farm. Their level of success recently led to the creation of national organic standards. Just as commodity groups have promoted their specific crop or animal products, organic growers and processors have pursued public education campaigns using grower fees and contributions. One result is that growers benefit from the premium price that organic products bring in the marketplace, thus increasing incomes by advancing their labeled products. Organic food sales have increased by about 20 percent per year for more than a decade in the United States and Europe (Lampkin, 2003). The U.S. Market for Organic Foods and Beverages report shows that annual sales soared from $1 billion to $11 billion during the period 1990- 2002 (http:// austin.bizjournals.com/austin/stories/2003/06/02/daily31.html). Groups that wish to promote sustainable agriculture in the marketplace can learn much from the successful outreach associated with the organic label. Perhaps one day some label such as SAFE will be seen on sustainable agricultural products. The word "safe" has a connotation that attracts consumers' attention and is logical, standing for Sustainable Agriculture: Farm Endorsed or possibly Food and Environment (Poincelot, 1997).

Emphasis on consumer education should include explanations of how ecological approaches used by sustainable farmers can produce food without pesticides and with alternatives to chemical fertilizers, while protecting the ecosystem and producing healthy, safe food for consumers. The farmer can be portrayed as an environmental steward who is also concerned about consumer health, an ally of the consumer who has made changes in farming practices to achieve these multiple goals. Customers can be affirmed when they make the choices to purchase food produced in this manner. Further education about how food is produced, where, by whom, and under what conditions will reinforce their decisions. Surveys in Norway have shown that consumers who regularly purchase ecological food are also concerned about overall pesticide and fertilizer use, about animal welfare, and about the conditions of farmers and families where food is produced (Torjusen et al., 2001). When agroecology studies include the entire food system, issues of consumer opinion and choice are part of the equation. Consumer education, if it is to lead to a successful marketing transition for sustainable agriculture, must be included in the new social contract.

RENEWING THE CONTRACT

In the following chapters, we explore contributions of SARE, land grant university programs, and the nonprofit sector to research and education related to both farming practices and systems and to consumer issues. Taken as a whole, the results in this book contribute to the creation of a new social contract with farming and ranching families and with rural communities.

Strategies to enhance the sustainability of farm practices are examined in Chapters 2 through 5. Pest management is discussed by George W. Bird and Michael Brewer, and soil fertility and management by Kent McVay. Intensive rotational grazing is a practice attracting growing interest, and this is presented by Terry Gompert. Whole-farm planning and the sustainable integration of enterprises are explored by Rhonda R. Janke in Chapter 5.

Foremost among the measures to determine the progress and success of the social contract with consumers, and of course farmers, is economics. John Ikerd is an advocate of diversity in farming and farm products, adding value on the farm and in the community, and finding innovative strategies to increase income and income stability on the family farm. He covers this topic in Chapter 6, "Economic Analysis and Multiple Valuation Strategies," discussing the considerable work that has been done over the past 17 years in SARE and other programs to enhance the viability of the family farm and the rural communities of the region.

Jerry DeWitt and Charles A. Francis describe the highly successful activities in Iowa as a case study in Chapter 7, to describe how the university can partner with farmer organizations and a high-profile sustainable agriculture center on campus to create effective programs that can reach a large number of farmers and consumers. In Chapter 9, "Regionalization of a Research and Education Competitive Grants Program," the administrative team in Nebraska of Steven S. Waller, Elbert C. Dickey, and Charles A. Francis discusses the elements of a successful regional effort that can backstop the state education and research activities in sustainable agriculture and enhance sustainable agriculture outreach to the farm community.

Workshops are a critical tool in the education process. Training others to carry the message forward, whether on the farm or in the community, can be initiated at the regional workshop level. Heidi Carter, Charles A. Francis, and Richard Olson have considerable expertise in this area which they share in Chapter 8, "Regional Training Workshops for Sustainable Agriculture." Workshops provide the opportunity for educators and specialists to share experiences, provide mutual support, and develop personal networking connections. Materials from regional workshops can have further educational outreach life when they are copied and distributed as educational support resources for local, state, and national programs.

The importance of rural communities to the success of sustainable agriculture is of first rank. Lorna Butler and Cornelia Flora examine this area from a sociological and anthropological perspective in Chapter 10, "Expanding Visions of Sustainable Agriculture." Their work on describing and applying the concepts of social capital and the importance of rural communities is seminal in linking production with economics and the broader social issues.

The interface between farms and communities is especially important to promote a consumer buy-in for products of sustainable agriculture. Bad relations at the interface as a result of different lifestyles and expectations can lead to misunderstandings and ripple effects further out in the consumer mainstream. Michele Schoeneberger, Gary Bentrup, Charles A. Francis, and Richard Straight cover this topic in Chapter 11, "Creating Viable Living Linkages Between Farms and Communities." Recognizing the interface as a zone of frequent conflict, the authors demonstrate positive solutions by using woody perennial plantings that separate the different activities, while providing multiple ecosystem and human benefits for those living at the interface.

Nonprofit farmer groups can play a vital role in practical research and education, especially when these activities are located on the farm. Credible on-farm research potentials for creating sustainable systems are discussed in Chapter 12 by Derrick Exner and Richard Thompson of the Practical Farmers of Iowa. Foundations can play an important role in promoting sustainable agriculture in the broader community. Oran Hesterman examines this role in Chapter 13, "Impacts of Private Foundations on Sustainable Agriculture and Food Systems." Foundation support can play a critical role in seeking reform in our institutions and supporting positive change through advocacy and support in the community. The motivations for change to more sustainable practices and systems are complex, and a careful study of change since the inception of the regional program is reported in Chapter 14, "Motivation Theory and Research in Sustainable Agriculture" by Shirley K. Trout, Charles A. Francis, and John E. Barbuto Jr.

Organic farming is certainly the best example of a sustainable agricultural system that has captured the hearts and minds of consumers. Fred Kirschenmann is an organic farmer and also a recognized authority in agricultural values and ethics, and George Bird has been active on state and national levels since SARE started. In Chapter 15, "Future Potential for Organic Farming: A Question of Ethics and Productivity," they use organic farming as a model system that deals with ethical issues, including the major current concerns about the industrialization of the organic food industry. This chapter describes a potential future for other sustainable agricultural approaches as well, and provides a good blueprint for planning ahead and avoiding pitfalls.

Of course the future is always challenging, and we provide a summary and vision by Charles A. Francis in Chapter 16. A future role for Cooperative Extension is explored in "Future Multifunctional Rural Landscapes and Communities."

In conclusion, we anticipate that this book will be useful to administrators, researchers, and educators in the North Central Region, where there have been many successful innovations, as well as across the United States. The editors and chapter authors invite your comments and suggestions on how we can further the agenda of sustainable agriculture in the U.S. food system.

REFERENCES

AFSIC. 2003. *Educational and training opportunities in sustainable agriculture* (15th ed.). Alternative Farming Systems Information Center, Beltsville, MD.

Alston, J.M. and P.G. Pardey. 1996. *Making science pay: The economics of agricultural R & D policy.* The AEI Press, Washington, DC.

Altieri, M.A. 1983. *Agroecology.* University of California Press, Berkeley, CA.

Bender, J. 1994. *Future harvest: Pesticide-free farming.* University of Nebraska Press, Lincoln, NE.

Bowman, G. (ed.) 1997. *Steel in the field: A farmer's guide to weed management tools.* Sustainable Agriculture Network, National Agricultural Library, Beltsville, MD.

Buck, L.E., J.P. Lassoie, and E.C.M. Fernandes (eds.) 1999. *Agroforestry in sustainable agricultural systems.* Lewis Publications, Boca Raton, FL.

Cavigelli, M.A., S.R. Deming, L.K. Probyn, and R.R. Harwood (eds.) 1998. *Michigan field crop ecology: Managing biological processes for productivity and environmental quality.* Michigan State University Extension Bulletin E-2646. Michigan State University, East Lansing, MI.

Cavigelli, M.A., S.R. Deming, L.K. Probyn, and D.R. Mutch (eds.) 2000. *Michigan field crop pest ecology and management.* Michigan State University Extension Bulletin E-2704. Michigan State University, East Lansing, MI.

Chaney, D., G. Feenstra, and J. Ohmart. 2004. *Direct marketing resource guide.* Sustainable Agriculture Network, National Agricultural Library, Beltsville, MD.

CIAS (Center for Integrated Agricultural Systems). 1991. *Toward a sustainable agriculture: A teacher's guide.* CIAS, Madison, WI.

Coleman, E. 1995. *The new organic grower: A master's manual of tools and techniques for the home and market gardener* (2nd ed.). Chelsea Green Publishing, White River Junction, VT.

Collins, W.W. and C.O. Qualset (eds.) 1998. *Biodiversity in agroecosystems.* CRC Press, Boca Raton, FL.

Daily, G.C. 1997. *Nature's services: Societal dependence on natural ecosystems.* Island Press, Covelo, CA.

Doran, J.W. and S. Peck. 2003. Editorial. *Am. J. Altern. Agric.* 18(4):173.

Edwards, C.A., R. Lal, P. Madden, R.H. Miller, and G. House (eds.) 1990. *Sustainable agricultural systems.* Soil and Water Conservation Society, Ankeney, IA.

El Titi, A. (ed.) 2003. *Soil tillage in agroecosystems.* CRC Press, Boca Raton, FL.

Flora, C. (ed.) 2001. *Interactions between agroecosystems and rural communities.* CRC Press, Boca Raton, FL.

Francis, C.A. 2000. How sustainable agriculture programs impact U.S. landgrant universities: An update. Sustainable Agricultural Workshop, North Central Region, University of Wisconsin, Madison, WI, November 2-4. 6 p.

Francis, C., C. Edwards, J. Gerber, R. Harwood, D. Keeney, W. Liebhardt, and M. Liebman. 1995. Impact of sustainable agriculture programs on U.S. landgrant universities. *J. Sustain. Agric.* 5(4):19-33.

Francis, C.A., C.B. Flora, and L.D. King (eds.) 1990. *Sustainable agriculture in temperate zones.* John Wiley and Sons, New York.

Francis, C., G. Lieblein, S. Gliessman, T.A. Breland, N. Creamer, R. Harwood, L. Salomonsson, et al. 2003. Agroecology: The ecology of food systems. *J. Sustain. Agric.* 22(3):99-118.

Gliessman, S.R. 1998. *Agroecology: Ecological processes in sustainable agriculture.* Ann Arbor Press, Chelsea, MI.

Gold, M.V. 2002. *Sustainable agricultural resources for teachers, K-12.* Alternative Farming Systems Information Center, Beltsville, MD.

Grant, M.J. 2002. *Down and out on the family farm: rural rehabilitation in the Great Plains,* 1929-1945. University Nebraska Press, Lincoln, NE.

Hildreth, R.J. and W.J. Armbruster. 1981. Extension program delivery—Past, present and future: An overview. *Am. J. Agric. Econ.,* December, 853.

Johnson, H. 1995. *Green plans: Greenprint for sustainability.* University of Nebraska Press, Lincoln, NE.

Lampkin, N. 2003. *Organic farming* (revised ed.). Diamond Farms Book Publications, United Kingdom.

Little, C.E. 1987. *Green fields forever.* Island Press, Washington, DC, 192 pp.

Magdoff, F. and H. van Es. 2000. *Building soils for better crops.* Sustainable Agriculture Research and Education Program.

Magdoff, F. and R. Weil (eds.) 2004. *Soil organic matter in sustainable agriculture.* CRC Press, Boca Raton, Florida.

McDowell, G.R. 2001. *Land-grant universities and extension into the 21st century: Renegotiating or abandoning the social contract.* Iowa State University Press, Ames, IA.

Poincelot, R. 1990. Statement of purpose. *J. Sustain. Agric.* 1(1): i.

———. 1997. From the editor. *J. Sustain. Agric.* 11(1): 1-2.

———. 2003. *Sustainable horticulture: Today and tomorrow.* Prentice Hall, Upper Saddle River, NJ.

Pretty, J., J. Ashby, A. Ball, J. Morison, and N. Uphoff. 2003. Editorial. *Int. J. Agric. Sustain.* 1(1):1-2.

SARE. 2001. *Managing cover crops profitably* (2nd ed.). Sustainable Agriculture Network, National Agricultural Library, Beltsville, MD.

Seyler, E. 1995. *The A to Z in sustainable agriculture: A curriculum directory for grades K-12.* Center for Sustainable Agriculture, University of Vermont, Burlington, Vermont.

Sooby, J. 2003. *State of the states, second edition: Organic farming research in the land grant universities.* Organic Farming Research Foundation, Santa Cruz, CA.

Soule, Judith D. and Jon K. Piper. 1992. *Farming in nature's image: An ecological approach to agriculture.* Island Press, Washington, DC.

Torjusen, H., G. Lieblein, M. Wandel, and C.A. Francis. 2001. Food system orientation and quality perception among consumers and producers of organic food in Hedmark County, Norway. *Food Qual. Prefer.* 12:207-216.

Youngberg, G. 1986. Why another journal? *Am. J. Altern. Agric.* 1(1):2.

Chapter 2

Integrated Pest Management, Ecoliteracy, and Unexpected Consequences

George W. Bird
Michael J. Brewer

INTRODUCTION

The roots of Integrated Pest Management (IPM) are deeply embedded within the land grant university system. IPM evolved during the last four decades of the twentieth century in response to the unexpected consequences associated with the Synthetic Pesticide Era of pest management. IPM has become both the national and global standard. The goal of this chapter is to describe the nature and current practice of IPM, and to illustrate its value in the enhancement of ecoliteracy. We also describe how the philosophy, process, and practices of IPM can be used to significantly reduce patterns of risk and unexpected consequences associated with ecosystem management decisions. An important part of the Sustainable Agriculture Research and Education (SARE)–funded research has dealt with weed and insect management using principles of IPM.

NATURE OF IPM

IPM is a philosophy and process used in the development, implementation, and evaluation of pest management practices that result in both favorable socioeconomic and environmental consequences. In a 1979 message to Congress, President Carter defined IPM as "a systems approach to reduce pest damage to tolerable levels through a variety of techniques, including

Developing and Extending Sustainable Agriculture
doi:10.1300/5709_02

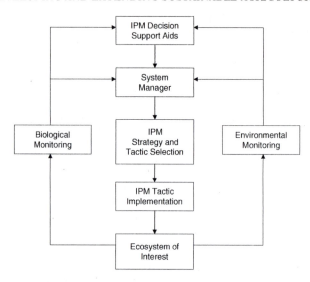

FIGURE 2.1. Conceptual model of the process of Integrated Pest Management (after Bird et al., 1989).

predators and parasites, genetically resistant hosts, natural environmental modifications and, when necessary and appropriate, chemical pesticides." The history of IPM in the United States is described in detail by Olsen et al. (2003). Successful IPM programs have been identified and documented in more than two dozen countries throughout Africa, Asia, Australia, Europe and the Americas (Maredia et al., 2003). Although development of IPM is continually being modified and improved, it can be conceptualized as a process (Figure 2.1) involving extensive ecosystem monitoring and implementation of various pest management strategies and tactics based on the concept of action thresholds (Figure 2.2).

MONITORING AND EVALUATION

Biological monitoring is a core component of IPM (Figure 2.1). It is often referred to as "scouting." This consists of standardized sampling procedures designed to estimate: (1) the developmental stages and population densities of pests, (2) the status of other organisms in the food web (usually referred to as beneficial organisms), and (3) the health of crops, livestock, or humans in the ecosystem of interest. Biological monitoring is a knowledge-intensive procedure requiring highly trained individuals. The system man-

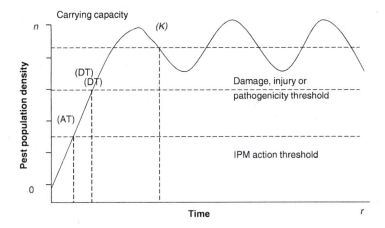

FIGURE 2.2. Relationships between the IPM action threshold (AT), damage-injury-pathogenicity threshold (DT), and ecosystem carrying capacity (K) (after Bird and Berney, 2000).

ager (e.g., professional agriculturist, certified crop advisor) is responsible for the current state of the system, and is frequently the best suited individual for monitoring and decision-making roles associated with IPM. Private sector scouts, IPM consultants, or pest control firms are often hired to do the biological monitoring. In the United States, most biological monitoring specialists are trained by land grant universities. This includes comprehensive Extension education programs and formal undergraduate and graduate degree programs in IPM. These programs have also evolved to include a major international emphasis (Maredia et al., 2003).

IPM literature is extensive and the majority of it is based on research at land grant institutions. It includes academic texts such as *Concepts in Integrated Pest Management* by Norris et al. (2003), and applied materials such as *IPM in Practice: Principles and Methods of Integrated Pest Management* by Flint and Gouveia (2001) and *Field Crop Pest Ecology and Management* (Cavigelli et al., 2000). There are numerous government-based assessments of IPM, including a National Research Council publication entitled *Ecologically Based Pest Management: New Solutions for a New Century* (Hardy, 1996). Other work ranges from mechanisms of biological interactions in nutrient cycling (Ingham et al., 1985) to the linkages between IPM and both sustainable-equitable development and organic agriculture (Benbrook, 1996; Bird, 2003).

Damage from pests occurs when population densities exceed specific injury-damage-pathogenicity thresholds (Figure 2.2). Population dynamics of pests and other food web organisms are governed by abiotic factors such as temperature, soil moisture, light, and relative humidity. This mandates that information about selected abiotic parameters must be readily available for use in predictive IPM decision making. Thus, environmental monitoring becomes another core component of IPM (Figure 2.1). The types of abiotic information needed are ecosystem specific, but usually include current weather data provided on a regional, local, or even microclimatological habitat scale. This information must be readily available, user friendly, and as close to real-time as possible. Dedicated computerized environmental monitoring systems are common in today's agricultural IPM programs.

Because of the large number of significant interactions among pests, beneficial organisms, crops, agricultural animals, and the environment, IPM decision-support aids are often necessary (Figure 2.1). These may be simple look-up tables or complex hardware-software systems. They can include simulation models, expert systems, or artificial intelligence technology. The associated human experience and wisdom is likely to always remain as an essential element of successful IPM programs.

The system manager or designated representative is responsible for pest management decisions. This aspect of IPM is perhaps the most knowledge-intensive part of the process. An individual within the enterprise may be assigned the IPM decision-making responsibility, or it can be delegated to an external private consultant or company. Individuals trained only in pest scouting often lack the breadth of experience needed to take on the responsibility of decision-making associated with IPM strategy-tactic selection and implementation.

A major difference between conventional pest control and IPM is that the latter is based on the use of damage-injury-pathogenicity, action, and economic and equilibrium thresholds (Figure 2.2). When a pest population density reaches the action threshold for a specific system, or is predicted to reach this level in the near future, it is time to select an appropriate IPM strategy and implement one or more tactics. It is imperative that this be done before the damage-injury-pathogenicity threshold is reached. IPM procedures should be implemented when the marginal revenue derived from the management tactic input is equal to or exceeds the marginal cost (economic threshold). The economic threshold is a dynamic concept that mandates information about the cost and efficacy of the management tactic, production system economics, nature of the pest and its population dynamics, and environmental variables. For instance, a field may have the exact same number of pests in two consecutive years, but due to a sharp rise in the cost of a management practice, the economic threshold may not be reached in the second year. Although IPM is designed to reduce environmental and human health risks, the economics associated with these factors are rarely available for inclusion in calculation of the economic threshold.

Integrated Pest Management is a dynamic and closed-loop process (Figure 2.1). After implementation of an IPM tactic, it is imperative that the biological and environmental monitoring programs be continued. This is done to determine if the desired management objectives were achieved or if there is a need for additional system evaluation, decision making, and action. The process and practices of IPM were described by James Kendrick Jr. as "an ecological approach to maintaining plant health. It is an attitude evolving into a concept of controlling pest and disease damage to plants. It is based on an understanding of the entire ecological system to which the host that we are interested in keeping healthy belongs." This concept of IPM is well suited for integration into the journey toward an overall equitable and sustainable future (Altieri, 1994; Potter et al., 1970; Rodale, 1984; UNCED, 1987). In this respect, the origins of IPM were well ahead of their time.

STATE IPM PROGRAM COORDINATOR OBSERVATIONS

All land grant institutions have an IPM Program Coordinator and the authors have both served in this role. State coordinators are responsible for the development and implementation of IPM programs that facilitate appropriate interactions among land grant university faculty, government, and the private sector. They are also responsible for interacting with the citizens of their respective states in the development of a future vision for IPM.

The concepts of IPM have been discussed in the agricultural literature since the late 1950s, mainly in response to the use of a multitude of broad spectrum synthetic chemicals that kill or impair the biology of a wide array of pests and nontarget organisms. The impacts of indiscriminant use of pesticides were revealed by Carson (1962) in her landmark book, *Silent Spring.* As an alternative, IPM encourages evaluating the economics of suppressing a pest or range of pests to below a level where short-term farm profitability is not adversely affected. The early constructs of the IPM concept emphasize the importance of using chemical controls in such a manner that adverse effects on biological control agents are avoided or minimized. Later writings emphasize the importance of minimizing as many negative environmental impacts as possible. Today, IPM systems often focus on resiliency of agroecosystems as farmers design them for the journey toward sustainable-equitable development. This extends the focus of IPM to more long-term protection from pests in a manner that does not disrupt other ecosystem structures, processes, and services. Although the focus of most IPM programs is on plant protection, there are excellent examples of successful IPM initiatives related to livestock and human living environmental issues. State coordinators have interacted closely with SARE to secure funds for research on experiment stations and commercial farms.

MANAGEMENT STRATEGIES AND TACTICS

Tactics chosen for IPM need to follow selection and implementation of an appropriate overall strategy. One or more tactics within that strategy are based on decision-making and use of biological monitoring, environmental monitoring, and threshold information. The decision maker has four basic strategic approaches that need to be considered before implementation of specific management tactics. These include the strategies of: (1) avoidance-exclusion, (2) containment-eradication, (3) control, and (4) no need to take action at this time. Following the choice of one of the first three strategic approaches, one or more tactics need to be selected for implementation. Five major categories of pest control tactics are generally available: (1) biological, (2) genetic, (3) chemical, (4) cultural, and (5) regulatory practices. Most IPM programs utilize multiple tactics. Although this is an ecology-based process, it is important to note that the language of IPM is a hybrid of elements associated with both conventional agriculture and fundamental ecology.

Biological control describes the actions of natural enemies (e.g., parasites, parasitoids, predators, and pathogens) that result in pest suppression (Barbosa, 1998; Bellows and Fisher, 1999; DeBach and Rosen, 1991). These organisms may be highly pest-specific or may have broad host ranges, and their interactions may result in the suppression of nontarget and potentially beneficial animal and plant species. There are three subcategories of biological control: natural, conservation biology, and manipulative practices. Naturally occurring biological control is not actively managed. Pest control results in resident organisms maintaining populations of pests to densities below the injury-damage-pathogenicity threshold. This pest–natural enemy interaction may be described as the innate capacity of the ecosystem to suppress pests. Conservation biological control functions through manipulation of land management practices at the site of interest and surrounding landscape. Properly implemented, this results in enhancement of resources that aid biological control organisms (e.g., nectar, alternate prey, refuge, and mating sites). Manipulative biological control involves direct manipulation of natural enemies to manage key pests. This may include importation of exotic biological control agents with the aim to permanently establish them for pest control in the site of interest (classical biological control) or augmentative-inundative biological control, which consists of the release of biological control agents with the aim of controlling the pest during critical periods of plant susceptibility to damage. In the latter case, biological control agents are not capable of becoming permanently established in the ecosystem. Both land grant institutions and the federal government have played major roles in the development of successful biological control, beginning with the introduction of the Vidalia Beetle for suppression of Cottony Cushion Scale in California in 1888 (Flint and Gouveia, 2001). In Michigan, highly successful biological control programs were implemented for the alfalfa weevil in the late 1960s and for the cereal leaf beetle in the early 1970s. Four native biological control agents were identified in the late 1970s for the control of the onion maggot (Haynes et al., 1980). In this case conservation biological control was necessary for both the identification of the control agents and their successful use.

Genetic control involves the use of plant cultivars with pest resistance or tolerance. Other practices can include the release of sterile males or very low dosages of chemicals that interfere with pest mating behaviors. Use of cultivars with resistance to pests increased significantly during the last decade of the twentieth century. One example is how the federal government identified, collected, and transported more than 100 different lines of soybeans with resistance to the soybean cyst nematode (SCN) to the United States. Seven of these have been used by private-sector seed companies for development of resistant varieties. The availability of SCN-resistant varieties and proof of their efficacy through extensive land grant institution variety demonstration trials and Extension education programs resulted in the elimination of the need for chemical controls for this potentially devastating pest. Another extremely important development for the Northeastern United States has been the release of scab resistant apple cultivars. For decades, wheat lines with resistance to stem rust have been provided to the seed industry by federal government and university breeding programs.

Second-generation pest resistant cultivars almost always have improved food quality traits, compared to first-generation cultivars. Use of biotechnology to develop genetically modified varieties has provided agriculture with new and widely adopted pest management tools. The most common of these are designed to be used with herbicides for weed control. Extensive use of resistant varieties has resulted in the need for new field management strategies and Extension education programs on resistance management. In a few cases, cultivars marketed as pest resistant have not been able to support this claim.

Cultural control involves the reduction of pest population densities or their effects through modification of production system management practices instead of through the use of additional system inputs (Bird and Berney, 2000). Examples of cultural control include the use of crop rotations and other spatial or temporal system manipulations such as time of planting or harvesting, cover crops, companion crops, tillage systems, or organic residue management. Cultural controls are usually highly site-specific. For example, no-till systems have been shown to be highly successful in the management of a pest species on one farm and have failed to provide satisfactory results on a neighboring field.

Chemical control involves the use of synthetic or natural chemical compounds to reduce pest populations through mortality or by disrupting various aspects of the pest biology. Broad-spectrum or high-risk compounds are toxic to a large array of target pests and other taxonomically related organisms in the ecosystem food web. One or more of these eliminated species are likely to be beneficial organisms for the farmer. This is a negative effect of pesticides. Some chemical compounds cross taxonomic boundaries and control multiple types of pests, such as nematodes and insects or fungi and mites. Methyl bromide has perhaps the broadest spectrum, providing control of insects, other arthropods, nematodes, fungi, and weeds. Narrow-spectrum or reduced-risk compounds are only toxic to a small array of organisms. Their effects are usually limited to the target organisms. It is important to note that these categories represent the extremes of a spectrum of effects of chemical compounds used in pest management, and most fall somewhere in between.

Regulatory control measures are used to prevent, avoid, contain, or possibly eradicate pests. Regulation can be implemented by federal and state governments, private-sector enterprises, or commodity groups. Regulatory practices may be either legally binding or voluntary. Examples of regulatory controls include quarantine restriction of agricultural products to prevent the transport of a pest into or out of an area and host-free periods that restrict planting and harvest intervals in ways that disrupt pest biology.

IPM PROGRAM IMPLEMENTATION

It is usually desirable to implement multiple pest management tactics that are compatible with each other, biologically regenerative or self-sustaining, nondisruptive to ecosystem health, and designed to minimize the use of nonrenewable resources. It is best to build these multifunctional tactics into the overall enterprise and whole-farm design. Ideally, when IPM is integrated into the system design, pest management tactics complement or enhance other aspects of the system, and pest management requires only modest maintenance after its initial implementation. Ironically, successful examples of this approach may lead to an underappreciation of the importance of the pest management program. One of the most frequently used tactics that attains this level of integration is the use of varieties that have multiple traits of resistant to pests while providing desirable commodity products.

Maintaining rotation systems, cover crops, and other accessory plants that suppress pests and also enhance various aspects of ecosystem quality are methods commonly used to reduce pest impact and stabilize enterprise income. The use of a continuum of different wheat cultivars with resistance to different races of stem rust is an example that has been in place for so long that the disease is rarely mentioned except in textbooks.

In addition to building IPM tactics into the enterprise design, identifying tactics that suppress pests on a district or regional level has a sound historical basis, and has regained popularity in recent years. Successful establishment of host-free periods and classical biological control agents have led to effective and resilient examples of pest management. Most biological·control texts cite impressive cases of biological control agents successfully controlling insects and weeds in citrus and other orchard crops, rangeland, and agronomic and vegetable cropping systems. Host-free periods are used as regulatory control techniques designed to restrict times of planting and harvesting to obtain extended periods of time when host plants are unavailable for pests. They can be effective for annual cropping systems that sustain damage from a key pest with few generations per year or a limited host range. More recently, expanding interest in mating disruption techniques and habitat management techniques designed to conserve biological control agents and disrupt pest population development have focused at a district or regional level. Development and implementation of innovative IPM systems always require an understanding of the pest management services provided by the overall ecosystem and adjacent landscape ecology. With this appreciation, new tactics can be integrated into the system to manage key pest disruptions, while maintaining the existing pest suppression strengths of the system.

IPM PROGRAM RESILIENCY

Although ecology is basically a post–World War II science, it has provided a comprehensive understanding of the structure and processes of ecosystems in relation to the transport and transformation of matter and energy. It has also become well known that it is not possible to simultaneously maximize two dependent variables. Development of IPM tactics has usually been addressed from the mechanistic view of targeting the suppression of a single specific pest species or pest group. There have been relatively few comprehensive efforts to assess the impacts of individual pest management tactics on other components and processes of ecosystems. Suggestions that the impacts of a pest management tactic are limited to one target pest should be suspect until sound documentation of its nontarget and broad environmental effects are obtained. The classical example of unintentional consequences is that many pesticides are broad-spectrum chemicals that impact a wide array of nontarget organisms, often resulting in the emergence of new pest problems or the resurgence of a known pest to levels exceeding the initial population density, and leading to crop damage.

The impacts of resilient IPM tactics are not necessarily limited to a single target. Host plant resistance traits may adversely affect multiple organisms, and in some cases traits that confer resistance to one organism may or may not benefit others. Bio-suppressive traits of soils and biological control agents often suppress, and in some cases benefit, organisms other than the targeted species. The mechanistic nature of developing IPM tactics lends itself well to the reductionist approach to problem-solving. The traditional list of IPM tactics (biological, genetic, cultural, regulatory, chemical) is frequently interpreted as categories of tools to be developed to address specific pest problems. It is highly likely that a more useful approach for the future will be based on a functional ecology approach. This ecological view is essential for the development of resilient IPM practices for integration into future enterprise systems.

The foundation of the ecological view of IPM is based on understanding the suite of pest management and affiliated ecological services available within the system, in addition to using pest management inputs and maintenance modifications. Development of resilient IPM systems requires recognition of the tendencies of a system to suppress or tolerate multiple pests, while introducing highly compatible tactics to address particularly disruptive species. Whenever possible, the IPM practices should be designed for self-maintenance in relation to existing ecosystem structures, processes, and patterns. Although land grant institutions are highly appropriate for meeting this mandate, new innovations in relation to human, fiscal, and institutional resources will likely be required.

The current understanding of the ecology of most systems is limited to selected interactions of only a few components. Though information about pest-plant and pest-biological control agent interactions and the impacts of abiotic factors on pest occurrence and abundance are very important, they fail to provide a comprehensive understanding of many important aspects of the functional ecology required for future IPM systems. Many components of systems do not function independently, and the only way that emergent properties can be studied is by understanding their multiple and complex interactions and through in-depth studies of model systems. It is highly likely that this will require use of the procedures of systems science (Bird et al., 1990). Systems research, education, and Extension projects have been supported by the USDA SARE program, National Research Initiative (NRI), and Agricultural Research Service (ARS). Bird et al. (1990) described in detail a number of such system studies and their relationships to both the evolution of IPM and the unexpected consequences of the synthetic pesticide era.

Studies of model systems need to be complemented by comprehensive biological inventories. IPM should take advantage of the new NEON (National Ecological Observatory Network) initiative sponsored by the National Science Foundation (NSF). This is likely to require integrated diagnostic laboratories that provide ecosystem food web analysis services. A new emphasis on both molecular and classical taxonomy will also be essential. The biological inventories will have to contain a functional ecological component to be of optimal value to future IPM initiatives. This will greatly enhance the predictive aspects of IPM and allow for improved strategic implementation of IPM practices designed to address particularly disruptive organisms, while maintaining the overall stability and resiliency of the system.

IPM TACTICS FOR RESILIENCY

IPM tactics should be multifunctional and regenerative. If they include the use of pesticides, narrow-spectrum or reduced-risk compounds should be emphasized and used in a limited strategic fashion that is designed to minimize ecosystem disturbance. Other IPM tactics (biological, genetic, cultural, regulatory) frequently tend to be more relevant to system stability and resiliency, compared to the use of individual chemical compounds.

Habitat modification is the foundation of conservation biological and cultural control. Information gained from studies of model systems will be helpful in the identification of habitat modification techniques designed to conserve biological control agents. For example, the structure of a crop production system such as plant spacing or the planting sequence and timing of planting accessory plants such as cover crops, field borders, and succession crops will usually play a role in the functioning of biological control agents. Such habitat manipulation may result in increased shelter and mating sites, nutrient sources (e.g., nectar and pollen sources), and alternative prey species that are essential to maintaining populations of beneficial organisms at levels necessary to maintain or suppress pests to a density below the associated damage threshold. A full cataloguing of these resources is difficult and rarely undertaken. It is imperative that comprehensive functional ecology information be available for a reasonable number of model systems.

Constraints to the model system approach must be recognized. For example, biological control agents can maintain themselves on alternate prey associated with accessory plants such as cover crops and those in field borders. When population densities of the pests and other prey are limited, the biological control agents may be forced to migrate to other locations. In some cases, alternate prey can be disruptive to the overall quality of the ecosystem of interest. Recognition of species that are rare, have special community function, or are linked to other sensitive environmental factors with special production system value needs to be integral to decision making in the biological inventory system.

Habitat manipulations through modification of standard practices such as cultural control are often well suited for IPM. An example is management of western corn rootworm through a two-year rotation of corn with soybean. The corn-soybean rotation suppresses western corn rootworm because progeny of western corn rootworm eggs laid into corn residues in the fall cannot develop in the next growing season if the field is rotated to a nonhost of western corn rootworm such as soybean. This tactic is multifunctional. It contributes to farm income by increasing crop yields on an average of 10 percent, reduces pest population density, and has the potential to enhance soil quality. It suffers from one tactical constraint associated with IPM development because it relies on the single attribute of egg-laying behavior for pest suppression. Recent behavioral shifts in the western corn rootworm have resulted in at least a portion of the population becoming capable of laying eggs in numerous crop residues, including soybeans. This results in larval hatch during the corn production year. The change in reproductive behavior is analogous to both pesticide and host plant resistance. Under these circumstances, additional management practices are required for the IPM system to be successful. The erosion of western corn rootworm suppression with a corn-soybean rotation can be forestalled or possibly prevented through use of longer rotations, such as corn-soybean-wheat, in addition to alternative suppression tactics that target other aspects of rootworm biology. Even then, it is possible that the pest of interest or an alternative one will adapt to the alternative ecosystem, and additional pest management changes will be necessary for the enterprise to thrive.

MULTIPLE BENEFITS OF IPM TACTICS

IPM tactics can have emergent system values beyond their original intent. A well-developed European system of maintaining strips of noncropped raised banks of vegetation in cereal fields promotes natural enemies that control cereal aphids (Wratten and Thomas, 1990). This system has benefited from a thorough study of natural enemy-prey-habitat interactions, which has led to recommendations on the frequency and width of non-cropped vegetation strips in cereal fields and the plant species composition of the strips. Adaptation of this European system appears to have occurred coincidentally in the North American Great Plains (Brewer and Elliott, 2004). Maintaining noncropped vegetation strips in large wheat fields is not economical because of low crop economic value. But alternating spring-sown, spatially rotated crops such as sunflower strips with wheat strips provides natural enemies with nectar and alternate prey sources when wheat is out of production. The addition of a spring-sown crop to the traditional wheat-fallow strip cropping system for the western Great Plains was designed for soil and moisture conservation and stability of farm income. Pest management research in this same system found that one of the spring-sown crops adopted in the rotation also enhances wheat pest management. This practice has gained acceptance where the multiple benefits of maximized soil moisture use, farm income stability, and pest management are realized. In contrast, habitat management to conserve and enhance biological control organisms has had lower rates of success when the application is only for pest management purposes.

Habitat management techniques to enhance wildlife populations in and around agricultural lands should be multifunctional in relation to pest management. For example, hedge-row reestablishment surrounding agricultural lands in England has increased bird populations and has contributed to increasing tourism to the countryside. Reestablishment of hedges around small agricultural fields also has been observed to improve pest management, thereby providing multifunctional benefits to agriculture, wildlife, and other community-valued activities such as tourism. In the United States, conversion of idled wheat fields in the Great Plains to grasslands has increased both bird and mammal populations. These grasslands can be successfully commingled with wheat fields. There is some concern that grass species approved for planting may harbor wheat pests and negate the positive effects of the more diverse system. Unfortunately, the original U.S. government-sponsored evaluation of grass species and cultivars planted in these managed grasslands was largely devoid of data on pest harborage. Subsequent work indicates that there is significant species and cultivar variation in pest harborage attributed to these grasses. Some studies have implicated the grassland conversion program in the rapid establishment of wheat disease incidence and increase in cereal aphid pest populations. Others have demonstrated that harborage of natural enemies in the grasslands is important and may offset the negative aspects of pest harborage. The debate about this system continues. This is one example where implementation of large government-sponsored resource conservation programs must be ecologically based and take into consideration both site-specific and regional attributes of the overall system.

ECONOMICS OF IPM

There are usually conversion costs associated with IPM. These may not immediately be offset by longer-term benefits in the production system and the environment. If long-term economic and other values can be established or some outside support provided to compensate farmers, even occasional continuing maintenance costs can be absorbed while farm managers maintain the competitiveness of the system. Unfortunately, the success of most IPM programs has been unpredictable and varied, particularly in large-scale systems where investors often focus on short-term economic returns.

When multifunctional IPM tactics are built into the system, many of the initial implementation costs can be publicly supported or amortized through time. This type of IPM has broad appeal. For example, the wheat production strip crop system in the North American Great Plains modified by incorporating a spring-sown crop into the system has gained acceptance because of the benefits of increased soil moisture availability and farm income stability. Although the initial costs of incorporating another crop into the system may have slowed adoption, many progressive farmers have made the conversion. Those planting sunflower and other crops also realize pest management benefits, including resilient cereal aphid management and reduced weed competition. In contrast, similar concepts in habitat management that are designed solely for pest management have had more limited adoption.

Whenever possible, early assessment of short- and long-term economic costs and benefits should include an analysis of the environmental and societal benefits of IPM systems. Frequent lack of societal investment in short-term economics is a significant constraint to optimal IPM systems. The long-term value of sustainable pest management deserves recognition as a societal benefit, and short-term costs in adopting such techniques deserve societal investment. Incentives to convert to IPM result in a good public investment that complements land grant research, Extension, and academic instruction initiatives. Such approaches in the United States include voluntary farmer adoption of sustainable agriculture practices and the conservation programs of the USDA Natural Resource Conservation Service (NRCS). Currently, NRCS programs provide only minor investments in IPM tactic implementation, even though program authorization language is well suited for the use of IPM practices. It is imperative that NRCS, the

Land Grant System, and local Conservation Districts work jointly to resolve this shortfall.

Agricultural production systems should be designed in ways that minimize ecosystem disturbance, and may function in ways that result in pest population densities that are maintained below their damage-injury-pathogenicity thresholds, except under infrequent and highly unusual circumstances. Innovative IPM initiatives should be founded on resilient and sustainable aspects of the system. While IPM must address particularly disruptive pests, it should always be multifunctional, regenerative, and create as little ecosystem disturbance as possible.

ECOLITERACY

Western civilization has thrived on a mechanistic and anthropocentric worldview based on the assumption that all components of the environment (air, water, soil, minerals, and all microbial, plant, and animal species) are resources to be used by humankind. This worldview assumes that natural resources are infinite, or that replacement technologies will be continually available, and that costs in terms of habitat and species loss are externalities in any rational economic evaluation. The concept is supported by government policy, multinational corporate development, economic theory, and the practice of science. The mechanistic worldview is based on linear relationships and the assumption that the whole represents the sum of the parts. It has relatively few direct feedback loops, and only a small number of system components with overlapping functions. This worldview does not mandate existence within a vibrant community of local ecological interdependence and partnerships (Capra, 1996). The technologies of the mechanistic worldview have resulted in major advances in food production, quality of life for many people, and a rapid increase in human population growth. There have also been a significant number of emergent properties, or unexpected consequences. Some of these have potential for major long-term detrimental impacts on society and our planet's biosphere. Existing evidence strongly suggests that this worldview is not sustainable for an era of 6.4 billion people, and even less appropriate for a population increasing to 10 billion during the twenty-first century. Alternative worldviews need to be evaluated in a serious and comprehensive manner.

One alternative worldview is an ecological one. Every fall semester, senior author George Bird teaches a course titled *Pests, Society, and the Environment* to 120 Michigan State University undergraduates. For a vast majority of the students, this is not only their initial introduction to the concept of alternative worldviews, but it is the first time they have ever been challenged to engage in a meaningful dialogue about the nature of food and farming systems.

An ecological worldview emphasizes cyclic systems based on the assumption that resources are finite and that the whole is greater than the sum of its parts. Under this model, systems have emergent properties that cannot be predicted by their components. They often require local ecological interdependence and partnerships, cyclic patterns of organization, system components with overlapping functions and multiple feedback loops, and existence within a vibrant community (Capra, 1996). This concept is based on self-organizing, interdependent, and interconnected networks of living organisms that are autopoietic (self-replicating), dissipative (requiring energy inputs and providing residual outputs), and cognitive (responsive to their environment). Since most individuals have familiarity with at least a few pests, case studies of pests and IPM can be used as tools for teaching the fundamentals of ecology, thereby enhancing ecoliteracy in a practical manner.

The mechanistic and ecological worldviews are fundamentally different. The synthetic pesticide era was based on the mechanistic worldview; in contrast, the origins of IPM were based on the fundamentals of ecological cycles and multiple feedback loops. For the most part, IPM was implemented into the existing mechanistic systems of the linearity associated with what is known as conventional agriculture (Bird and Ikerd, 1993). Conventional agriculture is a highly productive, industrial system of food, feed, and fiber production that relies heavily on external inputs. During the last half of the twentieth century, conventional agriculture produced large quantities of inexpensive commodities that are perceived by most as being of high quality, safe, and readily available throughout high-income nations, at least for those who have financial resources. Until recently, there has been limited serious critique of the long-term ecological sustainability or the social consequences of this dominant system. In addition, relatively few formal research, development, and education initiatives were related to management systems based on the ecological worldview (Bird, 1995; Lipson, 1997). The USDA SARE program was the first major exception to this on a national basis.

Sustainable development, a process for maintaining a system at a fuller or better state, is based on the ecological worldview (Meadows et al., 1991). It is possible that at some point in the future, society could shift from the industrial era to an era of sustainable development (Table 2.1). The practices, systems, and philosophies of IPM should be extremely useful in the development of the ecoliteracy essential for a successful journey towards an era of sustainable development.

TABLE 2.1. Major societal events and a potential future transformation.

Era	Time Frame
Tool revolution	2,400,000 years ago
Agricultural revolution	10,000 years ago
Industrial Revolution	250 years ago
Chemotechnology era	50 years ago
Electronic era	15 years ago
Biotechnology era	10 years ago
Sustainable-equitable development revolution	?
Industrial growth era	1750-xxxx
Sustainable-equitable development era	xxxx-xxxx

Although the issues of sustainable development and its challenges will be immense and critical in an era of rapidly expanding human population on a planet with limited resources, there are many similarities with the issues encountered in pest management during the last half of the twentieth century. In 1971, Barry Commoner indicated that "[W]e have become not less dependent on the balance of nature, but more dependent on it. Modern technology has so stressed the web of processes in the living environment at its most vulnerable points that there is little leeway left in the system" (Commoner, 1971). Commoner also described a fundamental aspect of ecoliteracy when he reminded us of the Law of Conservation of Matter by indicating that "everything has to go somewhere" and "everything is connected to everything else." In 1962, Rachel Carson illustrated the immense institutional impacts of a dominant worldview when she stated that "[t]he entomologist, whose specialty is insects, is not so qualified by training, and is not psychologically disposed to look for undesirable side effects of his control program" (Carson, 1962). Indeed, this interconnectiveness is the crux of the challenge. Capra's 1996 treatise entitled *The Web of Life: A New Scientific Understanding of Living Systems* is a truly interdisciplinary-holistic approach to an overall enhancement of ecoliteracy.

IPM is at once an interdisciplinary philosophy, process, and set of practices. It cannot be stronger than its weakest disciplinary component. The same is true for the emerging science of sustainable-equitable development. Both IPM and sustainable-equitable development must be based on the ecological worldview. People applying IPM have almost four decades of expe-

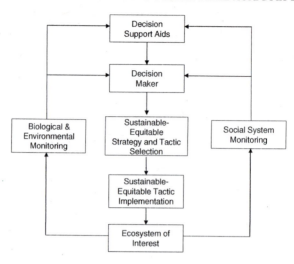

FIGURE 2.3. Conceptual model of the process of achieving a system of sustainable and equitable development (after Bird et al.,1989).

rience in dealing with the biological, economic, and other social aspects of pest management. Sustainable-equitable development has a significantly shorter but much broader formal history. The procedures and lessons of IPM can provide a sound foundation for the ecoliteracy mandated by a journey toward a future that is both sustainable and able to provide a high quality of life for current and future generations. The conceptual model of the process of IPM can be convertible into one for sustainable-equitable development (Figure 2.3).

LAW OF UNEXPECTED CONSEQUENCES

When a process (stimulus) interacts with a structure (system component), there will be a consequence or change in the system, as shown in Figures 2.4. Two important questions relate to this simple model. Is the result part of a predictable pattern? Is the result an unexpected consequence? The identification of predictable patterns resulting from ecosystem change forms the foundation of the IPM process. Without a comprehensive knowledge of the system, it is not possible to accurately predict system responses. The knowledge required for this type of prediction mandates a systems approach.

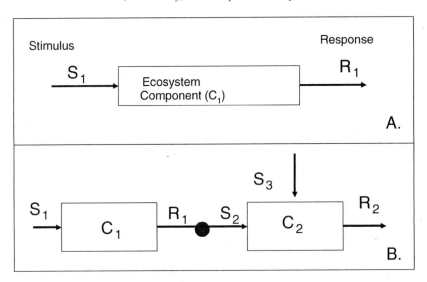

FIGURE 2.4. A. Illustration of the impact (R1) of a process (S1) on an ecosystem component (C1). B. Illustration of the joint impact (R2) of two processes (S2, S3) and a second ecosystem component (C2).

The initial development of IPM at Michigan State University was based on the principles of systems science and legendary interactions among engineers, biologists, economists, and resource development specialists (Bird, 2003; Olsen et al., 2003). This process was an outgrowth of the Macy Conferences of the 1940s that were designed to identify peace-time roles for the procedures of systems science, and the concept for the search for patterns of sustainability (Bird et al., 1990). Capra (1996) stresses that ecosystem patterns are the results of interactions between structure and process. The identification of ecosystem patterns is fundamental to IPM. Since most research during the twentieth century was based on the components of systems, there is a distinct lack of understanding of patterns associated with the emergent properties of systems. In the course *Pests, Society, and the Environment,* we use the classic example of NaCl to illustrate that there is currently no known way to predict the properties of table salt from a comprehensive understanding of each of its parts. It is only at the system level that the taste of NaCl, an emergent property, can be understood. This type of systems understanding is imperative for a reduction in the number of unexpected consequences associated with the design of future managed agroecosystems. The principles and procedures of systems science are both useful and highly appropriate for

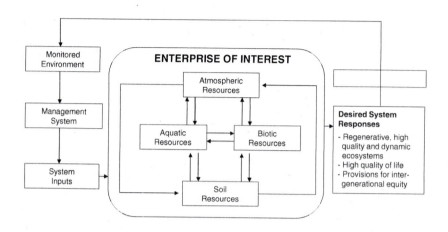

FIGURE 2.5. Conceptual model of a human-managed system.

evaluation of proposed future systems of IPM and sustainable- equitable development (Figure 2.5). The alternative appears to be an endless sequence of unexpected responses to our management decisions.

There are also many unexpected events associated with our institutional endeavors. The career of the senior author began in applied nematology, with academic training in plant pathology and entomology, a complementary set of perspectives and expertise that rapidly led to the area of IPM and then to sustainable agriculture. More recently, the author's focus has changed to the role of nematodes in soil quality and nutrient mineralization (Figure 2.6) and the overall process of organisms in the transport and transformation of matter and energy throughout soil ecosystems (Figure 2.7). This would not have been possible without the flexibility of the land grant system and its vast opportunities for interactions among research, Extension, and academic instruction. It is imperative that this highly unique institution be properly maintained as it evolves into a new leadership role for the twenty-first century.

Our ability to assign a high priority to understanding and mitigating humankind's long-term impacts on the biosphere will dictate the future success of human societies. Such an understanding requires a significantly longer temporal dimension that the short-term time frames that economists and politicians, and most professionals in agriculture, frequently use in their analyses. Geologists and anthropologists are accustomed to using decades, centuries,

Role of Nematodes in Nutrient Mineralization

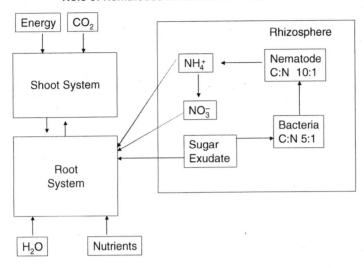

FIGURE 2.6. Role of nematodes in nutrient transport and transformation.

or even millennia as a temporal frame of reference. As a reference point, a time frame of at least a decade is necessary for understanding and preventing unexpected consequences associated with our management decisions. As indicated in a National Research Council report (NRC, 1996), it is necessary to conceptualize this process as a never-ending journey of moving toward a sustainable-equitable system, rather than attaining some narrow conceptual level of sustainability based on a few indicators.

Franklin D. Roosevelt's second inaugural address indicates that "[t]he test of our progress is not whether we add more to the abundance of those who have much; it is whether we provide enough for those who have too little." This illustrates both the need for identification of system boundaries, and the management practices designed to reduce risks associated with unexpected consequences of management. It also emphasizes the social realities that play a major part in determining for whom sustainability and equity are sought and for how long. In a survival mode at the edge of starvation, individuals are not likely to concern themselves with or take actions related to a high quality of life for the next generation. Preparing for the next day or next week is in itself a major task. Individuals in low-income nations and other stressed environments need and deserve the attention and support of high-income nations in order to move toward a sustainable-equitable future. Most high-income nations, however, currently operate with the philosophy

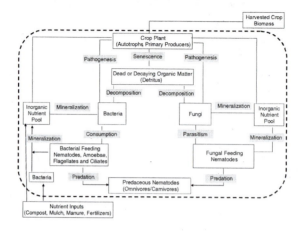

FIGURE 2.7. Role of soil fauna in nutrient cycling associated with crop production (concepts from R.E. Ingham et al., 1985).

and practices of the dominant mechanistic worldview. The system is designed on the principle of profit maximization. Since it is not possible to simultaneously maximize two dependent variables, this approach results in unexpected consequences that have negative impacts on other components of the system. Theoretically, IPM is based on the principle of optimization of all components of the system. In practice, this is rarely achieved because of a continued lack of understanding of the interactions among the structures, processes, and patterns of the system.

Sustainable-equitable development transcends agriculture. It is useless to argue for the sustainability of anything outside of the context of the total system within which it exists. This is one of the reasons that it is imperative for the science of sustainable-equitable development to rediscover the fundamentals of systems science as outlined at the Macy Conferences and implemented in various IPM programs. Although IPM is a small but very important subsystem, it must be compatible with the larger socioeconomic system.

The determination and will to develop and nurture sustainable-equitable development mandates both a goal for the present and a vision for the future. The future of IPM is highly objective-dependent and presents several significant challenges. Without major new developments in the prevailing attitude of technological dominance of nature, both agriculture and environment pest management will have to rely on increasingly knowledge-dependent systems. Such systems must be designed to use multiple strategies and tactics in ways that optimize the long-term sustainability of agroecosystems. Both sustainable-equitable development and IPM can be viewed as processes and described in terms of their temporal stability, inter- and intra-group equity, and ecological impacts across a multitude of time horizons. The problem quickly encountered is how, from a scientific view, one determines whether or not a practice or process will have unexpected negative consequences. What indicators need to be tested? It is highly probable that isolated components will never be suitable for use as indicators of system response. It is more likely that it will be possible to identify patterns of potential risk and patterns of potential success. These patterns result from the interactions among structures and processes. An approach recommended within IPM is to focus on process and evaluate elements of relevant processes in ways that assess their contributions to impacts on the nature and duration of subsystem cycles. For an ecological worldview to become dominant, global society must become ecologically literate. This should significantly decrease risks associated with the unexpected consequences of management. Both the lessons of IPM and its future have the potential to play key roles in this important global transformation.

REFERENCES

Altieri, M.A. 1994. *Biodiversity and pest management in agroecosystems.* Food Products Press, Binghamton, New York.

Barbosa, P. 1998. *Conservation biological control.* Academic Press, San Diego, CA.

Bellows, T. and T. Fisher. 1999. *Handbook of biological control.* Academic Press, San Diego, CA.

Benbrook, C. 1996. *Pest management at the crossroads.* Consumers Union, New York.

Bird, G. 1995. Sustainable agriculture: A case study of research relevancy classification. Proceedings of the 1995 Annual Meeting of the American Association Advancement of Science, Atlanta, GA.

————. 2003. Role of integrated pest management and sustainable development. In: *Integrated pest management in the global arena,* K. Maredia, M. Dakouo, and D. Mota-Sanchez (eds.). CABI Publishing, Cambridge, MA. pp. 73-85.

Bird, G. and M. Berney. 2000. Pest ecology and management. In: *Michigan field crop ecology,* M. Cavigelli, S. Deming, L. Probyn, and R. Harwood (eds.). Michigan State University Extension Bull. E-2624. East Lansing, MI.

Bird, G., T. Edens, F. Drummond, and E. Groden. 1990. Design of pest management systems for sustainable agriculture. In: *Sustainable agriculture in temperate zones*, C.A. Francis, C. Flora, and L. King (eds.). John Wiley and Sons, New York, pp. 55-110.

Bird, G. and J. Ikerd. 1993. Sustainable agriculture: A twenty-first century system. *Ann. Am. Acad. Pol. Soc. Sci.* 529:92-102.

Brewer, M.J. and N.C. Elliott. 2004. Biological control of cereal aphids and mediating effects of host plant and habitat manipulations. *Ann. Rev. Entomol.* 49:219-242.

Capra, F. 1996. *The web of life: A new scientific understanding of living systems.* Doubleday, New York.

Carson, R. 1962. *Silent spring.* Ballantine Books, Random House, New York.

Cavigelli, M., S. Deming, L. Probyn, and D. Mutch. 2000. *Michigan field crop pest ecology and management.* Michigan State University Ext. Bull. E-2704, East Lansing, MI.

Commoner, B. 1971. *The closing circle.* Alfred Knopf, New York.

DeBach, P. and D. Rosen. 1991. *Biological control by natural enemies* (2nd ed.). Cambridge University Press, New York.

Flint, M. and P. Gouveia. 2001. *IPM in practice: Principles and methods of integrated pest management.* University of California Extension Publ. 3418. University of California, Davis, CA.

Hardy, R.W.F. 1996. *Ecologically based pest management: New solutions for a new century.* National Research Council, National Academies Press, Washington DC.

Haynes, D., R. Tummala, and T. Ellis. 1980. Ecosystem management for pest control. *BioScience* 30:690-696.

Ingham, R., J. Trofymow, E. Ingham, and D. Coleman. 1985. Interactions of bacteria, fungi and their nematode grazers: Effects on nutrient cycling and plant growth. *Ecol. Monogr.* 55:119-140.

Lipson, M. 1997. *Searching for the O-word: Analyzing the USDA current research information system for pertinence to organic farming.* Organic Farming Research Foundation. Santa Cruz, CA.

Maredia, K., D. Dakouo, and D. Mota-Sanchez. 2003. *Integrated pest management in the global arena.* CABI, Cambridge, MA.

Meadows, D., D. Meadows, and J. Randers. 1991. *Beyond the limits.* Chelsea Green Publ. Co., Post Mills, VT.

Norris, R. F., E. Caswell-Chen, and M. Kogan. 2003. *Concepts in integrated pest management.* Prentice Hall, New Jersey.

NRC. 1996. *Ecologically based pest management: New solutions for a new century.* National Research Council, Board on Agriculture, National Academy Press. Washington DC.

Olsen, L., F. Zalom, and P. Adkisson. 2003. Integrated pest management in the U.S.A. In: *Integrated pest management in the global arena,* K. Maredia, D. Dakouo, and D. Mota-Sanchez (eds.). CABI Publishing, Cambridge, MA. pp. 249-272.

Potter, V., A. Baerreis, R. Bryson, J. Curvin, G. Johansen, J. McLeod, J. Rankin, and K. Symon. 1970. Purpose and function of the university. *Science* 167:1590-1593.

Rodale, R. 1984. *Our next frontier.* Rodale Press, Emmaus, PA.

UNCED. 1987. *World commission on environment and development: Our common future.* Oxford Press, New York.

Wratten S.D. and C.F.G. Thomas. 1990. Farm-scale spatial dynamics of predators and parasitoids in agricultural landscapes. In: *Species dispersal in agricultural habitats,* R. Bunce (ed.). Bellhaven Press, London. pp. 219-237.

Chapter 3

Soil Management in Sustainable Production Systems

Kent McVay

SOIL DEVELOPMENT IN MODERN HISTORY

Soils have developed over millions of years from the chemical, physical, and biological weathering of rocks. Through this process a rich medium for growing plants has evolved on the earth's surface, the biologically active and essential crust and interface of the atmosphere and land. Soil management describes how this soil medium is manipulated, rearranged, supplemented, and prepared so that we can purposefully place seeds and grow crops for food. We observe with interest that the natural system, which has developed over millennia, works fine without manipulation by farmers. In most natural settings, a certain balance has evolved that provides a great diversity of species and a certain buffering of resources that provides stability to the biological system.

Agriculture is not really a natural system. In our modern world of agricultural production, a monoculture of plants is grown that facilitates mechanical management and reduces the need for physical labor. Most of the grain crops grown in the Midwest are annual species planted and harvested over a period of four or five months for summer crops of corn, soybean, or grain sorghum, and around nine months for winter crops of wheat or canola. Until the advent of no-till systems beginning in the 1970s and 1980s, multiple tillage passes were conducted annually in nearly every field (Owens, 2001).

No-tillage, also referred to as direct seeding, is a soil management technique where farmers plant the seeds directly through previous crop residue. As with any new technology, there is a learning curve involved in making

Developing and Extending Sustainable Agriculture
© 2006 by The Haworth Press, Inc. All rights reserved.
doi:10.1300/5709_03

direct seeding work. But the widespread success achieved by many farmers and researchers across the Midwest from the 1970s to the present lead us to question why tillage is still deemed necessary by many in this region.

REASONS AND GOALS

The classic reasons for soil management include seedbed preparation, weed control, fertilization, improving soil structure, and erosion control. Let us look at each of these reasons a little closer.

Seedbed preparation sounds like a good idea. Nearly every farm field and everyone's backyard garden goes through a period of seedbed preparation. This could involve soil inversion followed by multiple passes with a disc or roto-tiller to break down clods and residue to produce a fine pulverized soil surface. On the other hand, a minimalist might use a hoe or rod on a small garden-sized scale to make a small opening just large enough to accept a seed. Is one method better than another? There are many interactions that complicate the answer. Soil moisture is affected by tillage, and any pass through the field can result in drying the tilled area. Soil temperature can be influenced by tillage by changing surface conditions such as residue cover, and amount of soil exposed to direct heating by sunlight. So emergence of a seedling under these two extremes of seedbed preparation is never just a function of the amount of soil disturbance, but in reality a result of a combination of many factors.

A better question would be, "Is it necessary to loosen the soil to assist the seedling to emerge?" The answer of course depends on the condition of the soil at the time of planting. If there is a compacted layer of soil directly above the seed, then emergence can be hampered, but that condition is not likely, since in both cases the soil directly above the seed was disturbed in order to plant the seed. Seedling emergence is not immediate, so crusting can happen to the soil surface after planting that could make seedling emergence difficult. The bottom line is, if the soil is in good physical condition with adequate organic matter, and the soil moisture levels are moderate, then crops can be planted successfully without the need to pulverize the soil to make a seedbed. Pulverization was primarily needed because of machinery limitations. Today's modern planting equipment can easily plant through high levels of residue into untilled soil, in a method quite similar to cutting the soil surface with a hoe and placing the seed into the soil slot.

Is not tillage necessary to control weeds? It is true that weeds can be effectively controlled by tillage. Tillage does eliminate the weeds currently growing. But think about where weed seeds are located. Unless the weed seed is a contaminant in our crop seed and gets planted when we plant our crop, the majority of weed seeds will remain at the soil surface. So each tillage pass kills growing weeds, but also plants a new set of weed seeds, and each operation results in a new flush of weeds about a week or so following the tillage. Weed seed is often hard-seeded and germination percentage is typically low, which enhances long-term species survival. When weed seeds are buried with a deep tillage pass, that same tillage pass is bringing up previously buried weed seeds that now may be ready to germinate following a necessary period of dormancy. Tillage for weed control is sort of a self-defeating exercise, although with enough passes, weed populations begin to subside. A moderate use of tillage for weed control can be justified if it is part of a larger weed management plan.

Most herbicides in the market today are quite environmental friendly. These chemicals are designed so that they have short half-lives, with a minimal off-target impact. Weed growth is best suppressed by a vigorously growing crop. Good crop management provides a dense population of plants and quick ground cover that will usually out-compete weeds. A good chemical weed control program can help bridge the fallow time between harvest of one crop and complete canopy closure of the next.

Good soil fertility is one of the keys to successful crop production. Crops require nutrients, and continued removal of grain results in a net loss of nutrients from the field. Producing crops without replacing these nutrients eventually leads to reduced levels of production as availability of one or more nutrients becomes limited. What role does tillage play in nutrient management? Soil fertility evaluation for nonalkaline soils in the United States has historically been based on soil sampling of the plow layer, a six- to eight-inch depth of topsoil. This zone has been considered homogenous due to mixing by tillage. As no-till becomes more prominent in management, nutrient stratification in soil begins to occur. This is really a function of the lack of complete mixing in that plow layer of soil. Over the past 30 years as tillage trends in the United States have changed, researchers have found that this plow layer has become stratified, with immobile nutrients such as phosphorus (P) being greatly affected when surface applied. Soil pH also stratifies in the upper layers which can affect herbicide effectiveness.

Is stratification a problem? Looking to the natural system, we see stratification in soil. Forest soils get annual inputs of leaf drop and input at the surface which develops an organic layer above the mineral layers of soil. The horizonation of mineral soils typically shows dark A horizons near the surface, underlain by B horizons that accumulate clay and have less organic matter content. So the natural world is stratified as well. Management can be used to minimize stratification. Fertilizer placement below the surface, below the residue, will help to prevent extreme values of nutrient stratification. Low soil pH values near the surface can be ameliorated with low application rates of lime. But homogenization of the surface soil is not necessary from a soil fertility point of view, nor is it preferred.

Crop yield responses are fairly well predicted for P and potassium (K) applications based on existing soil tests. Nitrogen (N) fertilizer, on the other hand is typically recommended, based on yield goal. Excessive N fertilization can result in groundwater contamination of nitrate (NO_3), yet under-fertilization of crops with N can result in severe yield limitations. Most research over the last few decades on N management has focused on soil NO_3 levels, either preplant or at side-dress time. Although this has improved management in some cases, the available soil NO_3 tested in early spring represents only a fraction of the total N that would be available and is necessary for sufficient crop yield. Mineralization of soil organic N complicates fertilizer N recommendations. Recent research in Illinois (Mulvaney et al., 2001) has identified a fraction of soil organic N that correlates well with yield response of corn. Amino sugar-N is less prone to temporal and spatial variability than NO_3, and therefore could become a more ideal soil N test.

Tillage is sometimes recommended to improve soil structure. First, soil structure must be defined. Soils consist of mineral and organic materials, and are home to microbial and terrestrial organisms. These constituents in total make up about 50 percent of the volume of a given soil. The other 50 percent, a critical part, consists of the soil pores, or void space within the soil where air and water move. Soil structure is basically a qualitative term describing the arrangement of all these particles and voids. A soil with good structure or tilth has an array of pore sizes from very small to quite large. For topsoil, good soil structure results in stable aggregates that are friable at intermediate moisture levels and are crumbly when held in the palm of your hand. Subsoils are typically higher in clay content and therefore good structure in these horizons tends to result in blocklike aggregates with irregular six-faced blocks. In this case, good structure is evident by the network of pores that flow over and around the faces of these aggregates. Soils that have poor structure can be naturally massive or become so following tillage, or under any induced pressure, such as wheel traffic or intense rainfall. Sandy soils tend to have less desirable structure as they inherently have lower clay content and typically have lower organic matter content, each of which helps to build soil structure.

A good way to quantify soil structure is by measuring water infiltration rates. Soils of good structure will maintain high infiltration rates indefinitely. Soils of poor structure tend to have poorer and less stable infiltration rates. During a precipitation event, as soil aggregates break down under the impact of raindrops, finer soil particles move with the water into soil pores, blocking them over time and reducing the rate of water transmittance.

How does tillage affect infiltration and soil structure? Immediately following a tillage event, soil structure can appear quite good, and infiltration rates are usually high. But these conditions quickly deteriorate under continued impact with rainfall, and sealing at the soil surface will begin to occur, reducing infiltration rates. Tillage mechanically breaks up soil aggregates, and destroys the connectivity of soil pores. For large pores to transport water to deep layers in the soil, these pores must remain both open and continuous to the soil surface. Tillage also reorganizes the soil constituents, making food sources more available to the microbial community. This mixing results in a flush of microbial activity which helps to rebuild soil aggregates, but similar to the example of weed control above, breaking up aggregates to create new ones is self-defeating. In the long-run, the continued use of tillage results in fewer total aggregates, and poorer overall soil structure.

Tillage can be a benefit to soils that have been previously damaged. Compaction occurs when a force large enough to cause a physical rearrangement of mineral particles is applied to the soil. This can be caused by traffic or by the tillage operation itself. Driving on soils and tilling them when they are too wet are the main causes of soil compaction. Even a disc that appears to break up clods and contribute to a better seedbed is really creating a dense layer right at the lower edge of the disk wheel. This physical damage to the soil can remain for many years, and when severe, can only be corrected by a proper use of tillage to break-up or shatter a compacted soil layer.

Tillage has been successfully used for erosion control. A rough cloddy surface has a fairly high storage volume, which can capture and hold precipitation in place, allowing enough opportunity time for water to infiltrate after the storm has passed. This practice works well for the first few storms, but over time, with exposure to rainfall, even cloddy surfaces will begin to seal, reducing infiltration rates and leading to runoff. Use of limited tillage practices to maintain crop residues on or near the soil surface can contribute substantially to the capture and storage of water, as heavy rain drops are intercepted and their energy dissipated when they strike surface residues. Maintaining higher residue levels tends to stabilize infiltration at higher rates. Thus more water infiltrates in soils with residue at the surface than when soils are maintained bare. Water that does not infiltrate runs off the surface, moving both nutrients and soil into our drainage and river system.

USE OF COVER CROPS

Cover crops can be managed in a way that helps control weeds. A cover crop is defined as a crop that is planted but not harvested. The purpose of a cover crop is to protect the soil surface from erosion, to scavenge nutrients left in the profile by the previous crop, to enhance activity in the microbial community, and to suppress weed growth. Weed growth is best suppressed by competition, and a vigorously growing cover crop has a competitive advantage over weeds. One of the trade-offs in using cover crops is the balance of moisture use. In regions of higher rainfall, cover crops can be grown easily in the winter fallow period between successive summer crops. In more arid regions, there is a definite risk that limited moisture will be used by the cover crop resulting in droughty conditions for the cash crop that follows (USDA, 1992).

The decision to use cover crops must be made locally and can be specific to field, soil type, and topography. The factors that need to be weighed in making the decision include the amount of expected precipitation to be received during the time the cover crop is to be grown; the risk of erosion, nutrient loss, or weed control cost that will occur if a cover crop is not grown; and the time and cost of terminating the cover crop. This last point is critical, because moisture used by the cover crop is most important, in terms of a following grain crop, right before planting. This time usually coincides with the time of maximum water use by the cover crop. In wetter regions, the cover crop might be allowed to grow right up until planting of the grain crop, so that soil moisture levels are not too high at planting. In drier climes, a cover crop might be killed a month or two prior to planting the grain crop to allow enough time for subsequent rains to refill the soil profile and for biological breakdown of the newly introduced plant material. If this time is not allowed, the organic material from the cover crop may attract biological activity and tie up nitrogen and other nutrients that are needed by the newly seeded crop plants.

Termination of cover crops can be achieved in many ways. In the early part of the twentieth century, tillage was used to turn under green manure crops. Tillage will provide a good kill of any crop, but for the reasons outlined earlier, tillage should be avoided if possible. Herbicides can be used to effectively kill a cover crop, and in most no-till programs, a spring burn-down is already planned so using a herbicide at this time is not an unexpected expense. Planting spring oats as a cover crop in the fall is an example of using winter temperatures to kill the cover crop at no added expense. This technique works well in areas of the Great Plains where winters are severe enough to provide 100 percent kill, and if moisture savings is needed prior to spring planting, using a summer annual in the fall as a cover crop can be a good choice.

A termination method used in South America, and being investigated now in the United States is mechanical rolling (Soil Quality Institute, 2002). The implement is basically a drum with blunt knives or bars that is rolled over a standing crop to knock the crop down, crimping the plants in sections, and either killing or greatly suppressing their growth. The grain crop is seeded directly into this rolled cover crop either immediately or within a few days so that the grain crop emerges and is quickly established. A roller will effectively kill dicotyledonous cover crops such as sunflower or sun hemp, but is less effective on grass crops. It can be effective on annual grasses that have reached the point in their development where they have switched from vegetative growth habit to reproductive development. Some U.S. studies have shown that rolling wheat or oats after these plants have reached jointing stage or have produced heads can provide as effective a kill as using glyphosate herbicide (Ashford et al., 2000).

CROPPING SYSTEMS THAT MIMIC A NATURAL ENVIRONMENT

When no-tillage was first attempted, very little in terms of management changed other than the fact that the discs and plows were left unused. After a couple of years, success was quite limited, especially in those situations where crop rotations were not practiced. Continuous mono-cropping, no matter what the crop, increases the incidence of weeds that have similar growth habit to the crop. Where the same herbicides were used repeatedly, selection of herbicide resistant weeds was the result. The crop residue provides material for disease organisms to persist, and insects to live. The challenges and experiences of producers from those early years have led to better understanding of the system and the development of more complex crop management systems. Successful cropping systems are strategies and techniques that diversify the cropping sequence to reduce disease and insect pressure, improve residue management, and improve the odds of acceptable yields from the crop being grown.

Observing growth of plants in a natural environment can provide insight into ways to manage crops without tillage and to help design cropping systems. For example, in native grass systems which evolved in a fire ecosystem, quick growth following fire suppresses competition from weedy species. In no-till systems, the colder soils associated with high residue levels lead to slow emergence of the crop. One way of countering the effect of cold soils is to use starter fertilizer to help these young seedlings get off to a quicker start. Like the native species following fire, quick growth helps crops compete against invasion by weeds. Another is to use a minimum tillage strategy that tills and opens a very narrow band, and with the dark color of exposed soil in the band, that soil will warm up more than the surrounding soil covered by residue, allowing seeds planted in that band to emerge more quickly as seedlings.

A native environment is typically species diverse, but managing diversity within a crop production field is difficult. Some of the benefits gained from this biological diversity can be obtained in production agriculture by rotating crops. By sequencing crops so that a field is out of any particular species for at least a year, and better if this lapse is at least two years, the negative impacts of disease and insect pressure can be minimized. Nearly all crops show positive yield responses to rotation as compared to continuous cropping of the same species, with an average expected advantage of 10 percent for Midwest grain crops (Crookston et al., 1991; Porter et al., 1997). The only negative impact to crop rotation is economic, usually in the short term, and this occurs only when one crop dominates from a profit standpoint. This is one reason for the dominance of continuous corn in the western part of the corn belt. In this situation, crop rotation appears economically counterproductive for the producer in the short term. But the farmer needs to realize that the lack of rotation will eventually lead to higher input costs to chemically control pathogens, weeds, and insect pests that will begin to build.

MAINTAINING A LIVING SOIL

Evelyn Balfour (1975), in her book *The Living Soil,* clearly described the link between soil organic matter and productivity. She points out that humus (soil organic matter) is "simply, yet accurately defined as the product of living matter and the source of it." For without decomposition, and without input of organic material, healthy plant life is not possible. So how does soil management affect this living system? It probably affects the system in ways we are just beginning to understand.

Realizing that soils are alive and understanding how they function are good first steps. Tillage can be seen as a catastrophic event to certain soil organisms. Soil aggregates and biological communities are shattered by the physical act of tillage. Some microscopic organisms such as bacteria may benefit from such an event. This can be especially true when food sources are scarce. Tillage typically introduces new food sources by breaking open soil aggregates and turning under organic debris, providing new energy to the system. But the growth of some organisms, such as soil fungi, is depressed by tillage. These macroscopic organisms produce mycelia, a fine fibrous network that spreads through the soil like plant roots. Fungi play an important role in building soil structure. For example, a shift from fallow of bare soil to more intensive cropping systems tends to increase large size aggregates (those >2 mm in diameter) and has been linked to increased levels of fungal glucosamine, a sort of glue that holds soil particles together (Chantigny et al., 1997). Role and dynamics of soil organic matter have been exceptionally well reviewed by Magdoff and Weil (2004).

EARTHWORMS

It is estimated that at least 5,000 species of earthworms inhabit the earth (Hendrix, 1995). Of these only a few dozen have been studied. Even with many questions unanswered, it is obvious that some general conclusions can be drawn. Where earthworms are abundant, they can have substantial effects on soil structure, turnover of organic matter, and nutrient cycling, and can also influence water infiltration rates and solute transport.

Earthworm populations have been seen to negatively correlate with tillage intensity. In many conventionally tilled fields, it is sometimes difficult to find any earthworms, while in adjacent native sites, earthworms can be plentiful. As some of these fields are converted to no-tillage management, earthworms soon migrate back into the barren fields.

Earthworms can impact soil fertility. As a direct result of feeding and burrowing, mineralization of organic matter is accelerated, increasing the availability of nutrients for plant uptake. Depending upon population, the amount of soil egested by earthworms has been estimated to be as great as 44,000 lbs/acre per year (Lavelle, 1988). The rate of decomposition of organic matter depends on the feeding behavior of the species. *Lumbricus terrestris,* which is a species found in temperate regions, lives in permanent vertical burrows that can extend in the soil to a depth of six feet. This species tends to feed on surface residues, pulling material back to their burrows and forming "middens," or piles of residue near the mouths of their burrows. Another species, *Lumbricus rubellus* also feeds on surface residues. But this species tends to burrow horizontally within the upper horizons of the soil, and does not make a permanent home.

Other genera found in North America such as *Diplocardia, Aporrectodea,* and *Octolasion* species feed less on surface residues, and have diets that contain more soil than residue. The result is that in soils dominated by these kinds of earthworms, surface residue remains on the surface longer, while homogenization of soils within horizons is accelerated.

There is also evidence that organic matter egested by earthworms is stabilized. Respiration studies of earthworm casts as compared to reference soils showed lower respiration rates, indicating that earthworm casts may resist decomposition. The overall effect helps soils retain organic matter (Lavelle and Martin, 1992).

VESICULAR-ARBUSCULAR MYCORRHIZAE

One of the clear differences between agricultural soils and native, or nonmanaged soils is the frequent occurrence of fungal associations in the undisturbed soils. Mycorrhizal fungi are common and likely to be the primary mutualistic organisms associated with grasses (Newsham and Watkinson, 1998). In the remnants of the tall grass prairie, a survey of over 90 grasses and forbs found that all species examined were colonized by vesicular-arbuscular mycorrhizal fungi (VAM) (Wilson and Hartnett, 1998). The warm-season perennial grasses generally show the highest growth response to colonization and the highest dependency on mycorrhiza for P uptake. The cool-season grasses usually are less dependent, while the forb species show a wide range of colonization levels.

This symbiosis between fungi and plants may be one of the least understood, but most important of biotic interactions. The symbiotic trade consists of carbohydrates for the fungi in exchange for nutrients and possibly water for the growing plant. It has been estimated that >90 percent of all plant species belong to genera that characteristically form mycorrhizae (Smith and Read, 1997), making associations the rule rather than the exception.

The benefits of this association can be increased growth and yield of a crop, or improved ecological function. In either case, the benefit is primarily due to mycorrhizal fungi expanding the volume of soil available to the plant. The hyphae of the fungi, which behave like the roots of plants, take up mineral ions from the soil solution and transport them to the plant. In nutrient-poor soils, or in soils lacking moisture, nutrients delivered to the plant result in greater plant growth and increase the plant's ability to tolerate environmental stress. As a result, mycorrhizal plants are typically more hardy than those that are grown without fungal association.

The greatest growth response to mycorrhizal association has typically been seen in soils low in fertility. In a pot study in West Africa, using a sterilized sandy soil, P application led to an 18- to 24-fold increase in pearl millet root and shoot dry matter independent of VAM. For sorghum and cowpea, the presence of VAM increased total uptake of P, K, Ca, Mg, and Zn, by 2.5- to 6-fold (Bagayoko et al., 2000). The researchers concluded that this significant interaction showed that early root growth enhanced by P applications was a prerequisite for mycorrhizal infection which then allowed VAM to enhance overall plant growth and nutrient uptake. A regional study in the United States evaluated six VAM fungal isolates on soybean and sorghum (Sylvia et al., 1993). Under the experimental conditions tested, the presence of VAM was the most important factor in determining plant growth response. The presence of the fungi was more important than the type of soil, or even the kind of host plant evaluated. These researchers also saw the greatest response in soil that tested low in extractable P.

There is also evidence that cropping system can influence mycorrhizal infection in a subsequent grain crop. Mycorrhiza cannot develop in the absence of host plants, and so populations decline during fallow periods. A cover crop of hairy vetch helped sustain VAM fungi in a study conducted in Maryland (Galvez et al., 1995). Kabir and Koide (2000) studied corn following cover crops of wheat and dandelion. They showed greater incidence of mycorrhizal colonization of corn roots following dandelion with significantly higher P content and greater shoot growth as compared to corn following winter wheat.

Tillage suppresses the influence of VAM. Vivekanandan and Fixen (1991) showed evidence of less colonization of corn roots in moldboard plowed soils versus those in a ridge till system. And with lower levels of mycorrhiza colonization, a greater response to P additions was noted. Galvez et al. (2001) compared colonization levels across different levels of tillage intensity, and found that chisel- disk systems reduced spore populations throughout the season as compared to no-till.

SOIL MANAGEMENT AND SOIL QUALITY

Soil quality is difficult to quantify, but one certain measure is biological diversity. Management practices such as lower tillage intensity tend to increase the activity of earthworms and fungi. Even with a limited knowledge of the specific response of crop plants to the presence of VAM, our observations of their importance in the natural environment gives reason to encourage their presence in production agriculture. Finding ways to provide plant nutrients such as P and N through means of the biological system helps reduce the need for input of these nutrients from inorganic sources (Doran and Jones, 1996). Magdoff and van Es (2000) review soil quality.

ENHANCING ENVIRONMENTAL QUALITY AND PRODUCTIVITY

Balancing agricultural production with protection of the environment does not have to be an either/or proposition. These two can actually be quite complementary. The greatest pollutant in U.S. waterways continues to be sediment when considered on a mass basis. There are arguments as to the main source of this pollutant, whether it comes directly from agricultural fields, urban construction sites, or from stream bank slumping. Most scientists agree that if agricultural fields are not the main source, they are certainly a primary source. When you multiply an acceptable rate of soil loss, say 5 ton/acre (NRCS specification) by the total acres of production (say 20 million acres for the state of Kansas), a rather large value for sediment contribution from agriculture is reached. Many fields currently erode at rates much greater than those considered acceptable, which further inflate that estimate.

Following sediment, the next greatest pollutants in the surface waters of the United States are nutrients. Excess levels of P and N in surface waters have led to algal blooms, and eutrophication of waters throughout the nation. Currently, a deadzone has developed along the coast of the Gulf of Mexico and has been linked to runoff water from agriculture in the U.S. Midwest. A large portion of this excess nutrient load can be tied to elevated sediment loads, and to runoff from agricultural fields. Best management practices to reduce nutrient losses include no-tillage soil management, fertilizer placement below the soil surface, and field and riparian buffers to help remove sediment and nutrients exiting the field. No-tillage cropping systems have been shown to reduce soil loss rates for most soils to levels less than one ton per acre per year across many different regions and ecosystems. Yields of crops vary for many reasons, tillage being one of them. The variability of yield due to tillage is far outweighed by the variability in variety or hybrid selection in nearly all locations. It is far outweighed by decisions that producers make in fertilizer and herbicide choice. The variability in yield is more dependent on climate and yearly fluctuations in rainfall, than on any decision a producer could make about tillage. So why do we still have a majority of our farmers practicing some type of conservation tillage? Good question. The answer lies somewhere in the fog of conflicting research results and corporate pressures to buy the latest new product or equipment. Add to this state of mind the farmer's reluctance to adopt new technology, or to manage in ways that may appear radical to the neighbors or to their predecessors. A third ingredient to this lack of adoption lies with the scientist's desire to remain as an unbiased source of information. Many scientists believe that once they express an opinion, they have crossed the line between science and salesmanship.

It appears that the argument for moving toward a new agriculture is stronger than for staying with the old. When viewed from the physical, chemical, and biological viewpoint, tillage is not necessary in our modern systems of production. In fact, there are more reasons not to till the soil than there are reasons to till it. Research in the future should concentrate on ways to improve on the no-till system. Direct seeding is a technology that provides a way to plant through high levels of residue, and maintaining high levels of residue is the key to conserving soil. We must concentrate our efforts on improving production using complementary best management practices. The benefits of mimicking the natural world and managing the soil as a living system should provide a more stable resource that will ensure that it remains viable for the next generation.

REFERENCES

Ashford, D.L., D.W. Reeves, M.G. Patterson, G.R. Wehtje, and M.S. Miller-Goodman. 2000. Roller vs. herbicides: An alternative kill method for cover crops. *Proceedings of the 23rd Southern Conservation Tillage Conference for Sustainable Agriculture*. Monroe, LA, June 19-21, 2000. pp. 64-69.

Bagayoko, M., E. George, V. Romheld, and A. Buerkert. 2000. Effects of mycorrhizae and phosphorus on growth and nutrient uptake of millet, cowpea and sorghum on the West African soil. *J. Agric. Sci.* 135:399-407.

Balfour, E. 1975. *The living soil.* Faber and Faber, London.

Chantigny, M.H., D.A. Angers, D. Prevost, L. Vezina, and F. Chalifour. 1997. Soil aggregation and fungal and bacterial boimass under annual and perennial cropping systems. *Soil Sci. Soc. Am. J.* 61:252-267.

Crookston, R.K., J.E. Kurle, P.J. Copeland, J.H. Ford, and W.E. Lueschen. 1991. Rotational cropping sequence affects yield of corn and soybean. *Agron. J.* 83:108-113.

Doran, J.W. and A.J. Jones (eds.). 1996. *Methods for assessing soil quality.* American Society of Agronomy, SSSA Spec. Publ No. 49, Madison, WI.

Galvez, L., D.D. Douds, L.E. Drinkwater, and P. Wagoner. 2001. Effect of tillage and farming system upon VAM fungus populations and mycorrhizas and nutrient uptake of maize. *Plant Soil* 228:299-308.

Galvez, L. D.D. Douds, P. Wagnoner, L.R. Longecker, L.E. Drinkwater, and R.R. Janke 1995. An overwintering cover crop increases inoculum of VAM fungi in agricultural soil. *Am. J. Altern. Agric.* 10:152-153.

Hendrix, P.F. 1995. *Earthworm ecology and biogeography in North America.* CRC Press, Inc., Boca Raton, FL.

Kabir, Z. and R.T. Koide. 2000. The effect of dandelion or a cover crop on mycorrhiza inoculum potential, soil aggregation and yield of maize. *Agric. Ecosyst. Environ.* 78:167-174.

Lavelle, P. 1988. Earthworm activities and the soil system. *Biol. Fertil. Soils* 6:237-251.

Lavelle, P. and A. Martin. 1992. Small-scale and large-scale effects on endogeic earthworms on soil organic matter dynamics in soil of the humid tropics. *Soil. Biol. Biochem.* 24:1491-1498.

Magdoff, F. and H. van Es. 2000. *Building soils for better crops.* Sustainable Agriculture Research and Education Program.

Magdoff, F. and R. Weil (eds.) 2004. *Soil organic matter in sustainable agriculture.* CRC Press, Boca Raton, Florida.

Mulvaney, R.L., S.A. Khan, R.G. Hoeft, and H.M. Brown. 2001. A soil nitrogen fraction that reduces the need for nitrogen fertilization. *Soil Sci. Soc. Am. J.* 65:1164-1172.

Newsham, K.K. and A.R. Watkinson. 1998. Arbuscular mycorrhizas and the population biology of grasses. In: *Population Biology of Grasses,* G.P. Cheplick (ed.). Cambridge University Press, Cambridge, U.K. pp. 286-312.

Owens, H.I. 2001. *Tillage: From plow to chisel and no-tillage, 1930-1999.* Midwest Plan Service, Iowa State University, Ames, IA.

Porter, P.M., J.G. Lauer, W.E. Lueschen, J.H. Ford, T.R. Hoverstad, E.S. Oplinger, and R.K. Crookston. 1997. Environment affects the corn and soybean rotation effect. *Agron. J.* 89:442-448.

Smith, S.E. and D.J. Read. 1997. *Mycorrhizal symbiosis* (2nd ed.). Academic Press, San Diego, CA.

Soil Quality Institute. 2002. *The knive roller (crimper): An alternative kill method for cover crops.* U.S. Department of Agriculture, NRCS-SQI. Agronomy Technical Note No. 13.

Sylvia, D.M., D.O. Wilson, J.H. Graham, J.J. Maddox, P. Millner, J.B. Morton, H.D. Skipper, S.F. Wright, and A.G. Jarstfer. 1993. Evaluation of vesicular-arbuscular mycorrhizal fungi in diverse plants and soils. *Soil Biol. Biochem.* 25(6): 705-713.

USDA. 1992. *Substituting legumes for fallow in US Great Plains wheat production: the first five years of research and demonstration, 1988-1992.* Sustainable Agriculture Research and Education Program. USDA-CSRS-SARE, Washington, DC.

Vivekanandan, M. and P.E. Fixen. 1991. Cropping systems on mycorrhizal colonization, early growth, and phosphorus uptake of corn. *Soil Sci. Soc. Am. J.* 55:136-140.

Wilson, G.W.T. and D.C. Hartnett. 1998. Interspecific variation in plant responses to mycorrhizal colonization in prairie grasses and forbs. *Am. J. Bot.* 85:1732-1738.

Chapter 4

Managed Grazing in Sustainable Farming Systems

Terry Gompert

INTRODUCTION

Grazing is a natural component of well-designed agroecosystems. Managed or planned grazing can increase profits, improve diets and welfare of cattle, contribute toward our quality of life, and provide a healthy and improved rural environment. One of the most practical and useful strategies developed over the past two decades is holistic management (Savory, 1999), an approach that parallels the implementation of SARE projects and a model that has been used in some regional training programs (see Chapter 8). Holistic management builds on ecological principles and clear goals to help farmers design systems that are resource conserving and efficient for long-term, sustained production.

Ken Schneider, coordinator of North Central Region SARE Producer Grants, reports that intensive grazing systems have been among the most popular projects funded for farmer research. They were a dominant activity in the early SARE grant awards in the region. Holistic management principles and trained educators were often called on to lead field tours.

Holistic Management™ practitioners consider four ecosystem processes: water cycle, mineral cycle, community dynamics, and energy flow. The water cycle (Figure 4.1) is most efficient when grazing or farm-

This chapter may be found in part in the UNL Extension Educator on Holistic Management™ In Practice, and guidelines in Gompert, 2004. The Allan Savory Center for Holistic Management can be located at 505-842-5252; http://www .holisticmanagement.org/. References to information presented at field days include name, place, and date, but these materials are not available in libraries.

The Water Cycle

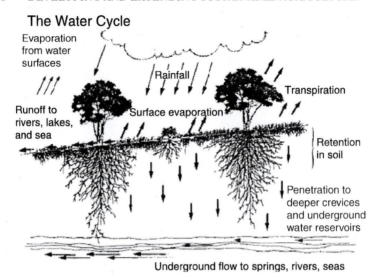

FIGURE 4.1. Ecosystem processes: The water cycle. *Source:* Reprinted with permission from Savory, 1999, Figure 12.2, p. 105.

ing systems maintain a year-long cover on the soil and the maximum possible soil biological activity. As described by Savory (1999),

> An effective water cycle requires a covered and biologically active soil. When effective, most water soaks in quickly where it falls. Later, it is released slowly through plants that transpire it, or through rivers, springs and aquifers that collect, through seepage, what the plants do not take. When soil is exposed and biological activity reduced, most water runs off as floods. What little soaks in is released rapidly from evaporation which draws moisture back up through the soil surface.

In pasture systems, rainfall soaks into the soil rather than running off the land, and snow is captured during winter months. In contrast, bare soil is subject to rapid erosion and evaporation of soil water.

Mineral cycles (Figure 4.2) also benefit from plant residues and the permanent green cover of a permanent pasture, since nutrients are cycled within the system and have less opportunity to be lost. Again from Savory (1999), "An effective mineral cycle requires a covered and biologically active soil. When effective (Figure 4.2, left side), many nutrients cycle between living plants and living soil continually. When soil is exposed and biological activity low (Figure 4.2, right side), nutrients become trapped at various points in the cycle or are lost to wind and water erosion."

The Mineral Cycle

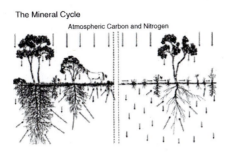

FIGURE 4.2. Ecosystem processes: The mineral cycle. *Source:* Reprinted with permission from Savory, 1999, Figure 14.1, p. 143.

Natural communities tend to develop greater complexity over time to maintain energy flows within the system, and thus develop more stability due to greater resilience (Figure 4.3). From Savory (1999),

> With few exceptions, natural communities strive to develop toward ever-greater complexity and thus stability. From unstable bare ground, where biological activity is low, stable range or forest communities develop over time. When humans reduce this complexity by planting monoculture crops or lawns, for instance, they so defy the principles of nature that they can only be maintained by unnatural means, and then only temporarily. As components within nature, humans cannot escape this principle any more than other organisms can.

The well-designed grazing system can emulate the natural system more than is possible with most cropping systems. Farmers who manage grazing need to also consider these ecosystem processes that represent the heart beat of a sustainable agroecosystem. These processes are summarized and put into a systems perspective in the holistic management model (Figure 4.4, Savory, 1999). Here are some of the reasons why ecological processes are important, and how grazing systems can contribute to sustainability.

WHY PROMOTE MANAGED GRAZING?

Managed, intensive grazing improves the health of soils and animals. These systems also increase cycling of nutrients, particularly with diversity

Community Dynamics

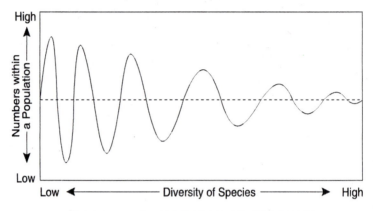

The more complex and diverse communities become,
the more stable populations within tend to be.

FIGURE 4.3. Ecosystem processes: Community dynamics. *Source:* Reprinted with permission from Savory, 1999, Figure 13.1, p. 127.

of plants and animals in the system. Harvesting is more efficient with grazing animals, compared to systems that need heavy use of equipment and fossil fuels. Grazing systems are more environmentally sound, since nutrients go directly back to pastures rather than concentrating in feedlots, making these systems more resource efficient for a long term. Finally, managed grazing systems can improve food quality and overall rural quality of life. These consequences and impacts of grazing systems are described in detail.

Soil Health

The ruminant animal is wonderful for the health of the soil. Across managed pastures, ruminants deposit mineral-enriched urine and feces that are full of beneficial micro-organisms. The deposits provide a source of enhanced food for the life underground that processes nutrients and makes them available for plant growth, and this in turn may become animal feed. Soil biological community dynamics among the earthworms, fungi, mycorrhiza, bacteria, nematodes, and other invertebrates are major factors in cleaning the earth and providing food for the rest of the species on the planet. There are more pounds of life and more types of living species in the soil as compared to those above ground as described by Pat Richardson at the Brush Creek Field Day in Nebraska in 2003. Grazing systems promote the health of this complex and living system.

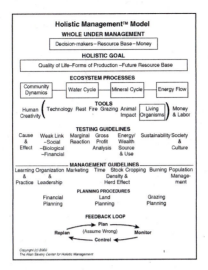

FIGURE 4.4. Holistic management is a framework for decision making. *Source:* Reprinted with permission from Savory, 1999, Figure 7.1, p. 51.

Animal Well-Being

An animal in confinement has less opportunity to be as healthy as one that is free to graze (Beetz, 1998). For example, a confined high-producing dairy cow on an average lactates only about 17 months of her life, because her body is quickly worn out. In contrast, a seasonal grass-based dairy cow can lactate up to 10 years, according to Alan Henning at the Mid Atlantic Dairy Grazing Conference in 2002. Confinement obviously lowers longevity and leads to larger costs for medications and for replacements in the milking herd. Kenneth King from Hutchinson, Kansas, noted the same findings in his SARE project report, *Profitable Dairy Options—Grazing, Marketing, Nutrient Management* (North Central Project FNC92-8/FNC93-57, 1992 & 1993).

Beef cattle are also healthier on grass. Cattle slaughtered from a feedlot commonly have more than 30 percent of the livers condemned, while pasture-raised and grass-finished beef animals rarely have a condemned liver according to William Sessions at the Henkel Grassfed Tour in Nebraska in 2004. It is healthier for an animal to be foraging or grazing in clean air with appropriate space and freedom to move.

Recycling Nutrients

There is a fixed supply of nutrients on earth. To be accessible for use in agroecosystems, these nutrients need to be available for managers to access them and apply or distribute according to soil types, and to crop or pasture needs. Pasture management will be most effective when designed to keep nutrients in place to the maximum extent possible. In a managed grazing system, the animal will consume, utilize, and return up to 90 percent of the nutrients consumed, leaving these materials where they can be utilized again and again with little or no added inputs or loss from the system other than what is extracted as beef or milk (Smith, 1986). In contrast, if we remove forage and feed livestock at some distance away from where hay or grain crops are produced, and we do not return the manure and urine to the land where crops were produced, the process results in mining the soil. The longer we harvest and haul away feedstuffs to a confined place, the less productive the crop and hay land becomes. Also, the more we fixate on the model of hauling feed to a confined area, the more the by-product of nutrients and minerals (manure and urine) is changed in our minds, from being a valuable resource to an expensive waste material to be disposed. The cost of mining from fields and concentrating nutrients near feedlots is externalized if this is not calculated as a part of the cattle enterprise. The manure can be distributed back to fields at a cost; however the larger cost or loss of soil productivity due to what is mined out of the soils is normally not calculated.

Diversity of Plants and Animals

A complex plant and animal community is more sustainable. In contrast, a monoculture has the highest level of environmental risk when the system is faced with quickly changing conditions. With continuous corn, the pattern has been an increase in purchased inputs every year; at the same time, soil health, soil organic matter, and farm profits have gone down. In contrast, a complex plant community that has dozens of plant species that include deep-rooted and shallow-rooted species, forbs, legumes, warm season, and cool season grasses and sedges tends to be more stable (Savory, 1999). In Larry and Judy Woodbury's SARE Project, *Incorporating Holistic Resource Management,* they saw warm season grass return to the plant communities along with an overall increase in desirable forage species in North Dakota, as reported in their final SARE report in 1999. This type of plant community will produce relatively better in either wet or dry years than a monoculture. It will be reasonably productive with little or no inputs. The deep-rooted plants bring up and recycle minerals for the other plants.

Managed grazing leads to more complex and stable systems that result in more profits. The more diverse the mixture of animals grazing the forage, the higher the overall productivity.

Harvesting Efficiency

The most energy efficient method to harvest in terms of fossil fuel use, is by grazing. Haying takes more energy than grazing; cutting, processing, and storing silage takes even more. It is important that grazing animals work for us, not vice versa. If we harvest the forage or feed and take it to the animal and then have to haul the manure away, we are in the hauling business and thus we are working for the animals. If we plan to graze and let the animals harvest feedstuffs, and than deposit the manure where we want it, we now have the animals working for us. That is a much more efficient strategy and better method of management.

Profits

Research data from various studies suggest that profit in the livestock business is a direct result of the amount of harvested forage fed to the cows according to specialist Rick Rasby of Nebraska Cooperative Extension. The less we have to feed harvested forage, while at the same time maintaining gains, the higher the profits. Grass-based dairies report higher net income per acre when compared to confinement dairies, according to the North Central Project FNC92-8/FNC93-57, a 1999 report on Profitable Dairy Options: Grazing, Marketing, and Nutrient Management in the North Central SARE office. Three factors that reduce profits on the farm are: rusting equipment, rotting of infrastructure, and depreciation. Many of these occur on haying and feeding equipment and facilities. If more attention is paid to grazing for a greater part of the year with fewer investments in equipment and buildings, profits can be increased.

Environment

Ecosystem processes and services improve when the soil is protected and complex plant and animal communities are present (Daily, 1997). Planned grazing allows and actively promotes the health of the agroecosystem. If plants are kept vegetative, healthy, and actively growing, the result is greater energy flow. If pasture plants are kept green and dense, with animals grazing the plants and redistributing their wastes in the same location, there is a more rapid and effective mineral cycling. If the vegetation is dense and the soil porous as a result of appropriate grazing management, the water cycle is kept at a high level of function. If there is abundant soil life, thriving wildlife, and people prospering from grazing animals, the health and economy of the nearby rural community are enhanced. This is a realistic and practical route to creating a green environment.

Food Quality

We are what we eat! If our soil is healthy, our crops will be healthy and so will our animals. If both crops and animals are healthy, then people will be healthy. It has been discovered that many organically grown foods contain two to three times more nutrients per pound of dry matter when compared to those from chemical-intensive farming (Beddoe, 1998). If our meat is produced by green grass rather than high carbohydrates grains, the type of fat is changed in the final product (Dhiman, 1999; Robinson, 2000). There are more good types of fats [conjugated linoleic acid (CLA) and omega 3 fatty acids] and sometimes less bad fats [such as saturated volatile fatty acids (VFA)], thus there is a potential for improving our health from products that are produced in a grazing system.

Quality of Life

Quality of life usually includes surplus income, time with friends and family as well as for yourself, good relations with family and community, good health and physical fitness, enjoyment of our surroundings, and using our talents and resources for ourselves and others. The author believes planned grazing can lead toward a desired quality of life, based on all of the details of other changes noted here. Those who work with nature are close to the land, they observe the natural functions of the immediate landscape, and react with quick and appropriate responses to any factors that would diminish the quality of life. Such a working situation leads to a better life on the farm in the rural landscape, and in the local community.

Long-Term Resource Sustainability

It is difficult to develop a sustainable production system without grazing animals. It has often been said that a beef cow is born to be a scavenger. A cow can eat and use fibrous plants and residues that plants or people cannot use directly. The reason they are able to utilize roughage is that the rumen is full of bacteria and other microorganisms that break down hard-to-use cellulose, lignin, and fiber and turn it into bacterial protein that is then digested by the animal. The ruminant in a well-planned grazing system will help the entire ecosystem to function well. Some farming systems in Argentina use

3-6 years of tame pasture (planned grazing) and then plant the same land to crops for 3-6 years according to Alan Nation who is editor of the Stockman Grass Farmer. The pasture builds enough organic matter and improves soil microbial population to produce crops for the first 2-3 years. According to Mr. Nation, in Uruguay this rotation system is reported to produce more total grain from three years of corn and three years of pasture, compared to six years of continuous corn. Organic matter is one of the great indicators of soil health and production. It is easier to maintain organic matter either on crop land or on pasture if the manager uses a planned grazing system.

Research at the University of Nebraska says corn yields can best be estimated within a given field by two factors: soil depth and percentage of organic matter (Schmidt, 2002). The deeper the soil and the higher the organic matter, the higher the yield potential. The life and health of the soil are also important. This is often referred to as our most important biological resource. The soil that is present in our fields and on our farms sets the limits on the production potential, and this resource needs to be protected and treasured.

WHAT DO WE GRAZE?

Through experience, the author has come to discover that all plants on earth appear to have a purpose in the large ecosystem, even if we do not understand their precise roles, and most have a potential grazing value. The niche for each species and uniqueness of its role is one principle of ecology, from which we can learn. The more we know about the plants, the better graziers or pasture managers we can become. Knowing when plants grow, how long they grow, under what conditions, which animals like them, how deep the roots are, the plant's quality and quantity at various stages of growth, the recovery period after grazing, and any toxic properties of the plants would help us to manage more effectively, and be better graziers. We need to be students of our land—always observing and accumulating knowledge and wisdom. Nearly all plants can be grazed. This grazing can be to our benefit and to the animal's, and the process can lead to improvement of the environment.

It is useful to focus on one plant as an example: Redroot Pigweed (*Amaranthus retroflexus* L.). Pigweed is invariably considered a weed. In *The Weeds of Nebraska and the Great Plains* (Stubbendieck et al., 2003) pigweed is described as: very common throughout Nebraska in cultivated and fallow fields, gardens, waste places, and roadsides; providing food and cover for upland game birds; causing bloat in cattle and accumulating nitrates; and serving as a source of seeds that were harvested by Native Americans and ground into flour. So we know that pigweed grows in most areas on its own, is wildlife food, has some toxic properties, yet can be eaten by humans. If we continue this study more in depth, we learn that the pigweed seed is 16.8 percent protein with total digestible nutrients (TDN) of 59.1 percent for cattle (National Academy of Sciences, 1971). The green plant has high quality, and in fact the green foliage of some species of *Amaranthus* is used for human food in parts of Africa and Latin America. Our observations show that most animals like to graze pigweed when the weed is at the immature stage. Organic farmers tell us that pigweed growth is an indicator of a soil imbalance—usually there is too much nitrate present in the soil. The pigweed plant is a healer of the land. It protects the earth and develops organic matter. It is productive, and does not cost anything to propagate in many circumstances, and can be used as a high-value and productive grazed forage. One does have to manage for the problems of bloat and toxic factors. In this example we have identified one plant that most people consider a weed, yet it is a wonderful resource that should be used and appreciated.

In nature, plant communities evolve toward greater complexity. We find deep and shallow rooted plants; annuals and perennials; forbs and grasses; warm and cool season grasses; bunch and sod forming grasses; tall and short plants; and plants with various levels of toxins. This complexity helps to protect the earth. Natural systems and their component plants have evolved to protect themselves under a wide range of variations in weather conditions.

The following ten personal observations based on experience provide useful generalizations about plants that can help guide our understanding and appreciation of what to graze and when.

1. The quality of a plant is higher at the more immature stages.
2. For cool season grasses to recover, three new leaves need to regrow before grazing.
3. Denser and greener forage indicates higher quality of forage.
4. High-stock density management increases utilization of available forage.
5. More complex plant populations give a longer growing and grazing season to the pasture.
6. Complex plant populations can handle weather fluctuations better.
7. Annuals are higher in nutritional value than perennials.
8. Deep-rooted plants and warm season plants grow best in hot, dry periods.
9. Deep-rooted forbs protect the soil and recycle deep minerals to the surface.
10. All plants of interest to graziers respond positively to grazing.

From the author's experience it appears that all plants may be available forage to graze, and as managers we must decide whether the existing plant mixtures in a field are optimum or if they present some opportunity costs compared to other mixtures that could be planted in that place. Mixtures may be planted as annual forages, annual or perennial crops, species adapted to waste areas, crop residues, complex native pastures, improved pastures, and even weeds. We must also consider the neighbor's drought-damaged or hail-damaged crops, a source of forage that may be available at very low costs.

USDA reports that grazing is the most efficient method of harvest (Gerrish, 2004). Allan Nation, editor of the *Stockman Grass Farmer,* has said often that the cow makes the most profit when it grazes what other animals won't graze. He said that the cow was born to be a scavenger. I think these points are extremely important and should be heeded—and we should consider grazing everything.

WHAT ECOLOGICAL PRINCIPLES DO WE CONSIDER?

In designing grazing systems, it is useful to consider a number of basic principles in ecology. Managed grazing systems, just as other agroecosystems, are established with goals that differ from those intrinsic goals of natural systems. Yet there are many lessons to learn from ecology that can be applied to the design of grazing.

Stocking Rate and Stock Density

If there are too many animals for the land available, your pastures are overstocked (stocking rate is too high). Stocking rate, the number of animals on the land either for the season or year, is definitely the most important management decision facing the grazier. Using too many animals on a given field is likely to reduce animal performance as well as damage the land. Stock density, on the other hand, is the number of animals or number of pounds of animals on a given amount of land at any one time. As a general rule the higher the stock density, the greater the forage utilization, the better the manure distribution, and the higher the herd effect and animal impact. Figure 4.5 illustrates stocking rate and stocking density. Herd effect and animal impact are important in brittle environments because capping of soil is reduced or broken, new seeds are established, and oxidized dead plants are recycled to the land. Stock density is likely to affect stocking rate. Stock density most likely will increase stocking rate. Larry Mason, in his

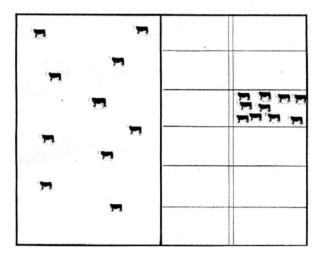

FIGURE 4.5. Stocking rates and stock density in two systems of grazing; stocking rate is the same in both examples, while stock density is greater in the right-hand side of the figure.

SARE Project on the TarBox Hollow Living Prairie (FNC95-89), used bison as the grazing ruminant and discovered the value of understanding stock density and stocking rate as he converted CRP grass to productive grazing land. His adjustments of both factors led to optimum production as well as healthy pastures.

Production, Utilization, and Dry Matter Intake

Production per acre can be measured in many ways and is a key factor in determining profitability. Forage production per acre per year is simply the total amount of forage available per acre (for example 5,000 pounds of dry matter yield). This is how we usually report harvested forage yields. Total forage available has meaning, but the quality of the forage and the percent utilized is even more important. Another measure is the total digestible nutrients, or TDN yield per acre. In the example above, a 50 percent TDN would provide 2,500 pounds per acre, and 60 percent TDN would provide 3,000 pounds of TDN per acre per year. Not only does the 60 percent rate produce more pounds of TDN per acre, it also is more efficient in producing animal gain since animals have to travel less distance to find feed. Thus, both quality and quantity are important factors.

Utilization is extremely important as well. In the same example, if one grazing system allows 25 percent utilization, that provides 1,250 pounds of forage; another system with 50 percent utilization will result in 2,500 pounds utilized forage from the production field. When it comes to what really counts, it is the combination of forage quantity, quality, and utilization that must be considered by the grazier.

Forage quantity is affected by many factors: adaptation of species, density of plants, complexity of plant population, ecosystem processes, plant and soil health, climatic conditions, grazing management, recovery period, fertility program, and many other management factors. Quality is primarily a result of maturity of the forage when grazed. Generally the more immature the forage, the higher the quality. Keeping the plant in a vegetative state at all times will increase quality. Utilization is mostly a function of the grazing method used, and generally the higher the stock density, the greater the utilization. The essence of managed grazing is putting these three factors together to accomplish your goals.

The other factor to consider in managed grazing is animal performance. Animal dry matter intake is the primary factor influencing animal performance, and is normally measured in gain per animal per day. For animals to gain weight there has to be enough of the correct forage of proper nutrition consumed each day. With a poor quality forage that is overmature or includes the wrong mixture of plants, the daily dry matter intake of forage could be as low as 1 percent of the animal's body weight per day. With such a ration there would be no gain. In contrast, if the forage is thick, leafy, immature, and actively growing with a high TDN, the daily dry matter forage intake could reach at least 3 percent of the body weight per day. That animal may gain up to 3 pounds per day if intake is high. Gain per animal per day is one of key factors in determining profit. Thus the forage factor of TDN production per acre, the utilization factor, plus the animal gain or performance per acre together are important in designing and evaluating a managed grazing plan. A moderately high quantity and quality forage production and a moderately high animal performance is usually more profitable than an extreme of either factor, just as natural systems are most sustainable when they are in balance.

Plant Diversity

Planted single annual species are the easiest to manage and are more likely to provide the highest quality forages. These same monocultures will have more yield variation from season to season and are more susceptible to drastic effects of extreme weather conditions. Some years the production is very high and some years very low, and distribution of yield through the growing season may fluctuate much more than a similar measure of the performance of mixtures. Monocultures of one forage species will produce well if conditions are just right. The author believes single annual species such as turnips, oats, rye, corn, and others have a place on every farm. These crops can be used to fill the gaps in the periods when pastures and other forages are not available, and thus are very important in the whole-farm managed grazing plan. These plants have the potential to respond to high inputs such as fertilizers, water, herbicides, and insecticides. Sometimes these annuals work well. It has also been discovered that combining several types of annuals often increases the overall system growing period and helps to reduce the great fluctuation in production of single species of annuals.

Planted single perennial species are relatively easy to manage and it is less complicated to know when and how to manage that forage—fertilize, graze, harvest. Perennials tend to be more consistent in production from one year to the next and they adapt to a particular niche in the growing season and management plan. Species such as alfalfa, wheatgrass, or brome have a specific period in the season when they produce very well, and thus should be part of the overall grazing plan. They often respond to inputs such as fertilizers. Perennial pastures are often planted into crop land that is converted to pasture. Monocultures of perennial forages are needed and often have a specific place in the grazing plan.

Complex plant communities in range and pastures take the most management, but have the greatest potential for capturing and converting sunlight into green plant material for the most possible days in the season. Mixtures can be the most functional systems for promoting effective water and mineral cycles. They can handle the greatest stresses in our variable and unpredictable climates, and can also provide excellent wildlife habitat. Species mixtures represent the most sustainable of all systems and require the least amount of outside inputs. Mixtures are the most stable types of plant communities that can be used for grazing. Many of these plants are symbiotic, and they compete less with each other than large monoculture plantings that have adjacent individual plants with the same growth cycle and nutrient/water needs. Plants in complex mixtures often live together and are supportive of each other, much like those in the native prairie.

Legumes

Legumes tend to have deeper roots, therefore are more drought tolerant than other species. They have the ability to capture nitrogen, which in turn can be used to produce more grass, thus they provide a built-in source of fertility to the neighboring grass plants. Legumes will also fit into a pasture among grasses—just filling in plant gaps without competing intensely with the grasses. They recycle minerals from deep in the soil profile for themselves and for neighboring plants. Legumes can assist is breaking up compaction, thus create conditions for better water percolation from rainfall or irrigation. In addition, legumes are very high quality, with high percent TDN.

The downside of legumes is they can lead to bloat and some are high in nonpalatable alkaloids and toxins. Legumes can also be damaged by broadleaf herbicides, so herbicide weed management options are more limited and difficult. Legumes have more insect pests than grasses. The author prefers to manage for 30-40 percent legumes in a pasture. With that mixture, the legumes can almost produce enough nitrogen for themselves and the neighboring grasses without any external input of nitrogen. This mixture of legumes with grasses also reduces bloat potential. A rule of thumb when deciding to increase or decrease legumes in a pasture is (1) to increase grass, graze high and (2) to increase legumes, graze short. Other forbs in the pasture, specifically curly dock *(Rumex crispus L)* that grows in lower, moist areas (Nelson and Burnside, 1975) contains antibloat factors. Again managing toward complex plant mixtures will tend to reduce production problems and stabilize availability of forage and its use. Sweetclover and red clover can be over-seeded into existing pastures with good success if the soil pH is between 6 and 8. Some success with over-seeding white clover has also been reported (Gerrish, 2004). Other legumes establish best if inter-seeded into the sod.

Matching Forage Quality to Animal Needs

A cow in mid gestation in good body condition can do well on low quality forage. On the other hand, a lactating grass-based dairy cow and a grass-fed and finished beef animal require extremely high quality forage in abundant amounts. Feeding very good feed to an animal with low nutrient requirement is often expensive, while using poor forage for an animal needing high quality feed will give disappointing results. We need to know the animal's nutrient needs and give them that diet, thus matching the quality of the grazed forage to the animal's needs.

Using low-cost cow/calf principles taught by Dick Diven (1998), we can learn how to match forage to animal needs. The three factors he suggests are: adjust calving period to a time of year when cows will require less feed; know the forage quality value and add appropriate nutrients; and know the effects of the photoperiod. With these concepts in mind, his conclusion is to calve as near as possible to June 21, add only appropriate minerals, observe nature, and only graze.

Year-Round Grazing and Filling the Gaps

Grazing is efficient! Profit tends to increase or at least the losses are reduced as days of fed forage and/or feeds are decreased. We can graze 12 months out of the year in Nebraska, if we can design and implement a good farm grazing plan, and this can be done in nearly any place in the United States. It just takes a well-organized plan that builds on the ecological principles already described. In holistic management they talk about developing one grazing plan during the open season when the grass is growing, and another appropriate grazing plan during the closed season when the grass is not actively growing. This strategy is called *filling the gaps*. Filling the gaps during the year requires knowing the forage needs of livestock and the forage production potential in every month, and then fillings the gaps when and where appropriate. Graziers who use good planning also take into consideration special conditions of grazing such as vacations for managers, lack of water, and wildlife needs. The main thing is to figure out what to use and provide some forage for animals to graze all year long. How to fill the gaps, considering different types of forage species, is illustrated in Table 4.1 for Center, Nebraska, a location in the north-central part of the state.

Multi-Species Grazing

Buffaloes prefer nearly 100 percent grass; cattle prefer about 70 percent grass; sheep prefer about half grass and half forbs; goats prefer about 70 percent forbs; and antelope and deer prefer mostly forbs. It is estimated that one can add 20 percent more goats (by weight) to a pasture grazed by cows without decreasing the number of cows while maintaining the same cow performance. This is a real increase of production per acre by 20 percent. Goats and some sheep will reduce populations of so-called *weeds* such as leafy spurge, cedar trees, buck brush, ragweed, and many others. Goats make better cow pastures and cows make better goat pastures.

Grass and Plant Growth

As our knowledge of plant growth increases we can understand the wisdom in grazing. Grass grows more grass! It takes leaves to capture the sunlight and absorb carbon dioxide to form carbohydrates and water through photosynthesis. We need the leaves to capture sunlight, so we don't want to graze too short or production will be reduced. The severity of grazing also affects regrowth in speed and amount. Leaves grow more grass! Over 50 percent of plant growth is underground in the root area. As the leaf is grazed, the living roots are reduced by approximately the same degree.

TABLE 4.1. Filling the gap: Dryland grazing plan for Center, Nebraska.

	Forage	Seeding rate	Planting date	Exceptional forage available	
Single-crop cool-season annuals or biennials	Oats	100-120	April 1-15	June 1-July 15	
	Oats	60-80	July 1-Aug 15	Sept 15-Nov 15	
	Triticales	60-100	Aug 1-Aug 15	Sept 15-Dec 1	April 15-July 15
	Triticales	60-70	April 1-15	May 15-July 15	Sept 1-Dec 1
	Rye	60-90	Aug 1- Oct 1	Oct 1-Dec 1	April 15-July 5
	Italian Ryegrass	10-25	Mar 15-April 15	June 1-July 15	Sept 1-Dec 1
	Wheat	45-90	Sept 1-Oct 1	April 15-June 1	
	Barley	45-90	Mar 15-April 15	May 5-June 10	
	Forage Peas	20-25	Mar 15-April 15	May 5-June 10	
	Hairy Vetch	10-15	Mar 15-April 15	May 5-June 10	
	Red Clover	8-12	April 15-May 15	June 15-Oct 15	
	Annual Alfalfa	8-15	April 15-May 15	June 15-Oct 15	
	Turnips	1-4	Mar 15-April 1	June 15-Sept 1	Oct 15-Jan 1
	Turnips	1-4	June 15-Aug 1	Oct 15-Jan 1	
	Essex Rape	1-2	Mar 15-April 1	June 15-Sept 1 (toxic if by itself)	
	Essex Rape	1-2	June 15-Aug 1	After frost by itself Oct 15-Jan 1	
Single-crop warm-season annuals or biennials	Corn	10-15	May 1-June 15	July 15-Sept 1	
	Corn	10-15	June 15-July 15	Sept 1-Nov 1	
	Sudangrass	10-20	June 1-15	July 15-Sept 1	
	Pearl Millet	5-20	June 1-15	July 15-Sept 1	
	Foxtail Millet	10-25	June 1-15	July 15-Sept 1	
	Soybeans	Thick	June 1-15	July 15-Sept 1	
	Crabgrasss	2-20	June 1	Aug 1-Sept 1	
Perennial	Alfalfa	8-20		May 15-Dec 1(bloat potential)	
	Cool Season Grass (single species or mixed)			April 15-Nov 15	

TABLE 4.1 *(continued)*

Forage	Seeding rate	Planting date	Exceptional forage available
Warm Season Grass (single species or mixed)			June 15-Aug 15
Legume, Cool Season Grass Mix			April 1-Dec 1
Legume, Cool Season Grass, Warm Season Grass Mix			April 1-Dec 1

There is a proper recovery period that the roots require after grazing, so if we regraze too soon and root recovery time is too short, the plant's regrowth is slow or it may die. Roots grow more grass!

Good resources that describe grass growth and management can be found in several university publications (Bishop-Burley, 2002; Blaser, 1986; University of Wisconsin, 1997).

Cool season grasses recover at a faster rate than warm season grasses. If growing conditions are good it takes about 30 days for cool season grass recovery, which is about the same recovery period as legumes. On the other hand, it takes about 45 days for warm season grass to recover to be ready for grazing. Of course it takes longer for all grasses and legumes to recover if moisture is limited and other growing conditions are slow and poor.

Cool season grasses grow in cooler conditions. Their seed germinates and the mature plant starts growing when the soil temperature is at 35-40°F. This compares to warm season plants that start growth at about 55-60°F. Another difference is that warm season plants have growing points in their sheath area, while cool season grasses do not. Therefore, it is important to avoid severe grazing of warm season plants in order to not damage some of the growing points.

In general the best time to graze is when the plant is in the stage of rapid growth and accumulation of dry matter. At this stage the plant roots are recovered and the plant still has relatively high quality. If a plant is kept from producing seeds, there is a tendency that the plant will grow actively and be more productive. Some plants, such as smooth brome *(Bromus inermis)*, can be affected greatly by the time of grazing. If smooth brome is grazed close to the ground when it is about 10 inches tall, after the growing point is above the ground and the seed head has not formed, grazing will more likely prevent the plant from forming a seed head during that year's growth. It will also form more tillers which result in the plant growing much longer into the season. Oats and other small grains react the same way to this type of grazing.

Rest

Many people confuse rest and recovery periods. Rest is only a management tool. Recovery is the time is takes a plant to recover from the last period of grazing. We previously discussed the principle of growth in the grasses. The roots need to be fully recovered prior to regrazing.

We overgraze plants and not pastures. If a plant is regrazed before it is fully recovered, its future growth will be reduced or the plant may die. We often see overgrazed plants in continuously grazed pastures, even when these are understocked. We can also undergraze plants. If the plants are not grazed, the top material will dry and oxidize, turning it into very low quality plant material. The center of the plant dies and the outside of the plant grows spindly shoots or tillers. This plant may eventually die and the pasture will naturally thin. This is easy to see in the Sandhills of Nebraska, especially with Little Bluestem *(Schizachyrium scoparium)*. In brittle environments it takes animal impact and herd action to break soil capping, break off oxidized plant material, and cause needed seed germination. Without grazing the pasture keeps deteriorating. Since Nebraska has moderately brittle conditions, herd effect and hoof action only help under certain conditions. The experienced manager will learn to get a feel for when and where it will not work.

Soil and Plant Health

Creating an environment or habitat for healthy soils and plants is a key to being a successful grazier. We need to consistently monitor conditions to see if our soil and plant health is improving. A healthy soil is active with a diversity of microbes, fungus, bacteria, earthworms, and other animal life. The number of earthworms may be one of our most easily seen and best indicators of soil life. The soil should be well covered with new litter ("litter one") and litter that is in the process of breaking down and becoming soil organic matter ("litter two"). A well protected soil needs to be totally covered with either plants or litter. Soil organic matter should be high and increasing. Our goal for most of Nebraska's soils should be between 4 to 6 percent organic matter, similar to the soil condition of the native prairie before cultivation. The depth of the topsoil should also be increasing. The soil should smell alive and be porous as well as friable.

A healthy plant will be more resistant to insects and pathogens. It will grow actively even in drought conditions (van Es and Magdoff, 2000). Ten days after grazing the ground cover should appear 100 percent green with actively growing plants. There will be a complex variety of plants. There will be plants that grow well during the longest possible period of the growing season (mid February to late November in Nebraska). Wildlife will be abundant and there will be no water loss due to run off or evaporation. The animals that eat the grass will be healthy and those who eat the animals will be healthy (Beddoe, 1998).

Yield Potential for Cropland Converted to Pasture

Soil, water, temperature, genetic plant potential, management, and other factors may affect the pasture yield potential. Many farmers are discouraged when there are inconsistent yields and limited or no profits where grains are harvested as a cash crop. Some are considering converting the land into pasture. They want to know the production potential of converted pastures from previous cropland. Of course, the answer to their question depends on their land quality and their management.

We should look at corn yields on a given field to estimate potential pasture yields on that same land. The top land in Nebraska under pivot irrigation, managed well, consistently will produce 200 bushels per acre of shelled corn. If that corn were fed in a feedlot to steers, that amount of feed would produce about 1,600 pounds of beef. If that same land were converted to pasture with a high level of management, the pasture yield would also approach 1,600 pounds of beef per acre. In contrast, land that would produce 65 bushels per acre yield of corn on a shallow soil and under dryland conditions would produce about 520 pounds of beef per acre, as calculated by the author and presented in an Irrigated Pasture Conference in Kearney, Nebraska, in 2002). In Table 4.2, the land yield potential is related to corn yields and also to pasture yields at several levels of production.

Irrigated pasture research in the 1970s by the University of Nebraska using a five paddock grazing system in North Platte produced more than 1,000 pounds of beef per acre, or 14 Animal Month Units (AMU) of grazing. That research was conducted on lands that would have produced about 160 bushels of corn per acre (Nichols, 1981). From an AMU and cost per acre stand-

TABLE 4.2. Hypothetical land yield potential for corn and pasture.

Corn yield potential (bu)	Beef produced in feedlot (pounds)	1,000# animal month units (AMU)
200	1,600	20
160	1,280	16
120	960	12
80	640	8
40	320	4

point, developing grazing systems on the most productive land first is normally the best choice.

Why are we surprised at these results? Argentina has some of the world's most productive grazing systems. They raise beef cattle to finished weight, and produce milk on their best, most productive lands. The soils are deep and fields are flat, and they are located in areas with highly desirable temperature and moisture. They rotate 3-6 years of crops with 3-6 years of pastures. This rotation is to build organic matter, control weeds, and have more flexibility with the cattle and grain cycles, as described by a group of visiting Argentinian graziers who visited Nebraska in 2000.

We can also look at the increasing number of grass-based dairies. Farmers who manage these dairies plant the most productive ground to grass and graze those areas with intensely managed rotations. Milkers on these dairies know immediately their milk yield shortly after their daily graze. One of the bright spots in U.S. agriculture is the grass based dairy—let us learn from them.

CONCLUSIONS

Intensive managed grazing is an innovation available at a relatively low cost to farmers and ranchers who want to substantially increase productivity of their pastures. The only requirements are fencing, water, time to move cattle, and especially the knowledge and skills needed to assess the pasture and know when to move animals. An intimate knowledge of the ruminant animal and the different types of forage that are available contributes immensely to the success of this approach. The skills are easy to acquire by following the guidelines given in this chapter and the references cited.

It is important to know the production potential of your several fields and areas within those fields. Yield potential is affected by soil quality, rainfall amount and distribution, temperature regime, genetic potential of the pasture species, and your management. One of the key concepts is to be in sync with nature and understand the lessons she is able to teach us. It is valuable to observe carefully and to plan our grazing patterns from the lessons learned in natural ecosystems. Nature is very forgiving, and will often help correct our bad management. She provides great potential for growing pastures, but also weeds, poisonous plants, pathogens that cause diseases, insects that cause infestations, and chances for erosion if the land is not managed well. But there is also a healing power in nature that can work for us with good management. Simple successes are possible when we copy nature and let the animal work for us by grazing, if we calve when the weather is warm, when we mix species in pasture polycultures, and when we eliminate the use of chemical pesticides.

Careful attention to setting goals, realistic planning, keeping good records, monitoring pastures, and replanning when necessary is also essential in our management plan (Savory, 1999). One thing the author has learnt from holistic management is to "first assume that you might be wrong." This mind set keeps us sharper, more proactive, and helps us reach our goals.

The author's conclusions about managed grazing were nicely summarized by Marcelo Fernandez, an Argentine veterinarian who works with large cow/calf, beef fattening, and grass-based dairies in the Pampa Area southwest of Buenos Aires. In June 2000, Dr. Fernandez visited ranches in Mead, Amherst, Callaway, Burwell, and St. Edward, Nebraska, and evaluated each stop as if consulting for them. He shared with us some techniques that Argentine stockmen use to produce high quality beef and dairy products in their country. He was a principal resource for the Nebraska/Ohio Beef Study Tour to Argentina in 1999 (Gompert and Scriven, 2000). Here are ten points from Dr. Fernandez' summary that are relevant to conditions in Nebraska and the North Central Region.

1. *Be friends with your grass.* "You must become friends with your grass. You must meet her often, appreciate her, be sensitive to her needs, understand her, care for her, and love her. Then she will give it all back." Monitoring the grass in all pastures was a big component listed for Argentine stockmanship. A weekly visit to each paddock was a minimum requirement.

2. *Know your grass yield—how much does it grow?* A practical way to collect, weigh and calculate dry matter available per acre uses a premeasured square. Dr. Fernandez collected all the grass within that square area by clipping and then weighing to determine the pounds of grass per acre using a formula based on the size of the

square. In the field, he estimated dry matter content to be approximately 30 percent, although dry matter content can actually be determined by drying the clipped sample in a microwave oven. After a few days of clipping and weighing, you will be able to estimate the dry matter yield within your clipped area. With experience and taking several clippings in a pasture, estimating is okay. He suggests doing a complete laboratory analysis each month to calculate energy levels and other details. All of this information will allow you to determine the grass yield in pounds of dry matter per acre. This is very important to help determine the carrying capacity or stocking rate of your pasture. You will also be able to use this accumulated information to determine the pounds of grass growth per day.

3. *Learn about "Animal Equivalents," the requirements for each animal species.* An animal equivalent is a measurement of an animal unit that is based on the weight of the animal and its expected daily weight gain or weight loss. An animal equivalent is based on the energy needed for a 900 pound cow and her calf for the year. There are animal equivalents for dairy cows and also for fattening cattle. It takes 21 mega-calories of energy per day to care for one animal equivalent. Because most high-quality grasses contain about 2.1 mega-calories of energy per kg, it takes about 10 kg or 22 pounds of dry matter material from the grass to supply the energy needed for one animal equivalent.

4. *Determine best stocking rate from feed budget.* Stocking rate is animal equivalents per acre. Harvesting efficiency needs to be taken into account, as described above. In the spring time, the harvesting efficiency may be 40 percent, but can reach much higher levels later in the season.

5. *Three leaf rule.* All grasses generally have maximum digestibility, maximum energy intake and minimum fiber at the three leaf stage, so this is a good and practical guideline for management. As the fourth leaf develops, the first leaf begins to deteriorate. A deteriorated leaf is a lost value. Graze at the third or fourth leaf stage.

6. *Intake is related to fiber content of the forage.* Forage intake by grazing livestock is directly determined by the fiber content. The lower the fiber content, the higher the intake of the forage available. Nitrogen detergent fiber, or NDF, is a measurement of the fiber content of the sample. Dry matter intake for each animal can be determined by dividing a factor, 120, by the NDF of the forage. For example, the grass sample containing 45 percent NDF on a dry matter basis would estimate an intake of 2.6 percent of the animal's body weight. Therefore, a cow weighing 1,100 pounds would be expected to consume

28.6 pounds of dry matter from the forage. If we estimate the dry matter content of the forage at about 30 percent, this cow would be eating about 95 pounds of forage per day.

7. *Quality of grass.* Quality of grass can be determined in several ways. One way is to compare the leaf to stem ratio. The higher the percent of leaf, the higher the quality. This can be done by separating leaves from stems, drying, and weighing the sample, or this can be estimated with experience.

8. *Three factors influence animal production.* The most important factors that determine production are height of grass at time of grazing, density of grass, and percent leaf tissue.

9. *Tillering.* The potential yield of pasture is partly determined by tiller density. Generally we can increase the tillering by grazing shorter. Short or close grazing is especially important in the fall in relationship to improved tillering.

10. *Grass fattening profits.* The one factor most responsible for the profitability of fattening cattle on grass is getting the highest or best utilization of spring grass. This must be followed by careful management of the warm season pastures and through the entire cycle of finishing, but the spring period is most important.

Grazing is a natural process, and well-managed and planned grazing should be beneficial to farmers and ranchers, to others in the rural community, and to the environment. Holistic Management (Savory, 1999) can provide many useful guidelines, but it is up to the individual grazier to adapt these ideas to each farm and ranch, and to each pasture. Managed grazing has been a key part of the SARE programs in the region, and it can help many farmers and ranchers achieve their holistic goals.

REFERENCES

Beddoe, A.F. 1998. *Biologic ionization as applied to farming and soil management.* S & J Unlimited Publishers, Washington, DC.

Beetz, A. 1998. *Grass-based and seasonal dairying.* ATTRA, Fayetteville, AR. Available from: http://www.attra.org/attra-pub/gbdairy.html [Accessed: October 20, 2004].

Bishop-Burley, G.T. 2002. *Dairy grazing manual.* University of Missouri Cooperative Extension, Columbia, MO.

Blaser, R.E. 1986. *Forage and animal management systems.* University of Virginia Cooperative Extension, Bulletin 86-7, Blacksburg, VA.

Daily, G.C. (ed.). 1997. *Nature's services: Societal dependence on natural ecosystems.* Island Press, Washington, DC.

Dhiman, T.R., E.D. Helmink, D.J. McMahon, R.L. Fife, and M.W. Pariza. 1999. Conjugated linoleic acid content of milk and cheese from cows fed extruded oilseeds. *J. Dairy Sci.* 82:412-419.

Diven, D. 1998. *Low cost cow calf program.* Agri-Concepts, Inc. Tucson, AZ. Available from: http://lowcostcowcalf.com/ [Accessed: November 24, 2004].

Gerrish, J. 2004. *Management intensive grazing: the grassroots of grass farming.* Green Park Press, Mississippi Valley Publishing, Corp., Ridgeland, MS.

Gompert, T. and B. Scriven. 2000. *Consulting graziers in Argentina.* University of Nebraska Cooperative Extension, 2000, Grazing Seminar for Extension Educators. Lincoln, NE.

National Academy of Sciences. 1971. *Atlas of nutritional data on United States and Canadian feeds.* National Research Council, Washington, DC.

Nelson, E.W. and O. Burnside. 1975. *Nebraska weeds, revised.* Nebraska Department of Agriculture, Lincoln, NE.

Nichols, J.T. 1981. *Grazing management of irrigated grass pasture.* NebGuide G81-563. University of Nebraska, Lincoln, NE.

Robinson, J. 2000. *Why grassfed beef is best.* Vashon Island Press, Vashon, WA.

Savory, A. with Jody Butterfield. 1999. *Holistic management: A new framework for decision making,* Second edition. Island Press, Washington, DC.

Schmidt, J.P. 2002. Nitrogen yield response to nitrogen at multiple in field locations. *Agron. J.* 94:798-806.

Smith, B. 1986. *Intensive grazing management: Forage, animal, men, profits.* The Graziers Hui, Kamuela, HI.

Stubbendieck, J.L., M.J. Coffin, and L.M. Landholt. 2003. *Weeds of the Great Plains.* Nebraska Department of Agriculture, Lincoln, NE.

University of Wisconsin. 1997. *Pastures for profit: A guide to rotational grazing.* University of Wisconsin Cooperative Extension, A3529, Madison, WI.

Van Es, H. and F. Magdoff. 2000. *Building soils for better crops* (2nd ed.). SARE Publications, University of Vermont, Burlington, VT.

Chapter 5

Whole-Farm Planning and Implementing Sustainable Agriculture

Rhonda R. Janke

BACKGROUND

There is a strong connection between whole-farm planning and the process of implementing sustainable agriculture. This connection is not always obvious to those doing sustainable agriculture research, especially when they are focused on component technologies. The history of whole-farm planning pre-dates the years when sustainable agriculture research began to be funded in the United States. This chapter reviews the history of both of these research directions, the connections between the two, some of my own work and that of collaborators who work in both areas, and our efforts to integrate the two. Specific examples from Kansas are used to illustrate these concepts.

First, a short history is required to understand the connection between whole-farm planning and sustainable agriculture. My college transcripts do

I would like to thank the following people who were collaborators throughout the development and implementation of the whole-farm planning tools described in this chapter. These projects would not have been possible except as a team effort; my co-author on the River Friendly Farms notebook, Dan Nagengast (KRC); the original Kaw Valley Heritage Alliance committee Kevin Licktieg, Greg McCabe, Chuck Otte, John Piscak, Danny Rogers, Don Snethan and Don Teske; research collaborators at KSU: Len Bloomquist, Stan Freyenberger, Hans Kok, David Norman, Morgan Powell, and Bryan Schurle; KSU administrators who initiated the extension trainings: Daryl Buchholtz and Bill Hargrove; and Clean Water Farms Project collaborators at the Kansas Rural Center, including Jim and Lisa French, Mary Fund (Project Leader), Dan and Mary Howell, Jerry Jost, Darrell Parks, and Ed Reznicek.

not contain a single class with the words "whole-farm planning" or "sustainable agriculture" in the title. In fact, I do not remember the subject matter being covered at all, with the exception of an international agriculture class at Cornell. In that class I studied aspects of agricultural systems in Mexico in the fall semester, spent two weeks touring farms and *ejidos* throughout Mexico during winter break, and then presented reports in the spring.

A typical agricultural graduate degree in the United States can include a wide variety of classes in agronomy, animal science, economics, rural sociology, and plant and systems ecology, but students are often encouraged to focus and to specialize. For example, as an undergraduate one starts by taking soils, but then moves on to more specialized courses such as soil chemistry, or soil and water relations. General courses in animal science are replaced by ruminant nutrition or forages. We can try to grasp the complexities of real world agriculture and whole-farm planning by participating in extra-curricular projects. My attempts to do this included an honors thesis as an undergraduate on plant-plant interactions, and participating in the "ecological agriculture research collective" as a graduate student. I also worked closely with organic farmers in upstate New York, first doing a survey, and then as a graduate student assisting with on-farm research. In the late 1970s and early 1980s, when I was completing undergraduate and graduate degrees, the terms ecological agriculture and alternative agriculture were being used only in discussion groups and guest lectures by those who were pioneers in the field.

At the Rodale Research Center in the late 1980s I was able to work with an interdisciplinary team, doing cropping systems research with collaborators with expertise in soil microbiology, nutrient cycling, and weed ecology. We also attempted to understand whole-farm processes by observing nearby fields, and later documenting practices used by our traditional farming neighbors. Colleagues at Penn State were beginning to study nitrogen and phosphorus dynamics (Bacon et al., 1990; Lanyon and Beegle, 1989) at the whole-farm level, and even at the county level, as nutrient loading in the Chesapeake Bay became a region-wide problem. Tools for researchers were being developed, and examples of farming systems research from the international agriculture community were being published, but practical whole-farm planning tools for farmers were not widely available and were scarcely used at all.

Historically, whole-farm planning was called "farm management," and included the study of the biological aspects of the farm, combined with the sociological and management dimensions. Farm management was multidisciplinary and involved the entire range of factors in running a farm. By the 1920s the term came to mean primarily the economics of operating a farm. By the late 1950s farm management in the United States became a subdiscipline of production economics, leaving the other disciplines out. In the 1970s, specialization had become the norm, and one simply took courses in "macro-econ" or "micro-econ," and farmers were on their own to integrate economic principles with those from crops, soils, animal science, and other disciplines.

IDENTIFYING AND FILLING THE GAPS

Beginning at Kansas State University (KSU) in 1994, I continued to work on various aspects of cropping systems, such as cover crops in rotations and soil quality. I also sought out collaborators in agricultural economics and rural sociology to explore the possibilities of whole-farm research. It was no accident that two of the collaborators brought to the group a strong background in farming systems research in an international agriculture context. The vital need for whole-farm planning and assessment tools was brought to the forefront by a series of conferences and proceedings published at about that time (reviewed in Freyenberger et al., 1997), and in a policy planning meeting in Washington, DC, hosted by a consortium of sustainable ag organizations attempting to interject sustainable agriculture language and concepts into the next Farm Bill. We talked at length in planning groups about the need to tie financial benefits to whole-farm planning . . . but where were the tools that would allow us to take this step?

Our group at KSU attempted to answer that question initially by taking on two projects. The first was a Sustainable Agricultural Research Education (SARE) grant called "Back to the Future," where we critically reviewed, cataloged, and then summarized research and extension literature published in Kansas. The goal was to determine if any of these publications, especially those written prior to the 1940s and 1950s, included credible information on sustainable agriculture practices and a whole-farm perspective that could be useful today. We were initially disappointed. Many of the older publications, even though they included more of a systems perspective, contained information that could not be used today without substantial editing and explanation. For example, a publication on apple orchards included good advice on cover cropping, site location, and soil quality, but also recommended the use of arsenic to kill insect pests! We managed to find concentrations of literature on topics such as cover cropping, alternative pest management, high value crops, and soil management, and summarized these on a Website, along with copies of the original documents, for others to use (www.ksu.edu/issa/sare). However, we did not find any particularly enlightening approaches to whole-farm planning or assessment.

Our second project took the question directly to farmers. In the fall of 1995 and 1996, focus group interviews were conducted with four groups of farmers, two with a sustainable ag orientation, and two that were simply local groups of representative farmers recommended by their county agents. We asked them to specifically address these questions:

1. How would you describe your approach to farming?
2. How did you develop your approach?
3. What kind of yardsticks do you use to evaluate your approach?
4. What other evaluation and planning tools would be helpful to you?

We learned that several of the farmers made some use of financial planning tools, including software programs to track farm expenses, and KSU Farm Management Association record keeping and tax preparation services (Norman et al., 2000). We also learned that those farmers who had attended one or more classes offered by "Holistic Management™," a private organization based in New Mexico, benefited greatly from that experience. They were more likely to have written farm goals, including family goals and resource management goals, as well as financial goals. The farmers in the sustainable ag focus groups were more likely to have taken these courses, and could articulate those goals most clearly. Other farmers spoke more of simply maintaining farming as a "way of life," while the sustainable ag farmers spoke of seeking "quality of life." One farmer even stated that his college degree in liberal arts gave him more of a systems perspective and tools for approaching sustainable ag than if he had received his degree in an agricultural discipline. Perhaps this reminds us that maybe whole-farm planning is an art, as much as a science? The sustainable ag farmers also described the importance of taking "thinking risk," in lieu of financial risk. Financial risk is incurred when farms take on the high debt load for farm expansion, but without significant changes in how the farm operates or in the array of farm enterprises.

When asked about yardsticks that they use to evaluate the sustainability of agriculture, the sustainable agriculture farmers generally agreed on the following: (1) reducing time required for farm operations, (2) ensuring community sustainability, (3) not making financial feasibility the only consideration, and (4) ensuring ecological sustainability. When asked about planning tools that allowed them to take into account these yardsticks and others, they had few suggestions. Beyond the financial planning tools and Holistic Management courses described above, farmers were not aware of any formalized analytical framework that allowed them to evaluate or plan for these goals. There was also some reluctance to enter into any planning exercise with a lot of time invested up-front without some significant benefit at the end. They also felt that an integrated index measuring changes had no particular utility. Farmer to farmer communication was seen as extremely valuable, especially when starting the transition to sustainable agriculture.

We began to wonder if the appropriate tools simply did not exist, or whether they did exist, but weren't widely available or known to this group of farmers. Our next project was to conduct an extensive review of the literature, including indicators of sustainability, how to measure them, and on planning tools, including both component tools and those for whole-farm integration.

LITERATURE REVIEW AND TESTING THE TOOLS

Our literature review on indicators of sustainability found some articles in published sources, especially on specific indicators such as soil quality (see summaries in Freyenberger et al., 1997). However, much of the work was still in the "gray" literature, in edited proceedings of recent conferences. Over 80 articles are summarized in this publication, and placed in categories such as overview papers, frameworks for classifying indicators, bio-physical indicators, socioeconomic and community indicators, policy papers, and indicators of sustainability that would be more user-friendly for farmers. Most of the literature was written by researchers, for other researchers, but a few of the articles summarized tools, indices, and ratios that could be useful to farmers and were written in a practical, accessible language.

In our search for planning tools developed specifically for farmers we found 11 (see Table 5.1) that presented some sort of overview or planning process (including a "Balanced Farming and Family Living" program that I had attended with my parents when I was two years old), two that could be used as decision-aid software, at least 19 that could be used for financial planning or tracking, and 14 that were designed specifically for nutrient or carbon tracking on-farm (Janke and Freyenberger, 1997). Clear guidelines for elements of a whole-farm plan were also described in the literature (Table 5.2), and though they differed in the specific number of elements, the lists had many concepts in common and illustrated consistency of ideas. The answer to our question as to whether planning tools existed was "yes." We now needed to determine if any of these tools were actually useful in a farming context.

TABLE 5.1. Summary of comprehensive whole-farm planning programs and tools.

Name of tool or program	Type of tool			Planning tool includes						Ease of use	
	Software	Work book	Text	Accounting or bookkeeping	Financial analysis	Environmental assessment	Social impact	Forward planning	Whole-farm perspective	Training required for use	Farmer friendly
Balanced farming			A	B	B	B	B	B	A	A	A
Crop rotations options program	A				A	A			A	B	B
Farm-A-Syst		A				A		A	B	A	A
Holistic management	A	A	A	A	A	A	A	A	A	A	B
Natural Resources Conservation Service		A				A		B	B		B
Ontario Environmental Farm Plan		A	B			A		A	A	A	A
Organic Certification		A				A		B	A		A
PLANETOR	A				A	A	B	A	A	A	B
Ranching for Profit		A	A		A	B	B	A	B	A	A
Western Integrated Ranch/Farm Education		A	A		A	A	A	A	A	A	A
Wisconsin Soil Health Scorecard		A				A		B	B		A

Source: Reviewed in Janke and Freyenberger, 1997, http://www.oznet.ksu.edu/sustainableag/Pubs_kcsaac/ksas3.htm.

Note: Since the 1997 publication, some of these tools have been expanded in their use, and others have been discontinued. Please see recent sources for current availability.

A = well developed; B = Evident, but could be improved; Blank = not evident.

TABLE 5.2. Key reference describing elements of a whole-farm plan.

Title	Reference	Elements
Successful Whole-Farm Planning: Essential Elements Recommended by the Great Lakes Basin Farm Planning Network	Kemp, L., 1996	1. Farm family goals 2. Economic viability 3. Water quality 4. Soil conservation 5. Nutrient management 6. Water management 7. Pest management 8. Soil quality 9. Crop rotations 10. Tillage
Holistic Management Course	Kroos, R.H., 1997	1. Define the whole that is managed, and define reasons for change 2. Identify the effectiveness of the ecosystem processes, and dependence on these ecosystem processes 3. Define a three-part goal for the future (people, finances, and land) 4. Brainstorm and select tools or actions, and test the ecological, financial, and social soundness of the actions 5. Plan, monitor, control, re-plan
An Evaluation of Tools for Whole-Farm Planning	Mulla, D.J., L.A. Everett, and J.L. Anderson 1997	1. Farm family goals 2. Inventory and assessment of farm resources 3. Action plan 4. Monitoring of progress toward the goals
Western Integrated Farm/Ranch Education	Hewlett, J.P., 1995	1. Strategic planning (establish goals, inventory resources) 2. Tactical planning (explore possible enterprises) 3. Develop enterprise plans, and develop the flow of resources 4. Operational planning (implement plans, monitor and adjust, and replan

TABLE 5.2 *(continued)*

Title	Reference	Elements
What is Comprehensive Farm Planning?	Kemp, 1996	1. Inventory farm resources, including soil tests and maps, cropping plan, economic data, and farm site information
		2. Develop goals for profitability, pollution prevention production and long-term ecosystem enhancement
		3. Analysis of management options, identifying problems and opportunities in the context of regulatory constraints
		4. A strategy for putting the plan into action, as well as to monitor and evaluate how the plan is working

A colleague and I offered a special topics course on whole-farm planning at KSU in the spring semester for two years. In 1997 we offered the course with a distance learning option (Telenet), and involved farmers and staff at the Kansas Rural Center as well as a few other farmers in rural Kansas. In 1998, the course was primarily taken by students on campus, but they were required to work with farmers known to them as they prepared the class assignments. Each spring the course students met for two hours per week, and one or two planning tools were discussed in each session. The homework consisted of trying out the planning tool, either on their own farm, or in collaboration with a farmer. This gave us a harsh reality check on the time investment required, and also the benefit or usefulness of each tool. About 12 to 15 different tools were presented to each class, chosen from the pool identified in our literature review as having the highest probability of usefulness. A third evaluation of this pool of possible tools was conducted with the help of county ag agents in a 2-h workshop at the annual KSU conference in the fall of 1997. Small groups of agents examined each tool, and then presented their evaluation of the tool to the larger group, based on their experience in working with farmers.

What we learned from these exercises was that no one tool really met all the needs for farm planning at this time. There were elements of several tools that seemed valuable however, and we decided that combining several tools into a package might be helpful. For example, aspects of goal setting contained within several of the tools seemed to be useful, and these ideas were new to most farmers. Also, resource and farming practice assessment tools such as those in "Farm-A-Syst" and the "Ontario Farm Plan" were easy to use, as they consisted of a checklist based on best management practices. They gave the farmers helpful information about how their farm was doing, and risks that they might be incurring, for example, when looking at farmstead safety or best management practices for conserving soil or enhancing soil quality. Financial planning tools were also seen as helpful. Even something as simple as an enterprise budget worksheet may be needed on some farms. One farmer who took the Telenet course, who already had over 40 years of farming experience, exclaimed that he didn't realize before how much money he was losing on wheat! He was not convinced that he should stop growing wheat, however, since "Kansas is the wheat state."

PUTTING TOGETHER A TOOL FOR KANSAS

Another SARE grant (collaboration between KSU and Kansas Rural Center [KRC]) at this time allowed us to put together a package of tools, and to work closely with four groups of farmers in a "study circle" or learning group format. Staff at KSU and KRC served as facilitators, and the goal setting and resource planning format from the Holistic Management course was used as the starting point. A certified Holistic Management instructor from out-of-state was brought in for two intensive weekend training sessions, and one person from KSU began his apprenticeship to become a certified Holistic Management instructor. The intensive course was well received by all, and the follow-up small group meetings were deemed very helpful by the groups that participated. However, once the initial broadbrush farm plans were created by participants, interest lagged in using any of the more specific tools, for example, nutrient management planning.

Training was also obtained for KSU and KRC staff in writing up decision case studies from lessons learned by farmers in these groups. A few case studies were written and used as part of the training materials for these four farmer study circles. However, each decision case included only a small snapshot of the overall farm, and an overview of the farming operation seemed to be more useful to other farmers, beginning a transition to a new enterprise or when adopting new sustainable ag practices. After this, farm case studies rather than decision cases were written up and published as a result of the exercise. These were used in subsequent agent trainings in the SARE PDP (Professional Development) programs and in Kansas Cooperative Extension programming (http://www.oznet.ksu.edu/sustainableag/farmer_profiles.html).

At about this same time, the agriculture committee within the Kaw Valley Heritage Alliance wanted to develop a "River Friendly Farmer" award. This committee represented a collaboration among KSU, the Kansas Rural

Center, state agencies, one of the commodity associations, and interested private citizens. The committee had a small amount of funding from an EPA grant, and our original intent was to model the award program after a similar program just beginning in another state. However, we had to develop our own criteria to choose the recipients of the award in a manner that was most relevant to Kansas conditions, climate, and rural culture.

We seemed to have two options available. The other state had simply used a one-page checklist. The Ontario Farm Plan was a 223-page workbook, with over 23 sections and an action plan. It had a high "thud factor"— the sound of a large notebook dropping on a table. We also had the Farm-A-Syst worksheets, with 12 sections, covering primarily farmstead health and safety issues. The Ontario Farm Plan had initially been adapted from Wisconsin Farm-A-Syst, and expanded to include fields, streams, dairy parlors, and other elements. Rather than use any of these tools in their present form, the committee, decided to adapt them to Kansas "best management practices" and to shorten the length of the tool if possible.

Initially, the 23 sections from the Ontario notebook were rearranged, and condensed into 4 sections, dealing with (1) natural resource quality (soil and water), (2) nutrient management, (3) pest management, and (4) livestock and manure management. In a later edition, two more sections were added, incorporating the Farm-A-Syst farmstead assessment sections, and a new section on irrigation management to specifically meet the needs of irrigators. The format of a self-rating of 1-4 was maintained, with 4 = exceeds best management, 3 = meets best management recommendations (at this time), 2 = does not meet best management criteria, and 1 = in need of serious improvement, or puts the farm at risk for health, safety, or environmental concerns. In an early version, we tried a 10-point scale rather than the 4-point scale, in order to capture some of the subtleties of real life, but found that the scale became confusing, and it was difficult to describe criteria at that many levels.

We also maintained the action plan portion of the Ontario Farm Plan, as it gives the farmer specific steps to take, rather than simply summarizing the results as a score or an index of sustainability. However, we realized that the REAL action plan is going to depend on farmers implementing the plan within the context of farm goals, so goal setting or "quality of life" worksheets were added to the front of the notebook. We added a field, pasture, stream, and pond inventory section in order to more precisely assess natural resource quality. This was preferable to coming up with a rating for the whole-farm for such indicators as soil erosion. Finally, a scorecard-prioritization section was added as a transition step between the worksheets and the action plan to allow farmers to determine for their specific situation, whether something that ranks "low" on the scorecard is really a high priority for change.

After about a year of writing, adapting, and reviewing by KSU subject matter specialists, the notebook was ready for field testing. Initial tests by members of the writing committee were eye-opening. For my own small farm of only 10 acres, 25 chickens, 15 sheep, and a garden, it took 8 h to complete the notebook! However, I learnt a lot about the farm in the process, even after helping to write the notebook. The goal setting section actually took the longest. How would other Kansas farmers react to the material? How would they respond to the time needed to complete the notebook?

TESTING, IMPLEMENTATION, AND OUTREACH

The first version of the notebook was pilot tested in 1998 with 25 farms, recruited by KRC staff. Farmers were reimbursed with $250 for their time and energy, and provided with the opportunity to apply for a larger grant on a cost-sharing basis for improvements on their farms. At that time, and continuing to the present, about half to three-quarters of the farmers who express initial interest end up completing the notebook. This range seems to reflect differing pressures farmers feel regarding regulation at the time they sign up, and also the possibilities for cost-sharing at any given time. Many of them also comment that the goal setting section is the hardest to complete, especially if there are several members of the family involved in the farm, landlords to keep happy, and many possible directions to go with the farm. The farm inventory section is also cumbersome, especially when some farms today involve several sections of land and dozens of fields. Farmers were encouraged to at least put a part of the farm in the inventory— either the "home place" or the part of the farm where they were seeking cost share for environmental improvements.

The checklist portion of the farm assessment goes fairly smoothly and quickly, but it is helpful if a project field assistant is nearby or if the notebooks are completed in a group setting. Questions occasionally come up regarding terminology, subtleties between categories, or explanations regarding what is listed as the best management practice. The score-card is straightforward and the action plans are not only helpful to the farmers, but also to KRC staff who are determining whether a particular cost-share request should be granted.

In this and other pilot tests, we asked participants if any questions should be dropped, added, or terminology clarified. So far, no one has identified a question that should be dropped. The wording was changed for a few questions, but not the content. The farmstead assessment (including well water, waste water disposal, other elements) was added for a couple of several reasons. First, the Kansas Department of Health and Environment (KDHE) staff on one of the project advisory committees suggested it. KRC field staff working closely with farmers in filling out the notebooks also noted that household waste disposal issues came up occasionally during farm visits. And finally, the KSU Extension staff responsible for the Farm-A-Syst assessment program felt that it would streamline the process and the publications to include Farm-A-Syst within the River Friendly Farms assessment notebook. Also, adding the irrigation management section makes the notebook more relevant to western KS farmers and other irrigators, as this material is not covered in other sections. In future versions, it is possible that new questions may be added to the livestock section, as more data become available on the risk of prophylactic antibiotic use and other livestock-human disease risk factors.

After two years of pilot testing and revision, another grant to KSU through the SARE PDP program allowed the county agent and agency staff training on whole-farm planning using the River Friendly Farms notebook. Five farm families agreed to host the workshops and become live case studies. The two-day workshop format included some orientation to the mechanics of the notebook and whole-farm planning, a farm visit to the host farm, and then sharing the scorecard, action plan, and farm goals developed by each host farm. The workshop participants' job on the second day was to explore options available to the farm and determine what, if any, technical assistance might be helpful to the farmer in implementing their action plan. Another important aspect of the farm planning and training was a FINPAK analysis of the farm's financial picture, and forward projection of various action plan scenarios using FINLRB. In some cases, "representative" farm financial data was presented to the workshop participants rather than actual farm data, in order to preserve confidentiality for the host farms' financial records.

Combining the financial analysis with the environmental assessment was the most eye-opening part of the entire exercise. It really drove home the point that these two planning processes *must* be linked in order for any action plan to be implemented on a farm. An Extension fact sheet is not now available (Janke, 2000), and specific examples have been provided (Janke, 2002). The range of options available to farmers for a particular environmental concern range from "do nothing," to "low-cost solution" (ranging from 100's to 1000's of dollars), to "high-cost solution" which may involve moving a feed lot, a barn, or other facility, with a price tag of up to $100,000. Some items on the action plan may be less expensive, such as "keep better pesticide application records" or "add legume to crop rotation."

Where are we now? We are currently in the "post pilot testing" phase, but not quite to the widespread implementation or adoption phase of whole-farm planning and assessment. Through an EPA 319 grant to the KRC, the $250 one-time payment is still available to farmers who fill out the notebook and some of them are also eligible for larger grants for farm improvements. The KSU watershed extension specialists occasionally work with farmers who are interested in filling out a notebook, but most are referred to the Rural Center and then to KSU and Natural Resource Conservation Service (NRCS) for technical assistance. Some farmers also take advantage of the opportunity for low-cost FINPAK analysis, but many do not. The current version of the notebook is available through KSU, the KRC, or on the web (Janke and Nagengast, 2002 or http://www.oznet.ksu.edu/library/h20ql2/S138.pdf). We never did develop an awards program based on the River Friendly concept, but we did make the notebook shorter than the Ontario version. Ours has only 116 pages, not counting the appendices. A unique video explaining the whole-farm planning process using the River Friendly Farm plan notebook was created under the leadership of Jim French, with a small grant to the Kansas Rural Center. This should help with our outreach efforts during this next phase.

Informal feedback from farmers indicates that they learn plenty from filling out the notebook, both about recommended best management practices and about their own farms. They also report that the notebook is a real time commitment, and that yes, the goal section is still the hardest to complete. We hesitate at this time to either shorten the notebook or to take out the goal setting without more formal assessment of the value of these sections. We are concerned about the downside to the overall planning if these were left out. We also talked about eliminating the wildlife assessment exercise, since it is important, but not central to environmental planning. However, at a recent workshop, two of the three farmers presenting their experience with farm planning spoke highly of the wildlife assessment, and said it was one of their favorite sections of the notebook. Reasons for coming to the program range from "wanting to do the right thing" to "would like access to cost-share" to "my neighbor reported my livestock operation to KDHE because I shot his dog."

Next steps involve getting more buy-in from the various farm organizations, and access to more cost-share dollars through state and federal farm programs to make it worth farmers' time to complete the notebook. The simple satisfaction of doing a farm plan will not be enough for most people, until its benefits are widely accepted. The notebook could become a required step to obtaining farm loans or government subsidies. Financial planning needs to be either highly encouraged or required, in addition to the intrinsic and practical value from the environmental assessment offered in the River Friendly Farms notebook.

Another needed project is a formal survey or set of structured interviews to determine exactly what farmers are getting from the planning process now and to follow up on the farmers who completed notebooks three years ago. We need to see how having this information influenced their farming practices. Both participants and nonparticipants would be interviewed, and their attitudes and actions compared.

Soil and water testing recommendations, field test kit sources, and interpretation guidelines are also being developed to complement the River Friendly Farm notebook. This is funded through KDHE and Environmental Protection Agency (EPA) grants, and is currently being pilot tested with high school agricultural classes and selected farmers from the Clean Water Farms program (http://www.oznet. ksu.edu/kswater/).

OTHER FORMS OF WHOLE-FARM PLANNING

A couple of years ago, a SARE grant to the KRC sponsored business planning training using the NxLevel curriculum, adapted by Nebraska for agricultural producers. Jerry Jost and Claire Homitsky became certified trainers, and conducted several classes across Kansas. I was fortunate to be a participant as a small farmer, in one of those classes. In this situation as in many others, I found that my extension work and role as a small farmer complement one another (for more info, see Janke, 2003). Again, it's amazing how much time it takes to write a plan, especially a business plan for a 10 acre tract and a few animals, and I cannot even say I did a particularly good job on the financial sections. It did reinforce the need for ALL farms to have a business plan. The NxLevel curriculum complements the environmental assessment found in the River Friendly Farms notebook, and also the record keeping and forward planning offered by FINPAK (2005) and FINLRB [http://www.cffm.umn.edu/Software/FINPACK/]. As a side benefit from the class, I now have two new business partners (my carpool companions) in a start-up CSA. The farmer-to-farmer interaction in the class was a real strength, along with the quality of the material, the instructors, and the guest speakers.

Currently, harvesting equipment with yield monitors and computer programs allow farmers to integrate this information and practice in what some call "precision farming." These tools allow farmers to get a better feel of the within-field variability in yield, and sometimes for soil nutrients, water content, and other factors if the proper data has been collected and entered into the program. Precision farming does not help farmers integrate that crop field with livestock or other enterprises that may be on the farm, nor with any plans for the future. It is one more tool that allows fine-tuning of some production inputs, but is not a prescription for sustainable or even efficient farming.

The demand for organic food has continued to increase at a rate of 20 percent per year in the United States for over two decades. As more farmers become certified to meet this demand and as some farms become larger, organic certification becomes a form of whole-farm planning. Certainly the crop rotation documentation, field maps, production, and storage records lend themselves to the "monitoring" aspect of whole-farm planning. However, organic certification does not require goal setting or financial planning—two other critical aspects of whole-farm planning.

In a similar vein, new "eco-labels" or green labels are springing up. Some don't require any documentation, and are little more than marketing gimmicks, but the better labels require documentation and set certain strict standards for participation in a program. For example, Wegmans has developed an IPM label in collaboration with Cornell University that requires adherence to specific pest management guidelines and protocols. A comprehensive whole-farm plan is not required (www.eco-labels.org).

A quick survey of the recent literature, to see if a lot had progressed since our survey in 1997, revealed that nutrient management is still an environmental concern and model development continues to be a widely used approach. As a recent reviewer of USDA/ARS systems projects, I came across many that proposed the use of some sort of tool to integrate the results and present them to farmers. One of them proposed many high-tech monitoring techniques—spatially precise geographic information systems, satellite images, plant reflectance to monitor stress—in great detail, but their integrative tool was left up to the imagination of the reviewers, and "would be developed later by the team."

In looking at where Ontario has gone with its program, after which ours is modeled, the news is encouraging. In the first 10 years of the program, over 20,000 Ontario farmers have completed farm plans. The Ontario government just gave them another 4 year grant for over $100 million to continue the program. Alberta and Saskatchewan began promoting the program in February and April of 2003, and Prince Edward Island (PEI) has had a program in place since 1996. The PEI program is particularly interesting because having a farm plan is now a requirement for taking part in virtually any provincial government program. Many financial institutions are also requesting plans from their agricultural clients. Over 1,100 farmers have participated in workshops in this small province, and some counties have a participation rate as high as 80 percent (The Guardian [Charlottetown, Prince Edward Island], Dec. 23, 2003). Perhaps U.S. banks and government programs, especially conservation programs, should require having an environmental farm plan in place? Similar programs exist in Newfoundland, New Brunswick, Nova Scotia, and Quebec. A search using LexisNexis to access articles in the popular press reveals that similar whole-farm planning efforts are underway or available in Tasmania–Australia, New Zealand, and Scotland.

CONCLUSIONS

In the past, whole-farm planning was common only in extremely large agri-business operations, where bank loans or other forms of external financing required it, or where operations were so large (e.g., corporate hogs), that one would be foolhardy to proceed without a plan. The sustainable ag community has promoted the idea of whole-farm planning for small and medium sized family farmers through political discussion, research, and extension activities. Environmental as well as financial sustainability have both been emphasized, along with goal setting, to meet personal goals as well as community responsibilities. It may be ironic, but fitting, if the tools initially developed to promote environmental stewardship end up "saving the family farm" financially. We have always maintained in the sustainable ag community that these goals are not diametrically opposed, but can work together. It would be a nice thing if 10 years down the road we can look back at some of these tools and efforts to promote whole-farm planning, and say that this work contributed to preserving farms and rural quality of life, as well as environmental quality.

REFERENCES

Bacon, S.C., L.E. Lanyon, and R.M. Schlauder, Jr. 1990. Plant nutrient flow in the managed pathways of an intensive dairy farm. *Agron. J.* 82:755-761.

FINPAK. 2005. *Center for Farm Financial Management.* University of Minnesota, St. Paul, MN. Available from: http://www.cffm.umn.edu/products/finpack.aspx. Accessed March 28, 2006.

Freyenberger, S., R. Janke, and D. Norman. 1997. *Indicators of sustainability in whole-farm planning: A literature review.* Kansas Sustainable Agriculture Series. AES Contr. No. 97-482-D. Sustainable Agriculture Series, Publication No. 2.

Hewlett, J.P. 1995. *The Wyoming WIRE (Western Integrated Farm/Ranch Education) team, and the Western Regional WIRE-SARE Coordinating Committee.* Department of Ag. Economics, University of Wyoming, Laramie.

Janke, R. 2000. *whole-farm planning.* Fact Sheet MF-2403, Kansas Cooperative Extension, Kansas State University, Manhattan.

————. 2002. Composing a landscape. In: *The farm as a natural habitat,* D.L. Jackson and L.L. Jackson (ed.). Island Press, Covelo, CA. Chapter 14, pp. 209-219.

————. 2003. Pedagogical farming: One experience combining horticultural research, extension education, and organic farming. Chapter 17 in: *Community and the world: Participating in social change through local to global connections,* T. Dickinson (ed.). Nova Science Publishing Inc., New York.

Janke, R. and S. Freyenberger. 1997. *Indicators of sustainability in whole-farm planning: Planning tools.* Kansas Sustainable Agriculture Series. AES Contr. No. 98-124-D. Sustainable Agriculture Series, Publication No. 3, Kansas State University, Manhattan.

Janke, R. and D. Nagengast. 2002. *The River Friendly Farm environmental assessment tool—revised version.* Publ. No. S-138, Kansas Cooperative Extension, Kansas State University, Manhattan.

Kemp, L. 1996. *Successful whole-farm planning: Essential elements recommended by the Great Lakes Basin Farm Planning Network.* The Minnesota Project, St. Paul, MN. Available on line: http: //www.mnproject.org/publications/successfulWFP.pdf. Accessed 28 March 2006.

Kroos, R.H. 1997. *Your comprehensive guide to the study and practice of holistic management.* Crossroads & Company, Belgrade, MT.

Lanyon, L.E. and D.B. Beegle. 1989. The role of on-farm nutrient balance assessments in an integrated approach to nutrient management. *J. Soil Water Conserv.* 644:164-168.

Mulla, D.J., L.A. Everett and J.L. Anderson. 1997. An evaluation of tools for whole-farm planning. *Amercian Society of Agronomy Abstracts.* p. 33.

Norman, D., L. Bloomquist, R. Janke, S. Freyenberger, J. Jost, B. Schurle, and H. Kok. 2000. Sustainable agriculture: Reflections of a few Kansas practitioners. *Am. J. Altern. Agric.* 15:129-136.

Chapter 6

Economic Analysis and Multiple Impact Valuation Strategies

John Ikerd

INTRODUCTION

The economics of sustainable farming has been a controversial issue since the emergence of sustainable agriculture as a significant public issue during the late 1980s. Opponents of United States Department of Agriculture's (USDA) original Low Input Sustainable Agriculture (LISA) program dubbed it as Low Input "Subsistence" Agriculture, with the indictment that LISA farmers would never be able make a decent economic living. Others wanted to define sustainable agriculture as profitable agriculture, with claims that sustainable farming must be profitable farming, period. More serious questions were raised concerning whether or not farming systems that were ecologically sound and socially responsible, as sustainable farms must be, can also be economic viable. So the emergence of sustainable agriculture during the late 1980s brought with it a public mandate for agricultural economists to assess the short run economic feasibility and long run economic viability of agricultural professionals (in this chapter called farmers), shifting to more ecologically sound and socially responsible systems of farming.

Agricultural economists responded to this new public mandate in various ways. In reality, most economists essentially ignored it. Sustainable agriculture, with its emphasis on the potentially positive ecological and social

---

Editors' Note: John Ikerd recently published a book, *Sustainable Capitalism:a Matter of Common Sense* [Kumarian Press, Bloomfield, Connecticut, 2005], after he wrote this chapter. The book expands on many of the themes included here.

aspects of farming, somehow seemed to threaten the status of economics as the driving force for change in American agriculture. Historically, the environment and society had been viewed by economists as "constraints" to greater profits and growth, not as sources of renewable capital or direct contributors to quality of life. Economics, as a profession, had gained new credibility during the 1980s during an era of global "free markets" and "privatization." Economists were not inclined to allow questions of sustainability diminish this newfound respectability.

The new sustainable paradigm suggested that contemporary approaches to economic development were extractive and exploitative, and perhaps economists should reexamine the very foundations of their discipline. As most economists in positions of influence were simply unwilling to accept the risks of starting to think all over again, they quickly joined the attack on sustainable agriculture, without seriously examining the legitimacy of the questions of sustainability. Most without influence were unwilling to risk their professional advancement by challenging their profession; they chose to pursue other less controversial issues.

However, a small core group of agricultural economists did accept the challenge. Some stayed above the fray over whether sustainable farming meant going back to farms of the 40s, with horses and hoe handles, whether farmers were already the country's greatest environmentalists, or whether today's farms were simply too large or too specialized to be sustainable. Instead, they focused their time and efforts on evaluations of the comparative economics of conventional versus alternative farming practices and methods, assessing whether more sustainable farming might be more or less profitable. Others synthesized sets of alternative farming practices and methods to evaluate alternative economic enterprises, while others combined enterprises to simulate and compare the economic performance of *conventional* and *sustainable* whole-farm systems.

Thankfully, agricultural economists have been joined by agronomists, animal scientists, soil scientists, rural sociologists, anthropologists, farmers, and others who have recognized the need for economic analysis of sustainable farming systems. No scientists who are serious about understanding and addressing the important issues of sustainability can afford the luxury of staying within the narrow bounds of their respective disciplines. Noneconomists have made valuable contributions in addressing the economics of sustainable farming, just as economists have made significant contributions in addressing ecological and social dimensions of sustainability.

In addition, a growing number of sustainable agriculture researchers, educators, and practitioners seem willing to challenge the adequacy of contemporary neoconservative economic thinking in addressing the most important issues of sustainability. Ultimately, economic analysis of sustainable farming systems must be made within the context of multiple impact evaluation strategies, integrating the economic, ecological, and social implications of each management decision. Economists and others will continue their uninformed attacks on the economic performance of sustainable agriculture until they are forced to confront the inherent exploitative and extractive nature of contemporary economic development strategies—including conventional approaches to farming.

COMPARISONS OF ALTERNATIVE FARMING PRACTICES, METHODS, AND ENTERPRISES

Farming *practices* are specific farming activities such as mulching, plowing, planting, feeding, cultivating, spraying, threshing, loading, and shipping. Farming *methods* are combinations of farming practices selected to perform some basic function of crop or livestock production, such as genetic selection, germination or birth, nourishment, protection, and marketing. Farming *enterprises* include all of the farming practices and methods associated with a particular crop or livestock production process, usually but not necessarily resulting in a marketable product. Farming *systems* are integrated combinations of farm enterprises, organized according to some set of principles to achieve an overall purpose.

Sustainability is a characteristic of whole farming systems, not of farming practices, methods, or enterprises. Thus, economic analyses of specific farming enterprises and methods must be based on assumptions, explicit or implicit, that the practices and methods being evaluated are either more or less likely to be associated with more sustainable whole-farm systems. For example, conservation tillage practices, such as ridge tillage and direct seeding, being more ecologically sound than conventional tillage, might be logical components of sustainable farming systems. Integrated pest management (IPM) is a logical alternative method of managing pests in sustainable farming because its reduced reliance on chemical pesticides reduces risks to the environment. Many of the early economic analyses in sustainable agriculture focused on comparing the economic implications of conventional farming practices and methods with alternatives considered good candidates for inclusion in more sustainable whole-farm systems.

Low input farming required less reliance on off-farm inputs, such as pesticides and fertilizers, and greater reliance on management of resources internal to the farm, such as utilization of crop rotations and cover crops. Thus, farming practices or methods that reduced reliance on fertilizers, pesticides, fuel, or borrowed money were logical subjects for economic analysis. Integrated pest management, below-label rate application of pesticides, and reduced-rate nitrogen application also were popular subjects for economic analysis during the 1990s. Soil testing, mulching, composting, and other nonchemical approaches to maintaining soil fertility were also subjected to economic analyses. Economic analyses of alternative farming practices and methods typically focus on relative costs, as it is difficult to evaluate the contribution of a specific practice or method to the overall value of production.

The reference list includes a small sample from hundreds of such studies that have been carried out since sustainable agriculture emerged as a research topic in the late 1980s. The list also includes references to reports of on-farm trials and demonstrations, which lack the statistical validity of research trials, but nonetheless provide some interesting insights. Obviously, the different studies led to somewhat different conclusions regarding the economic feasibility of farmers' adopting alternative practices and methods in their quest for sustainability. Some studies showed economic advantages for the alternatives while others showed some sacrifices in cost or effectiveness. In general, most studies led to conclusions that the alternatives were at least competitive with conventional farming practices and methods, in terms of both cost and effectiveness in maintaining productivity. And equally important, there was reason to believe the alternatives were more ecologically sound or socially responsible than their conventional counterparts.

Farm enterprises provide a convenient level of aggregation of farming practices and methods for purposes of economic analysis because enterprises typically result in a saleable product. Typical farm enterprises include corn, hogs, vegetables, soybeans, beef cattle, milk, fruit, and berries. Farms may also include intermediate enterprises such as pastures, cover crops, and green manure crops, typically valued economically as inputs to other enterprises that produce saleable products. Regardless, the relative ease of assessing value of production has contributed to the popularity of economic analysis of conventional and sustainable enterprises.

The differences between enterprise analysis and the analysis of practices and methods are subtle and may not seem too important to agricultural entrepreneurs. When the purpose of an analysis is to compare the economic performance of one enterprise identified as conventional, and another identified as alternative or sustainable, attention must be given to all of the farming practices and methods used in each enterprise. For example, an economist can evaluate conventional and no-till cultivation in corn production, focusing on the costs of reducing soil loss, without assessing other differences between corn enterprises that might logically utilize the two different tillage systems. However, if an economist compares a no-till corn enterprise with a conventionally tilled corn enterprise, the economic implications of all practices and methods associated with each enterprise must be considered. For example, the researcher must consider differences in types and amounts of herbicides, fuel and equipment costs, labor costs and timing of activities, and farm size, as well as differences in yields, costs, and returns.

An analysis of differences in costs may be adequate when the focus is on economic evaluation of a specific farming practice or method, even when the practice or method is used for a specific enterprise. But, when the focus is on one enterprise, the economic implications of all practices and methods utilized for the enterprise must be given consideration. The reference list includes studies conducted over the past fifteen years evaluating the economics of sustainable farming at farm enterprise level of aggregation. As with practices and methods, different studies have led to different conclusions regarding the economic feasibility of conventional and alternative or sustainable farm enterprises. For a wide range of farm enterprises, previous economic analyses have shown that more ecologically sound and socially responsible alternatives can at least be as profitable as their conventional counterparts.

COMPARISONS OF ALTERNATIVE FARMING SYSTEMS

The lowest practical level of aggregation for assessing overall agricultural sustainability is at the whole-farm level. Realistically, economic viability is associated with economic organizations as wholes, not with specific practices, methods, or enterprises. A profitable farming operation may include one or more unprofitable enterprises, which nonetheless have valuable attributes that support other enterprises or the farm as a whole, and thus contribute to profitability of the whole-farm system. Historically, hogs and corn have been complementary enterprises within diversified farming operations. In some years, losses in the hog enterprise, resulting from high-priced corn, were offset by profits in the corn enterprise. In other years, profits from hogs, resulting from low corn prices, offset losses in the corn enterprises.

Diversification among enterprises within whole-farm systems provides opportunities to create whole-farm systems that enhance sustainability through complementary relationships not limited to economics. Synergisms among enterprises can enhance the ecological integrity of farming operations, as when waste from one enterprise becomes a productive input for another. Complementary enterprises can also improve the social or quality of life dimension of farming by spreading labor requirements across seasons or creating opportunities for all family members to participate in appropriate activities. A diversified farm is a complex living organization of soils, plants, animals, and people. Organizations that function in harmony are inherently more resistant, resilient, regenerative, and thus more sustainable.

Obviously, farm enterprises often are competitive rather than complementary—competing for limited planning time, labor, land, and capital. However, the overall complementary or competitive nature of relationships among enterprises depends on the nature types of specific farming practices and methods used in carrying out the various enterprises. So evaluation of whole-farm systems must take account of the relationships among the various farming enterprises, methods, and practices that contribute to or detract from economic, ecological, and social sustainability.

It is difficult to evaluate whole-farm systems within design constraints imposed by accepted scientific methods. For example, comparisons of conventional and sustainable farms would require detailed observations from two sets of farms that were alike in all respects, except for those characteristics used to classify them as either conventional or sustainable. Each set would need to be identical, or at least similar, in size, soil types, production history, and management expertise. The performance of paired farms, differing only in their conventional and sustainable characteristics, would need to be observed over a number growing seasons—with each farm experiencing the same external conditions such as weather, markets, and farm subsidies. A few studies have focused on comparisons of actual farming operations that were considered similar in most respect other than specific characteristics logically associated with sustainability. Most researchers avoided the inherent difficulties of whole-farm comparisons by focusing their attention on logical subsystems of conventional and alternative farming systems. Thus, most farming systems research has focused on subsystems or multi-enterprise components rather than whole-farm systems.

Alternative crop rotations have been studied extensively. Mono-cropping of corn or soybeans has been compared with corn-soybean rotations, showing yield advantages for both crops grown in rotation. Extended rotations, adding clover and wheat to corn and soybeans, have shown further economic and ecological benefits. Similar comparisons of alternative vegetable rotations have been studied extensively by sustainable agriculture researchers, showing economic and ecological advantages for diversification over mono-crop systems. While not a simple task, conventional and alternative cropping systems can be replicated both within years and over time on a number research farms, including those of universities, the USDA, and nonprofit organizations.

Obviously, the multienterprise subsystems studied on research farms may be similar to the total complement of enterprises that comprise the whole-farm system on some farms. Thus economic analyses of subsystems may represent analyses of whole-farm systems, at least for some farmers. Yet the context within which replicated trials take place on research farms typically differs significantly—for example, plot size, labor skills, management flexibility, site specificity, personal commitment—from the context of a typical family farm. Information from such cropping systems trials may be useful to farmers, but the results are not directly transferable from research plots to whole farms.

Farming subsystems can be identified as organic versus conventional. In the former, reliance on multiple crop and livestock enterprise systems allows farmers to eliminate commercial chemical fertilizers and pesticides. Using livestock manure instead of commercial fertilizers allows researchers to simulate organic systems, even if no livestock enterprise is otherwise included in the study. A significant number of economic analyses have been completed comparing conventional and organic cropping subsystems that include a wide range of farm enterprises.

Research focusing on management intensive grazing, grass-fed beef or poultry, and organic meats of all types is similar in concept to cropping systems research. In each case, a number of different methods and practices— pastures, feeds, animals, manure, marketing—are integrated into farming subsystems, or wholes, which have emergent qualities not present in any of the individual elements. For example, grass-based livestock and poultry systems may be far more profitable than would be indicated by simply summing the expected profitability of each individual enterprise evaluated in isolation from the others.

Several studies of farming systems and subsystems are included in the list of selected references. The full potential, including economic potential, for sustainable agriculture begins to emerge from research at the farming systems level of aggregation. The potential economic synergy from alternative "arrangements" of crop and livestock enterprises—across space, over time, among individuals—becomes more apparent when integrated crop and livestock systems are compared with specialized single-species or monocrop farming operations. The synergy is not limited to economic gains, but also arises in the form of ecological and social benefits. The "wholes" of more sustainable farming systems are fundamentally different from the simple sums of their parts.

COMMUNITY, REGIONAL, AND NATIONAL IMPACTS

Several farming systems research projects have evaluated the economic impacts of alternative farming systems on farming communities, in addition to comparing profitability at the farm level. Conventional input-output models, such as IMPLAN (an economic impact modeling system), and the estimates of economic impact multipliers have been of limited usefulness in analyses involving sustainable agriculture. Sustainable agriculture is significantly different in many important respects from conventional agriculture and most other commercial ventures. Many of the data used in estimating critical parameters in these models are not appropriate for assessing the economic impacts of sustainable agriculture. For example, at the community level of aggregation, the basic nature of economic costs and benefits from farming is perhaps more important than the magnitude of costs and benefits. Low input farmers reduce their reliance on off-farm commercial inputs by using more on-farm resources, including management and labor. So even if conventional and low-input farming systems were to result in the same level of production and profits, the low-input systems would typically contribute significantly more to the economic viability of the local community.

Commercial farm inputs are manufactured in urban agribusiness centers. The only economic contribution commercial inputs make to local farming communities is the proportion of value added by retailing and delivery, which may amount to ten percent or less of the total purchase price. Economic returns to farm management and labor may accrue almost entirely to farmers and farm workers from the local community. Thus, a dollar of economic return to the farm management and labor might return ten times as much to the local community as a dollar spent for fertilizers or pesticides. Conventional input-output models were not designed to detect such differences. Lower-input, sustainable farms also can be smaller farms, in that they rely more on management and labor and less on land and capital. The land resources of a given community will support more low-input farm families, and create more consumptive spending in the local community. Studies have shown that operators of smaller farms tend to make a greater proportion of their purchases in their local communities, thus supporting more local nonfarm employment. The potential economic benefits of a more sustainable agriculture are increasingly apparent at higher levels of systems aggregation.

Several analyses of sustainable farming systems have assessed economic, ecological, and social impacts at regional and national levels of aggregation. Such studies must rely on broad assumptions regarding general characteristics of conventional and sustainable farming systems and the potential extent of adoption of alternative systems by farmers at regional or national levels. Once the necessary assumptions are made, alternative farming systems can be simulated for representative crop and livestock farming operations, and then models developed to aggregate individual crop and livestock impacts into regional and national estimates for impacts in changes in agricultural systems overall.

The results of regional and national impact studies depend heavily on initial assumptions. For example, if researchers assume dramatically lower yield levels for sustainable farming systems based on "expert opinions" of conventional agriculturalists, the simulation model will generate lower aggregate production levels and higher consumer food prices for the sustainable agricultural alternative. On the other hand, if researchers assume little, if any, reduction in yields for sustainable farming systems based on actual yields of experienced organic farmers, the simulation model will indicate little difference in agricultural production or consumer food prices between conventional and sustainable systems. All credible regional and national assessments of aggregate impacts of shifting to more sustainable farming systems show that any changes in overall production or prices of food would be relatively small and affordable by most consumers.

Any credible aggregate model of the U.S. food system must reflect the well-documented fact that the average American consumer spends only about ten percent of their disposable income for food. Farmers receive only about 20 percent of the amount that consumers spend for food. The rest goes to pay for processing, packaging, transportation, advertising, and other marketing services. More than half the amount that American farmers receive is spent for purchased inputs—seed, fertilizers, pesticides, and fuel. Thus, on average, farmers have the ability to affect less than ten percent of the total food costs by the efficiency of their farming operations. In other words, of the dime of each dollar of disposable income consumers spend for food, farming accounts for less than a penny. So, if farming were perfectly efficient, if farmers required nothing for their contribution, food costs would be only 10 percent lower. Conversely, if the efficiency of farming were cut in half, if farming required two cents rather than one cent of each dime spent for food, food prices in grocery stores and restaurants would only be 10 percent higher—and a consumer would spend eleven cents of each income dollar for food.

Most national studies by researchers actively involved with sustainable agriculture have shown farm level costs of production from 0 to 10 percent higher for sustainable and organic farming systems. This translates into increases in consumer prices of up to one percent or one cent for each dollar spent for food. Some studies funded by those opposing sustainable agriculture have assumed that farm level production would be cut by more than half and farm level costs of production would more than double if farmers shifted from conventional to sustainable agriculture systems. However, estimated food price increases at the consumer level were still only five to six percent. Researchers cannot ignore the fact that what happens at the farm level has relatively little impact on retail food prices.

WHOLE-FARM CASE STUDIES

Case studies allow researchers to avoid the potential pitfalls of generalized assumptions and complex simulation models by dealing directly with reality. Whole-farm case studies were among the earliest form of inquiry into the fundamental nature of differences between conventional and alternative or sustainable farming systems. With case studies, no attempt is made to pair conventional and alternative farms or to discover or define quantifiable differences in performance through comparisons of randomly selected representative farms of each type. Instead, individual farms are identified as being likely representatives of sustainable farming—based on their economic, ecological, and social attributes. The selected farmers are interviewed and case studies are developed to bring out the salient characteristics of each farm. The objective is to gain an intuitive understanding of sustainable farming in general through comprehensive descriptions of individual "sustainable" farms.

Case studies have been used for decades by business schools to prepare students to cope with real world management situations. There is no expectation that the real world situations will be identical, or even similar in detail, to the case study situations addressed in the classroom. The purpose of case studies is to help the student grasp basic management principles that can be applied to a wide variety of management situations. The student learns general principles by studying specific examples. Whole-farm case studies of sustainable farms are developed for this same basic purpose. Decision case studies focus on providing information relevant to a course of action. Farm profiles differ from decision case studies; the focus of a profile is understanding the basic nature of the overall farming operation. As the numbers of individual farm case studies and profiles grow, our understanding of the fundamental principles of sustainable farming will grow as well.

Case studies have a number of advantages over replicated research trials. whole-farm case studies allow the researcher to address the goals and aspirations of individual farmers. Farmers may be asked, "Is farming a commercial venture, a commitment to stewardship, a preferred lifestyle, or all three?" Case studies often validate the proposition that sustainable farming is not just a simple matter of economics. Sustainable farmers care about the land, they care about families and communities, and they are willing to accept a lower standard of living, if necessary, to achieve a higher quality of life as they define it.

Case studies also tend to validate the importance of sustainable farmers reconnecting with people beyond the farm gate—with their neighbors and their customers—to help maintain economically viable farms. Farmers who market directly to customers who want to know where their food comes from, how their food is produced, and the farmer who produced it tend to be most successful economically. Farmer profiles also validate that sustainable farming is as much a way of life as a way to make a living. Sustainable farmers are "quality of life" farmers, and quality of life has personal, interpersonal, and spiritual dimensions. In addition, the success of the farm may be as much a matter of personal commitment as inherent competency. These intangible but potentially critical dimensions of sustainable farming cannot be derived from complex sets of statistical data, but are readily apparent in case studies based on actual interviews with farmers.

The reference list of case studies has been compiled from interviews with farmers who identify with sustainable agriculture. The numbers of case studies available have grown rapidly in recent years, with many states committed to developing sets of farming cases. State extension sustainable agriculture program coordinators, located at most state universities, should be able to guide the reader to case studies available for their states.

DECISION SUPPORT METHODOLOGY

Sustainable farming requires a fundamentally different approach to farm decision making. Specifically, decisions driven by the pursuit of greater sustainability require simultaneous consideration of the inherent economic, ecological, and social dimensions of all sustainable systems. Some of the earliest research and educational work carried out under the LISA program focused on the development of microcomputer-based farm decision support systems for sustainable farming. The objective was to find practical means by which farmers could balance considerations of soil loss, water quality, productivity, and profitability in developing multiyear, whole-farm plans. After several generations of computer programs, the program was largely abandoned because of a lack of incentive for farmers to go through such a formalized planning process. Perhaps the whole-farm option of the new Conservation Security Program will rekindle interest in a resource-based, multiyear decision support system for whole-farm planning (see Chapter 5).

Some economists have focused on developing farm decision models, not for use by farmers, but for use in analysis of alternative farm policies. Policy analysts wanted to model farmers' likely responses to various conservation and environmental regulations and incentives, such as those related to soil erosion, water quality, wetland protection, carbon sequestration, and wildlife habitat. Policy support programs can rely on much more comprehensive data bases and complex models. With a few notable exceptions, these models have been funded, supported, and used in policy debates by those opposed to resource conservation and environmental protection policies.

Other researchers have explored fundamentally different approaches to decision making and have applied new analytical tools, which can more easily accommodate the multidimensional, holistic nature of sustainable farming. Several references addressing decision support methodology are provided. Much work remains to be done in developing a better understanding of how to manage holistic, diverse, dynamic, decentralized, sustainable farming systems—as well as in developing better decision support systems.

ISSUES OF SUSTAINABILITY

Much of the attention given to sustainable agriculture by economists has addressed the potential implications of the sustainable agriculture movement on the economy, environment, and society as a whole. Most such studies articulate a particular perspective on the issues of sustainability and provide a particular point of view regarding their importance and implications. Research results and other data are used more as means of supporting particular points of view than as means of searching for truth. It is not that such research is somehow inappropriate or useless, but the fact is that each study reflects a particular bias.

The issue of sustainability ultimately is rooted in a worldview that is fundamentally different from the mechanistic worldview that has dominated the modern era of science and industrial development. Sustainability is rooted in an organismic worldview, which ultimately will be reflected in a broader, fundamentally different, concept of science and a different paradigm for development. Most economists still hold the old mechanistic worldview, similar to most other scientists, and thus fail to grasp the full relevance of sustainable agriculture (see Chapter 15). They don't understand the importance of holism, synergy, or interdependence within a specialized, standardized, centralized economy. Thus, most economists who have written on the general subject of sustainable agriculture tend to treat it is a simple matter of internalizing ecological and social externalities. Those who understand that sustainable agriculture represents a different agricultural paradigm tend to view it as idealistic or nostalgic and are skeptical that sustainable agriculture will occupy more than a small niche within a rapidly industrializing, high-tech agriculture. Science is incapable of proving that one worldview is superior to another, because science as we know reflects a particular worldview and uses a specific set of methods. One's worldview is a matter of personal belief, and reflects how we believe the world works and what we believe about our place within it.

Some critics argue that sustainable agriculture is far less productive than conventional agriculture, and with sustainable agriculture half the people of the world would starve, even after all of the forests, plains, and other wild places were plowed up and put into farmland. Others respond that sustainable systems can be just as productive as conventional agriculture, that conventional agriculture is destroying the very resources on which it must depend for long run productivity, and sustainable agriculture is the only logical means of feeding a growing world population. Some argue that sustainable agriculture will require more farmers on smaller farms, because this approach is inherently management intensive. Others are of the opinion that larger farms are inherently more capable of addressing environmental concerns, and thus larger farms will be required to achieve ecological sustainability.

The primary value of studies addressing the broad conceptual issues of sustainability is to help inform the debate, not that such studies will ever be capable of proving that sustainable agriculture is or is not feasible or necessary. Ultimately human society must choose one agricultural paradigm or the other, either conventional or sustainable, to provide for the food and fiber needs of humanity. If the world works like a big complex machine, perhaps the conventional, industrial paradigm of agriculture is the appropriate choice. But if the world works more like a big complex living organism, the regenerative or sustainable paradigm of agriculture will be essential to the survival of humanity. No one is capable of presenting a truly balanced perspective on such issues, because each person holds one particular worldview—whether they are willing to admit it or not. So the best science can do under such circumstances is to inform, rather than resolve, the debate.

Economics of Sustainability

A number of sustainable agriculture advocates have openly identified contemporary economics as a fundamental threat to the sustainability of agriculture, and thus a threat to the future of humanity. At its very disciplinary core, economics is about finding the optimum means of using things up. There is nothing in contemporary economic theory that addresses how to restore, renew, or regenerate the economic resource base.

The environment and society are viewed as constraints to economic development that must be reduced or eliminated to allow maximum economic progress. Scarcity is the source of all economic value—if it is not scarce, it has no value. Thus, economics will not value biological diversity, healthy soil, clean air, or clean water until these things become sufficiently scarce to command a price in the marketplace. The economy will not ration use of nonrenewable resources through higher market prices until the years of proven reserves are less than the planning horizon of global industry— which is probably somewhere between five and ten years. The economy will not ensure equality of economic opportunity among people, or any other aspect of a humane, civil, and just human society, because economics doesn't deal with interpersonal relationships. Economics is about the pursuit of short-run, material self-interests of individuals.

A few economists have begun to rethink their discipline in search of an approach to economics capable of addressing the long run ecological, social, and economic interests of human society, as well as the short-run self-interests of individuals. Such economics are quite likely to require a return to the philosophy of economics as a means for the pursuit of happiness or quality of life—which has social and ethical or moral as well as individual dimensions. Economics will have to be broadened to include the social economy as well as the individual economy if it is to address the social responsibility dimension of sustainability. Economics will have to be deepened to include the moral economy as well, if it is to address the long run ecological integrity of sustainability.

References are listed that relate the need to rethink the economic foundations of modern society, if economists are to be relevant in addressing the issue of sustainability. A few of these works go beyond documenting the nature of the problem to suggest solutions. However, the long, difficult task of developing an economics of sustainability has hardly begun.

Multiple Impact Valuation Strategies

The economic, ecological, and social dimensions of sustainability are inseparable aspects of the same whole. Thus, researchers must learn to integrate multiple criteria when evaluating the performance of sustainable farming systems or assessing their impacts on families, communities, regions, and society as a whole. Economics will continue to play an important role in such activities, but economists must be willing to address the social and ecological dimensions of sustainability as well, if they are to remain relevant. Likewise, ecologists and sociologists must be willing to take active roles in helping to inform the economic dimension of sustainability.

Internalizing ecological and social externalities is not the answer; this simply transforms things that are inherently social and ethical in nature into economic goods and services. Once ecological and social goods have been brought into the economy, they can be sold to the highest bidder in the marketplace. If issues are fundamentally social or public in nature, everyone must have an equal voice in the decision making process—regardless of their wealth or income. Long run ecological issues are fundamentally ethical or moral in nature. Such issues cannot be decided in the marketplace or by majority vote; they must be resolved by consensus. Ecologists and sociologists cannot simply defer to the economists—it's not simply a matter of economics.

The impacts of sustainable farming on communities, regions, and nations must be evaluated using measures of economic, social, and ecological costs and benefits. Sustainable farming operations must be managed using these multiple indicators of performance. Overall sustainability must be assessed in terms of balance and harmony among the economic, ecological, and social issues. Effective indicators must be capable of detecting imbalance and disharmony within farming systems. Thus, sustainable agriculture researchers and educators must become competent in the use of multiple impact valuation strategies.

Economic Indicators and Measures of Sustainable Farming

Economic indicators of sustainability might seem the most straightforward of the three. Farming, like most other business endeavors, has been driven for decades by the economic bottom line. "If it's profitable it's sustainable, if it's not profitable, it's not," according to early skeptics. But, it's not quite that simple. A farm that is profitable in the short run may not be able to sustain profits over time; it may not be economically viable, even if it appears to be profitable. Also, while a farm must be profitable over time to be sustainable, it need not be profitable at all times. And a farm that maximizes profits in the short run most likely will not be economically viable over the long run. Economically sustainable farming is not simply a matter of profitable farming. Nonetheless, profitability is an important indicator of economic sustainability.

Robert Rodale, an early leader in the sustainable agriculture movement, frequently expounded the fact that all sustainable systems must be resistant, resilient, renewable, and regenerative. While these characteristics may be more readily understood in reference to biological systems, they are equally relevant to economic systems. Sustainable economic systems, like biological systems, must be managed as living systems. All living systems are vulnerable to injury or death, and thus a farm's capacities to resist economic adversity and to rebound from economic misfortune are critical to its sustainability. An economic system must be capable of self-renewal and regeneration if it is to be sustainable over time. Thus, a sustainable farm must be capable of maintaining its productivity and profitability from generation to generation, while relying primarily on renewable sources of energy and capital.

A farm that relies less on external sources for its production inputs and external markets for its products distribution is less vulnerable to inevitable volatility in market prices. This does not imply that a farm must be internally self-sufficient to be sustainable; the economic sacrifices of total self-sufficiency are simply too high. But neither can a sustainable farm be completely dependent on outside markets or sources of inputs over which it has no control. The more a farmer can internalize inputs and markets, either through integration of enterprises within the farming operation or through collaborative relationships with neighbors, input suppliers, and customers, the greater the ability of the farm to resist economic adversity. Degree of economic self-reliance is an important indicator of economic sustainability.

A farm that relies on economically diverse enterprises not only is more resistant to economic adversity, but also is more resilient when struck by inevitable economic misfortune. Economic diversity is not simply a matter of total number of enterprises in a whole-farm system. If the economic risk associated with one enterprise is negatively correlated with the risks associated with the other, as in the case of hogs and corn, the two enterprises are economically diverse. Enterprises that are uncorrelated economically also add diversity—when one suffers a loss, the odds that the others will show a profit are unaffected. The economically diverse farm is less affected by economic adversity and recovers more quickly from economic misfortune. On the other hand, if two enterprises are vulnerable to the same production and market risks, as in the case of corn and soybeans, the two enterprises lack economic diversity, and thus, adding the second enterprise does very little to improve economic sustainability. Economic diversity is an important indicator of economic sustainability.

A farm that relies less on production facilities, machinery, and equipment—things that rust, rot, or wear out—and relies more on living organisms has greater self-renewing and regenerative capacity. The fundamental purpose of farming is to convert solar energy into energy forms more useful to humans. Fertilizers, pesticides, fuel, and most other off-farm inputs are manufactured using fossil energy, which is not renewable within any reasonable time frame. However, a farm is a naturally regenerative living system if it is allowed to function as such. Living organisms in the soil feed living plants, which capture and store renewable energy from the sun. People eat or otherwise use the energy stored in plants and in animals. Many animals used for food by humans eat plants or residues that humans cannot digest. A diversity of other living species thrives within the living ecosystems of farms. The wastes from soil organisms, plants, animals, and people become resources to feed future generations of soil organisms, plants, people—in a self-renewing cycle of life.

Farms that rely on continuing infusions of nonrenewable energy eventually lose and in some cases destroy the regenerative capacity of living systems on the farm. Farms that rely heavily on buildings, machinery, and equipment, which inevitably depreciate, eventually must restore or rebuild those assets using capital generated by the farm. Such investments inevitably divert resources from ecological investments in living systems that are necessary to maintain the farm's regenerative capacity. The more a farm relies on its self-renewing, living systems for productivity and profitability, the more economically regenerative it will be. Reliance on living systems is an important indicator of economic sustainability.

The purpose here is to suggest a logical approach to developing appropriate criteria, indicators, and measures of sustainable farming, rather than develop an exhaustive listing. The indicators of economically sustainable farms include profitability, self-reliance, diversity, and reliance on living systems. Each of these indicators suggests a number of different farm-level measures of sustainability.

Profitability may be interpreted in a number of ways. For some purposes, profit may be defined as a return over cash costs or variable costs, implying that costs of land, management, and other committed factors of production must be paid out of profits. A more common interpretation of profit is return over all costs, fixed and variable, implying that previous investments in land, buildings, machinery, and management are more than repaid. In the purest sense, profit is defined as a return to risk or a return over opportunity costs, implying a greater return is being earned from one strategy than by committing land, labor, capital, and management to any less risky alternative venture.

For assessing sustainability, it would seem logical to measure profit as a return over replacement costs, including the cost of replacing both variable inputs and fixed resources. The key questions in the economic sustainability of farming relate to whether the farm is capable of continuing to operate over the long run. To do so, it must be able to replace anything that is used up, worn out, or wasted in the production process. A profitable farm, in this sense, is a farm capable of doing more than just sustaining its productivity over time.

Economic self-reliance can be measured using the same type of economic data required to calculate profitability. However, in measuring self-reliance it is important to distinguish between fixed and variable costs and between costs associated with off-farm purchases and opportunity costs of inputs produced and used on the farm. Fixed costs are those already committed to the production process, while variable costs are uncommitted and thus can be avoided if nothing is produced. Cash costs, variable or fixed, are those that require cash outlays during any given production period. In general, external input costs include all variable cash costs, although some external inputs might be obtained through barter. Internal opportunity costs reflect the value foregone by using something on the farm rather than selling it. Fixed cash costs include annual taxes, insurance, and installment payments on land, buildings, and machinery.

A farm that fails to cover its variable costs plus fixed cash costs will suffer a direct loss, which it may or may not be able to survive financially. A farm that covers its variable costs and fixed cash costs, but fails to cover total fixed costs, can simply accept a lower return to committed resources—land, operator labor, management—for the current production period, and hope to make up the deficit later on. Thus farms with lower variable costs and fewer cash costs are less vulnerable to economic adversity, even if they have no advantage with respect to total cost. Also, if two farms have the same level of variable costs, but one has lower external cash costs, the farmer with lower external cash costs will be more self-reliant, and thus more resistant to economic threats.

Economic diversity can be estimated by statistically analyzing yield and price data for the enterprises included in a farming system. Farm level yield and price data are likely to be less correlated, either positively or negatively, than county, state, or national yield and price data. However, estimates of correlations at any given level of aggregation may provide valuable insights into economic diversity at other levels. In measuring economic diversity, enterprises that are highly positively correlated may be combined and treated as a single enterprise.

In general, if enterprises are uncorrelated or independent, most of the economic diversity possible will be achieved by four to six comparable size enterprises. In rolling dice, the average after rolling the dice four to six times will generally be close to the same average after rolling the same dice 20 times. If enterprises are significantly different in size, it will take more enterprises to achieve the same level of economic diversity. Good reasons may exist for having a far larger number of enterprises on a farm, such as providing a variety of vegetables for a Community Supported Agriculture (CSA) operation. However, most of the economic stability and resilience is achieved with a relatively small number of independent enterprises.

If enterprises are negatively correlated, meaning that losses in one are generally offset by gains in another, most of the possible economic diversity is achieved with even fewer enterprises than if enterprises are independent. Thus, three or four such enterprises may reflect a high level of economic diversity and sustainability. Again, if enterprises are significantly different in size, it may take several small enterprises to offset the risks associated with one larger one. Although detailed statistical analysis may be justified in assessing economics in research, most of the relationships involved in economic diversity are intuitively obvious to many farmers.

A farm's reliance on living systems may be clearer in addressing ecological indicators, but some obvious economic indicators of reliance on living systems also exist. One such indicator is the amount of fixed costs committed to purchased physical elements that do not replace themselves, thus reducing the natural regenerative capacity of the farm. A farm that houses livestock outside or in low cost shelters, stores grass in the field or in outside stacks, and harvests crops by using livestock or with old equipment is utilizing renewable capital.

Reliance on nonrenewable capital is also reflected in different types of variable input costs. Variable expenditures for fertilizers, pesticides, fuel, and other such inputs reflect a farm's lack of ability to create productivity from its living systems—from healthy soils, plants, animals, and people. Conversely, the proportion of total costs committed to hired or family labor and management reflects a farm's reliance on renewable human energy.

Ultimately, the economic sustainability of a farm depends on its regenerative capacity. Profitability, self-reliance, diversity, and reliance on living systems are all indicators of the economic regenerative capacity of a farm, and thus its long run sustainability.

Ecological Indicators and Measures of Sustainable Farming

Ecological indicators and measures of sustainability have received a great deal of attention in sustainable agriculture research and education. Much of the work in this area could have benefited from a greater emphasis on sustainable farming systems as living systems and on defining indicators and measures of the ecological health of these living systems.

Sustainable systems must be regenerative systems, capable of reproducing and renewing themselves and maintaining their productivity and vitality from generation to generation, indefinitely. Living systems are distinguished by their self-making and repairing ability. Nonliving or dead systems cannot remake themselves. Bacteria, insects, plants, animals, and humans are examples of living systems. Clocks, bicycles, automobiles, machines, and factories are examples of dead systems. All systems, both living and dead, can be characterized by their pattern, structure, and process. But the process by which the structure is created and recreated is fundamentally different for living and dead systems.

The key to ecological sustainability in farming is to maintain the ecological health of a farm as a living system, and the capacity of the system to remake itself. If the ecological health of the soil, plants, animals, and people of a farm remain healthy, the farm will remain ecologically regenerative and sustainable. Thus, our knowledge of human health should provide some valuable insights into the nature of appropriate indicators and measures of ecological sustainability.

The most basic indicator of human health is general physical appearance. The skins of healthy people have the glow of life; they also have clear, bright eyes, and their walk and posture reflect physical stamina and energy. The characteristics of a healthy farm may not be as well defined, but a healthy farm will have a healthy appearance. Its soil, fields, and landscape will appear alive with a diversity of plants and animals. Its streams will run clear, even after a hard rain. Its plants will grow quickly and its livestock will look slick and full of life. The family who lives on a healthy farm also will be hopeful, if not optimistic, and full of life. It may be difficult to define a healthy appearance in scientific terms, but for the keen observer, health is relatively easy to see in both people and in farms.

The more objective routine measures of human health include body temperature, blood pressure, pulse, respiration, and physical weight and height. A high body temperature indicates the presence of an infection or some other invasion of the body's immune system. High blood pressure indicates stress or some chronic malady that restricts blood flow. The rapid pulse rate indicates the heart has a limited ability to increase blood flow if more is needed. Shallow, rapid breathing indicates a limited ability of the body to process more oxygen if needed. Abnormal physical dimensions of the body can be indicative of a whole host of problems.

Researchers and farmers alike need a similar set of readily observable, objective indicators of health for farming systems. Farmers need indicators to warn them immediately when their farm is under attack from some unseen and unexpected threat or risk. They need indicators that tell when their farming system is under stress or when it is not processing and regenerating the amount of energy necessary to maintain its productivity. They need indicators of the ability of the farm to respond to stress, of its capacity for resistance and resilience. They need indicators to tell them if their farm is too large to be managed with ecological integrity or too small to be managed for economic viability. They need indicators that will let them know they are imposing ecological or social costs on their neighbors and society as a whole. Farmers, researchers, and educators need simple, objective indicators of ecological sustainability.

Many of the common indicators of ecological sustainability are good candidates for basic indicators of the ecological health of a farm—such as soil loss, soil structure, water infiltration, biological diversity, pest pressures, and drought resistance of crops. The challenge is to develop a few reliable indicators and to explain how these indicators relate to the health of a farm. Such indicators of ecological health will not pinpoint the specific nature of a problem any more than a high body temperature, high blood pressure, or obesity pinpoint the source of human health problems. But they will indicate to farmers that something is wrong—that their farm may no longer be sustainable. Knowing that something is wrong increases the odds that farmers can find the problem and restore the health of the farm before it is too late. A healthy farm is an ecologically sustainable farm. As farmers and researchers find better indicators and measures of farm health, society will have better indicators and measures of ecological sustainability.

Social Indicators and Measures of Sustainable Farming

Social responsibility is no less essential to sustainable farming than economic viability and ecological integrity. However, the social dimension of sustainability appears to be less well understood, and thus less appreciated than the other two factors. Social responsibility has three distinct aspects: responsibilities to people as consumers, to people as producers, and to citizens of civil society. While some aspects of the three are more important at higher levels of aggregation, all are relevant to sustainable farming.

The first is reasonably straightforward. In agriculture, social responsibility to consumers typically is defined as providing adequate quantities and qualities of safe and healthful food at a reasonable cost. In other words, agriculture has a social responsibility to provide for the food and fiber needs of people. Thus, a socially responsible farm must be productive.

The typical and logical indicators of social responsibility to consumers include food taste, flavor, freshness, variety, safety, nutrition, and cost. Specific measures are readily available because some these are indicators of economic viability as well as social responsibility. Specific measures of food safety and nutrition also exist, though not as readily available as quantities and prices. Sensory traits generally have been ignored in agriculture in recent years, as economic efficiency through specialization and standardization has taken precedence over quality. Together, these indicators define product quality—those attributes of a product that determine its acceptability to consumers.

Social responsibility to consumers is fundamentally different from profitability. Economic viability results from providing for the food and fiber needs of people who are willing and able to pay—which means food production and distribution are driven by economic returns from the marketplace. Social responsibility deals with providing for the basic food and fiber needs of all, regardless of their willingness and ability to pay—it is about food equity, not just abundance.

A socially responsible farm must ensure the safety and healthfulness of its food products, regardless of whether safety and nutrition maximize the short run economic bottom line. A socially responsible farm must put human health above short run economic efficiency. A socially responsibility farm must consider food quality as one aspect of the basic right of all consumers to good food, regardless of their income or wealth. A sustainable farmer must give priority to producing quality products for all.

The second dimension, social responsibility to producers, is easy to understand, but it is not as easy to reach an agreement concerning its relevance to sustainability. It is easier to understand the relevance of social responsibility to people as producers at the level of the aggregate economy, rather than at the level of the individual farm or community. But the responsibilities are the same, in principle, at all levels of aggregation.

First, if people are to be able to participate in society as consumers, people must have opportunities to participate in society as producers. Otherwise, they won't have money to spend as consumers. Even a strong and growing private economy doesn't always provide opportunities for people to participate fully as producers or workers. In a sustainable society, each sector of the economy, each community, and each farm or business should provide employment opportunities for people in proportion to its productive capacity. When one sector fails to provide employment opportunities, the rest of the society must accept the added burden of either creating additional employment opportunities or suffering the social consequences of increasing overall unemployment. Thus, each sector of a sustainable economy must accept its social responsibility to provide opportunities for people to be useful, productive members of society.

The same is true for each community or farm. When one community fails to provide continuing employment for its people, it places an additional burden on other communities to provide employment. When one farm fails to provide quality employment opportunities for the farmer, farm family, or farm workers, it places an additional burden on other farms and businesses. An agriculture that diminishes employment opportunities and drives people from the land is neither socially responsible nor sustainable.

The opportunity to be useful and productive is not just a matter of having a job; it is also a matter of having an opportunity to reach one's full productive potential. If a society is to be sustainable over time, it must rely on its renewable resources. Among the most potentially productive of these renewable resources is the uniquely human capacity to think—to solve problems, innovate, create, imagine, and dream. There is dignity in all types of work, but a sustainable society must provide "quality employment opportunities" where people can fully express their capacities as productive human beings.

Among such quality employment opportunities, nothing more fully empowers the individual to reach his or her full productive potential than self-employment. Lacking the aptitude or desire for self-employment, a farm worker should be allowed the maximum degree of self-determination and individual responsibility consistent with the economic and ecological necessities of the farm. A socially responsible farm must provide quality employment opportunities for people—farmers, farm families, farm workers, and others in rural communities.

Indicators of the social responsibility of a farm to people as producers include total employment, employment per unit of production, farm wages and salaries, farm income, self-employment versus contract production, and empowerment of workers. With the exception of empowerment, these measures are readily available for most farming operations.

As with social responsibility to consumers, social responsibility to producers is fundamentally different from economic sustainability. Economic viability is derived from the marketplace, where farmers and farm workers are valued at whatever they are worth in dollar and cent terms, regardless of the long run consequences for society as a whole. Farms are expected to compete for their survival, regardless of the nature of the competition. If a farm needs to fire half of its workers and replace them with machines to remain competitive, those workers are simply left without a job. If it is easier and cheaper to *de-skill* or *dumb-down* a job than to help someone learn how to do it, then the job and its associated pay are down-graded. But each of these market-driven actions diminishes the productivity of the human resource needed to sustain the farm and society over the long run. Each time a farm diminishes the quality of employment opportunity it provides, it places an added burden on the community and the rest of society.

The final type is social responsibility to people as citizens of civil society. Quality of life is not just about consuming or producing—we humans are created as social beings. We need positive relationships with other people—regardless of whether such relationships return individual or economic benefits. A sustainable society must provide a desirable quality of life, and quality of life is inevitably linked to the quality of human relationships.

The existence and nature of human relationships is referred to as *social capital*. A family, community, or nation that has more social capital is a more desirable place to live—other things equal. A family, community, or nation without social capital is not a very desirable place to live, regardless of its economic or ecological assets. A civil society is one that values relationships among its members. An uncivilized society is a society lacking in social capital.

Although different social scientists may define social capital a bit differently, a few basic indicators of social capital—trust, local rules and sanctions, respect and reciprocity, links and networks—are common in socio-logical literature. The four obviously are related, in that people often make connections that make them willing to do favors for another person, and if the favor is reciprocated then it leads to trust. Each of the four elements both nurture and are nurtured by the other three. Indicators and measures of so-cial capital are readily available in sociological literature (see Chapter 10).

Finally, social capital is fundamentally different from economic capital. Investing in social capital to strengthen personal relationships is fundamen-tally different from forming relationships as a means of achieving economic benefits. When people build and maintain relationships for economic purposes, the purely social aspects of those relationships are invariably strained and fre-quently are broken—by economic necessity. In fact, business relationships tend to degrade or deplete social capital, so that it has to be constantly re-plenished by the rest of society. Thus a sustainable farm, family, commu-nity, or society must consciously conserve and regenerate its social capital to offset this inevitable loss. Any sector of society, community, or farm that fails to build and conserve social capital puts an added burden on the rest of society and is not socially responsible. Quality food, employment, social capital, and food security all could be enhanced by local food systems.

Indicators and Measures of Sustainability

No objective, quantifiable means exist for ensuring balance and harmony among the economic, ecological, and social dimensions of sustainable farming. No single scale of measurement exists that is capable of integrat-ing economic viability, ecological integrity, and social responsibility into a common unit of measure. Ecological and social costs and benefits cannot be converted into dollars and cents, nor can their measures be converted into votes. Social and economic issues need not be resolved through ethical or moral consensus. Thus, no single indicator or method can be devised for measuring sustainability.

Sustainable farming depends on balance and harmony among the eco-nomic, ecological, and social aspects of any given farming operation. En-hanced sustainability is not a matter of increasing performance in one di-mension by more than enough to offset decreasing performance in the other two, or increasing one while holding the others constant. The three dimen-sions of sustainability are inherently interrelated; as in natural ecosystems, we can never do just one thing. For farming systems that are already in balance, increased performance in one dimension will require comparable increases in the other two to avoid detracting from sustainability. If a farm becomes more productive and profitable, it must simultaneously be made more ecologically sound and socially responsible, if it is to sustain the higher levels of production and profits. If a farm becomes more ecologically sound, it must be made more economically viable and socially responsible in order to sustain the higher ecological investments. If a farm becomes more socially responsible it must be made more ecologically sound and economically viable in order to regenerate higher levels of social capital. If a farm is out of balance, the key is to identify the weak links, and then to do whatever is necessary to improve harmony and balance among the economic, ecological, and social dimensions of the farm. Once a farming system is in balance, sustainability at a higher level will require improvements in all three—economic, ecological, and social performance.

Finding better indicators and measures of sustainability will not allow researchers to prove that some farms are sustainable and others are not, or even that some farms are more sustainable than others. One's evaluation of sustainability ultimately is a reflection of their worldview, what they believe about how the world works, and where we people fit within it. Better indicators and measures of sustainability will simply allow people more accurately to assess how well a farm fits their perceptions of the balance and harmony needed for sustainability and then to draw whatever conclusions or take whatever action they feel is justified. Ultimately, sustainability is about people and their associations with each other and their relationships to the earth. So perhaps sustainability ultimately depends not on proving that some systems are sustainable and some are not, but instead upon people having the courage to explore the world around them and to act on their beliefs. Better indicators and measures of sustainability should at least improve the process of exploration. Perhaps this is enough to justify our continuing efforts to find better indicators and measures of sustainability.

SELECTED BIBLIOGRAPHY

Economic Aspects of Sustainable Agriculture

Schneider, K.R. 1987. *Alternative farming systems-economic aspects, 1970-1986.* National Agricultural Library. U.S. Dept. Agriculture, Beltsville, MD. Available from: http://www.nal.usda.gov/afsic/agnic/agnic.htm#print. Accessed March 29, 2006.

———. 1996. *Alternative farming systems: Economic aspects July 1993-June 1996.* AGRICOLA Database Quick Bibliography Series no. QB 98-08 Updates QB 93-17 National Agricultural Library. U.S. Department of Agriculture, Beltsville, MD. Available from:http://www.nal.usda.gov/afsic/agnic/agnic.htm#print. Accessed March 29, 2006.

————. 1998. *Alternative farming systems: Economic aspects July 1996-June 1998*. 451. AGRICOLA Database Quick Bibliography Series no. QB 98-01 Updates QB 96-08 National Agricultural Library. U.S. Department of Agriculture, Beltsville, MD. Available from: http://www.nal.usda.gov/afsic/agnic/agnic.htm#print. Accessed March 29, 2006.

Farming Practices and Methods

Benbrook, C.M. 1995. *Healthy food, healthy farms: Pest management in the public interest*. National Campaign for Pesticide Policy Reform (U.S.), Washington, DC. 44 p., September 18.

Calkins, J.B. and B.T. Swanson. 1996. Comparison of conventional and alternative nursery field management systems: Tree growth and performance. *J. Environ. Hort.* 14(3):142-149.

Carr, P.M., G.R. Carlson, J.S. Jacobsen, G.A. Nielsen, and E.O. Skogley. 1991. Farming soils, not fields: A strategy for increasing fertilizer profitability. *J. Prod. Agric.* 4(1):57-61.

Doll, J., R. Doersch, R. Proost, and T. Mulder. 1990. Weed management with reduced herbicide use and reduced tillage. *Proceedings Progress in Wisconsin Sustainable Agriculture*, March 1990, University of Wisconsin. pp. 7-16.

Gleason, M.L., M.K. Ali, P.A. Domoto, D.R. Lewis, and M.D. Duffy. 1994. Comparing integrated pest management and protection strategies for control of apple scab and codling moth in an Iowa apple orchard. *HortTechnology* 4(2): 136-141.

Hall, C., T. Longbrake, R. Knutson, S. Cotner, and E. Smith. 1994. Yield and cost impacts of reduced pesticide use on onion production. *Subtropical Plant Sci.* [Weslaco, Texas]: *Rio Grande Valley Hort. Soc.* 46:22-28.

Johnson, W.G., J.A. Kendig, R.E. Massey, M.S. DeFelice, and C.D. Becker. 1997. Weed control and economic returns with postemergence herbicides in narrow-row soybeans (Glycine max). *Weed Technol.* 11(3):453-459.

Knutson, R.D. 1993. Economic impacts of reduced pesticide use on fruits and vegetables: Executive summary. In: *Reduced pesticide use on fruits and vegetables*, R.D. Knutson (ed.). American Farm Bureau Research Foundation. 21 p.

Merwin, I.A., D.A. Rosenberger, C.A. Engle, D.L. Rist, and M. Fargione. 1995. Comparing mulches, herbicides, and cultivation as orchard groundcover management systems. *HortTechnology* 5(2):151-158.

Mugalla, C.I., C.M. Jolly, and N.R. Martin Jr. 1996. Profitability of black plastic mulch for limited resource farmers. *J. Prod. Agric.* 9(2):283-288.

Farm Enterprises

American Farmland Trust. 1991. *1990 Indiana on-farm demonstration project results*. American Farmland Trust. Indiana Sustainable Agriculture Association, West Lafayette, Indiana. 35 p.

Brumfield, R.G., F.E. Adelaja, and S. Reiners. 1995. Economic analysis of three tomato production systems. *Acta Hortic.* 340:255-260.

Creamer, N.G., M.A. Bennett, B.R. Stinner, and J. Cardina. 1996. A comparison of four processing tomato production systems differing in cover crop and chemical inputs. *J. Am. Soc. Hort. Sci.* 121(3):559-568.

Ennis, J., R. Klemme, and B. Rajhandary. 1990. Economic analysis of low-input and conventional dairy cropping systems. *Proceedings Progress in Wisconsin Sustainable Agriculture,* March 1990, University of Wisconsin, Madison, WI. pp. 41-47.

Ferguson, J.J., M.E. Swisher, and P. Monaghan. 1995. Commercial organic citrus production in Florida. *Proc. Ann. Meet. Florida State Hort. Soc.* 107:26-29.

Gliessman, S.R., M.R. Werner, J. Allison, and J. Cochran. 1996. A comparison of strawberry plant development and yield under organic and conventional management on the central California coast. *Biol. Agric. Hort.* 12(4):327-338.

Hasey, J.K., R.S. Johnson, R.D. Meyer, and K. Klonsky. 1997. An organic versus a conventional farming system in kiwifruit. *Acta Hortic.* 444:223-228.

Honeyman, M.S. 1995. Vastgotmodellen: Sweden's sustainable alternative for swine production. *Am. J. Altern. Agric.* 10(3):129-132.

Jacobsen, J.S., S.H. Lorbeer, H.A.R. Houlton, and G.R. Carlson. 1997. Reduced-till spring wheat response to fertilizer sources and placement methods. *Commun. Soil Sci. Plant Anal.* 28(13/14):1237-1244.

Klonsky, K., L. Tourte, and D. Chaney. 1994. *Production practices and sample costs for organic processing tomatoes in the Sacramento Valley.* University of California Coop. Extension, Davis, CA. 24 p.

Paudel, K.P., N.R. Martin Jr., N. Kokalis Burelle, and R. Rodriguez Kabana. 1996. Economic & environmental evaluations of peanut rotations with switchgrass and cotton. *Highlights Agric. Res.* 43(1):4-7.

Smith, L.L. 1990. Comparison of rotational to intensive rotational grazing of yearling cattle. *Proceedings Progress in Wisconsin Sustainable Agriculture,* March, 1990. University of Wisconsin, Madison. pp. 50-55.

Taylor, D.C., D.M. Feuz, and M. Guan. 1996. Comparison of organic and sustainable fed cattle production: A South Dakota case study. *Am. J. Altern. Agric.* 11(1):30-38.

Farming Systems

Dobbs, T.L. and J.D. Smolik. 1996. Productivity and profitability of conventional and alternative farming systems: A long-term on-farm paired comparison. *J. Sustain. Agric.* 9(1):63-79.

Ewbank, M., C. Dobbins, and D. Mengel. 1993. *Cropping systems research—July 1993.* Dept. Agric. Econ., Agric. Exper. Station, Purdue Univ. Station Bull. 667. 32 p.

Klonsky, K., L. Tourte, D. Chaney, P. Livingston, and R. Smith. 1994. *Cultural practices and sample costs for organic vegetable production on the Central Coast of California.* Giannini Foundation of Agricultural Economics. Giannini-Found-inf-ser. Davis, California. March. No. 94-2. 87 p.

Mends, C., T.L. Dobbs, and J.D. Smolik. 1989. *Economic results of alternative farming systems trials at South Dakota State University's Northeast Research*

Station: 1985-1988. Economics Dept., South Dakota State Univ., Brookings. 95 p. August.

Painter, K.M., D.L. Young, D.M. Granatstein, and D.J. Mulla. 1995. Combining alternative and conventional systems for environmental gains. *Am. J. Altern. Agric.* 10(2):88-96.

Pfeifer, R.A., M. Rudstrom, S.N. Mitchell, and O.C. Doering. 1995. Evaluation of alternative cropping practices under herbicide use/soil loss restrictions. *Proceedings of the Clean Water, Clean Environment, 21st Century Team Agriculture, March 5-8, Kansas City, Missouri /. St. Joseph, MI.* ASAE. 1:145-148.

Salamon, S., R.L. Farnsworth, D.G. Bullock, and R. Yusuf. 1997. Family factors affecting adoption of sustainable farming systems. *J. Soil Water Conserv.* 52(4): 265-271.

Sellen, D., J.H. Tolman, D.G.R. McLeod, A. Weersink, and E.K. Yiridoe. 1995. A comparison of financial returns during early transition from conventional to organic vegetable production. *J. Veg. Crop Prod.* 1(2):11-39.

Smolik, J.D., T.L. Dobbs, and D.H. Rickerl. 1995. The relative sustainability of alternative, conventional, and reduced-till farming systems. *Am. J. Altern. Agric.* 10(1):25-35.

Community, Regional, and National Impacts

Batie, S.S. and D.B. Taylor. 1991. Assessing the character of agricultural production systems: Issues and implications. *Am. J. Altern. Agric.* 6(4):184-187.

Bischoff, J.H., T.L. Dobbs, B.W. Pflueger, and L.D. Henning. *1995.* Environmental and farm profitability objectives in water quality sensitive areas: Evaluating the tradeoffs. *Proceedings of the Clean Water, Clean Environment, 21st Century Team Agriculture,* March 5-8, Kansas City, Missouri /. St. Joseph, MI. ASAE. 3:25-28.

Chase, C. and M.Duffy. 1991. An economic comparison of conventional and reduced- chemical farming systems in Iowa. *Am. J. Altern. Agric.* 6(4): 160-173.

Diebel, P.L., D.B. Taylor, and S.S. Batie. *1993.* Barriers to low-input agriculture adoption: A case study of Richmond County, Virginia. *Am. J. Altern. Agric.* 8(3):120-127.

Dobbs, T. L. and J.D. Cole. 1992. Potential effects on rural economies of conversion to sustainable farming systems. *Am. J. Altern. Agric.* 7(1/2):70-80.

Hammer, J. and K. Foster. 1996. *Land use regulations supportive of sustaining agriculture and natural resources in urbanizing rural communities: Evaluation criteria and municipal officials' perspectives.* Rodale Institute, Kutztown, PA. August, 1996.

Ikerd, J., G. Devino, and S. Traiyongwanich. 1996. Evaluating the sustainability of alternative farming systems: A case study. *Am. J. Altern. Agric.* 11(1):25-29.

Ikerd, J., S. Monson, and D. Van Dyne. 1993. Alternative farming systems for U.S. agriculture: New estimates of profit and environmental effects. *Choices* 8:37-38.

Karlen, D.L., M.D. Duffy, and T.S. Colvin. 1995. Nutrient, labor, energy, and economic evaluations of two farming systems in Iowa. *J. Prod. Agric.* 8(4):540-546.

Lee, L.K. 1992. A perspective on the economic impacts of reducing agricultural chemical use. *Am. J. Altern. Agric.* 7(1/2):82-88.

Levins, R.A. 1996. *Monitoring sustainable agriculture with conventional financial data based on the work of the Biological, Social and Financial Monitoring Team.* U.S. Cooperative State Research Service and Land Stewardship Project, White Bear Lake, MN. 29 p.

Lockeretz, W. 1975. *A comparison of the production, economic returns, and energy-intensiveness of corn belt farms that do and do not use inorganic fertilizers and pesticides.* Center for the Biology of Natural Systems, Washington University, St. Louis, MO. 62 p.

O'Neill, K. 1997. *Emerging markets for family farms: Opportunities to prosper through social and environmental responsibility.* Center for Rural Affairs, Walthill, NE. p. 61. May.

Whole-Farm Cases and Profiles

Berton, V. 2001. *The new American farmer—profiles in agricultural innovation.* Sustainable Agriculture Research and Education Program, USDA, Washington, DC. Available from: http//www.sare.org/publications/naf.htm. Accessed March 29, 2006.

Chan Muehlbauer, C., J. Dansingburg, and D. Gunnink. 1994. *An agriculture that makes sense: Profitability of four sustainable farms in Minnesota.* Land Stewardship Project, St. Croix, MN. 43 p.

Madden, J.P. 1989. Farms that succeed using LISA. In: *Farm management: How to achieve your farm business*, D.T. Smith (ed.). Yearbook of Agric. U.S. Dept. of Agric., Washington DC. pp. 220-225.

Richards, K.S., J. Bachmann, and H. Vinton. 1994. *Farming more sustainably in the South: Nine farmers' stories.* Southern Sustainable Agriculture Working Group. 30 p.

Thrupp, L.A. 1996. *New partnerships for sustainable agriculture.* World Resources Inst., Washington, DC. 136 p. September.

Farm Decision Support Systems

Buzby, J.C., R.C. Ready, and J.R. Skees. 1995. Contingent valuation in food policy analysis: A case study of a pesticide-residue risk reduction. *J. Agric. Appl. Econ.* 27(2):613-625.

Dunn, E.G., J.M. Keller, and L.A. Marks. 1998. Integrated decision making for sustainability: A fuzzy MADM model for agriculture. In: *Proceedings of the Modeling for Agriculture*, S.A. El-Swaify and D.S. Yakowitz (eds.). Lewis Publishers, Boca Raton, FL. pp. 313-322.

Ikerd, J.E. 1990. A farm decision support system for sustainable farming systems. *J. Farm. Systems Res. Exten.* 1(1):99-107.

———. 1993. The need for a systems approach to sustainable agriculture. *Agric. Ecosyst. Environ.* 46:147-160.

Jones, J.W., W.T. Bowen, W.G. Boggess, and J.T. Ritchie. 1993. Decision support systems for sustainable agriculture. In: *Technologies for sustainable agriculture in the tropics*, J. Ragland and R. Lal (eds.). American Society of Agronomy, Madison, WI, Spec. Publ. 56:123-138.

Mawampanga, M.N. and D.L. Debertin. 1996. Choosing between alternative farming systems: An application of the analytic hierarchy process. *Rev. Agric. Econ.* 18(3):385-401.

Issues of Sustainability

Allen, P. 1993. *Food for the future: Conditions and contradictions of sustainability.* Wiley Interscience, New York. 328 p.

Avery, D.T. 1994. *The organic farming threat to people and wildlife.* Hudson Institute. Indianapolis, IN. 12 p.

Bird, G. and J. Ikerd 1993. Reinventing U.S. agriculture for the twenty-first century. *Ann. Am. Acad. Polit. Soc. Sci.* September, 529:92-102.

Breimyer, H.B. 1991. Science and scientific principles in agricultural economics: An historical review. *Am. J. Agric. Econ.* 73(2):243-254.

Council for Agricultural Science and Technology. 1990. *Alternative agriculture: Scientists' review.* CAST, Ames, IA. 182 p.

D'Souza, G. and J. Ikerd. 1996. Small farms and sustainable development: Is small more sustainable. *J. Agric. Appl. Econ.* 28(1):73- 87.

Faeth, P. 1995. *Growing green: Enhancing the economic and environmental performance of U.S. agriculture.* World Resources Inst., Washington, DC. 81 p.

Francis, C.A. and J.P. Madden 1993. Designing the future: Sustainable agriculture in the U.S. *Agric. Ecosyst. Environ.* 46(1/4):123-134.

Hewitt, T.I. and K. Smith. 1995. *Intensive agriculture and environmental quality: Examining the newest agricultural myth.* Henry A. Wallace Inst. Alternative Agriculture, Greenbelt, MD. 12 p.

Ikerd, J.E. 1993. The question of good science. *Am. J. Altern. Agric.* 8(2):91-93.

Novak, J.L. and W.R. Goodman. 1994. Profitable and environmentally sound agriculture: A sustainable approach to the future. *J. Agric. Food Inf.* 2(4):43-62.

Ruttan, V.W. 1996. Research to achieve sustainable growth in agriculture production: Into the 21st century. *Can. J. Plant Pathol.* 18(2):123-132.

Sustainable Agriculture Network. 1997. *Exploring sustainability in agriculture: Ways to enhance profits, protect the environment and improve quality of life.* Sustainable Agric. Res. Education Program, USDA. Washington, DC.

Tweeten, L.G. and W.A. Amponsah. 1996. Alternatives for small farm survival: Government policies versus the free market. *J. Agric. Appl. Econ.* 28(1):88-98.

Vasavada, U., J. Hrubovcak, and J. Aldy. 1997. Incentives for sustainable agriculture. *Agric. Outlook* 238:21-24.

Economics of Sustainability

Allen, P. 1993. Connecting the social and the ecological in sustainable agriculture. In: *Food for the future: Conditions and contradictions for sustainability,* P. Allen (ed.). John Wiley & Sons, New York.

Berry, W. 1977. *The unsettling of America: Culture and agriculture.* Sierra Club Books, San Francisco, CA.

Brewster, J.M. 1970. *A philosopher among economists.* J.T. Murphy Co, Philadelphia, PA.

Cochrane, W. 1996. *The troubled American economy—an institutional policy analysis.* Department of Applied Economics, University of Minnesota, St. Paul, MN, Staff Paper P96-9.

Common, M. and C. Perrings. 1992. Toward an ecological economics of sustainability. *Ecol. Econ.* 6(1):7-34.

Daly, H.S. and J.B. Cobb. 1989. *For the common good: Redirecting the economy toward community, the environment and sustainable future.* Beacon Press, Boston, MA.

Ikerd, J.E. 1990. Agriculture's search for sustainability and profitability. *J. Soil Water Conserv.* 45(1):19-23.

Kirschenmann, F. 1991. Fundamental fallacies of building agricultural sustainability. *J. Soil Water Conserv.* 2:165-168.

Levins, R.A. 1989. On farmers who solve equations. *Choices* 4(4):8-10.

McCloskey, D. 1984. The rhetoric of economics. In: *Appraisal and criticism on economics: A book of readings,* B. Caldwell (ed.). Allen & Unwin, Boston. pp. 320-356.

McNaughton, N. 1988. Sustainable agriculture: Farming that lasts. *Agric. For. Bull.* 11(4):3-7.

Rodale, R. 1983. Breaking new ground: The search for a sustainable agriculture. *Futurist* 1(1):15-20.

Chapter 7

Transformation in the Heartland: Emergence of Sustainable Agriculture in Iowa

Jerry DeWitt
Charles A. Francis

INTRODUCTION

A major transformation in agriculture has occurred over the past two decades in the U.S. heartland. Sustainable agriculture has become a recognized and successful alternative to conventional industrial approaches to crop and animal production, particularly in Iowa. With a unique array of cooperating farmers, nonprofit farmer organizations, and university faculty the state has responded to the farm crisis of the 1980s and more recent changes in federal farm bills to innovate in new directions. From lowered nitrogen applications to hoop houses for swine, and from marketing specialty pork products to community supported agriculture, Iowa farmers have found themselves on the frontiers of innovation in an agriculture that supports families and rural communities. This chapter chronicles the sweeping changes that have occurred in Iowa in recent years and how these have been supported by the regional Sustainable Agriculture and Research Education (SARE) program.

BACKGROUND FOR CHANGE

Iroquois County, Illinois was on the front line in a battle in the mid 1950s between the USDA and the Japanese beetle. As detailed in her seminal book *Silent Spring,* biologist Rachel Carson (1962) described the eradication program undertaken, using broad spraying of dieldrin on thousands of acres

Developing and Extending Sustainable Agriculture
© 2006 by The Haworth Press, Inc. All rights reserved.
doi:10.1300/5709_07

around Sheldon, Illinois, to prevent the spread of the beetle to more farmland. The results on wildlife were predictable, since dieldrin was more potent than DDT. In the path of this chemical onslaught were also domestic cats, sheep, and other livestock. Dead muskrats, rabbits, and foxes littered the area. And most affected were the birds.

One of the authors grew up in the 1950s near Sheldon. Jerry DeWitt remembers that one of his most important chores on the farm was to gather the dead robins off the lawn each Saturday morning. He also remembers the loss of wildlife in general, and can attest to what he experienced and what was described in Carson's book. Could anyone imagine that he would eventually be a key administrator in Iowa, shaped by the events of childhood, and thus predisposed to help lead an effort to explore alternatives to a chemical-intensive agriculture? The well-known irony described in *Silent Spring* was that the Japanese beetles were suppressed for a time, but still continued their march to the West. DeWitt moved west to Ames in 1972 to his job with Iowa State University (ISU) Extension. Japanese beetles moved west to Iowa in the mid 1990s. Their confrontation continues yet today.

Another factor in promoting awareness of the negative consequences of industrial farming methods in Iowa was the growing number of communities with nitrate in the water supply. Virtually all of the potable water for people and livestock comes from wells or surface streams, and Iowans were surprised to learn that many of these sources had more than the allowable 10 parts per million of dissolved nitrate. Since agriculture and especially corn production accounted for application of some 145 lbs of nitrogen/acre across more than 11 million acres of corn per year in 1980, it was soon obvious that there was a connection between these two pieces of data. Long-used guidelines of a pound and a half of N fertilizer for every bushel of corn harvested were then challenged. Soil scientists calculated careful nitrogen budgets of where that important nutrient came from in the field system, and found that in fact excessive application of N fertilizer just to be sure the crop would not be short was both an expensive strategy and a disaster for water quality.

Finally, the farm crisis of the 1980s was a wake-up call for many in Iowa agriculture. The well-designed and smoothly running system was clearly not providing enough income for many families that had been cultivating the land for decades. This was a gut-wrenching experience for farm operators whose land had been in the family for several generations, through depressions and the dust bowl of the dirty thirties, when they came face-to-face with losing the farm. Suicide rates climbed among farmers. The situation was vividly described by former ISU faculty member, Jane Smiley, in her book *A Thousand Acres* (Smiley, 1991) and later a movie by the same title. Although there was widespread concern in Iowa, the university community was perceived by many rural people as oblivious to the problems grow-ing all around them. Only those in Cooperative Extension who were in the field, close to their clients, were aware that all was not well. These are the key people who emerged in leadership roles in the early days of the SARE program to cause programs to begin to change. The stage was now set for some major first steps that would lead to an exemplary state program in sus-tainable agriculture that is now the envy of their midwestern neighbors and other agriculturists across the United States.

KEY EARLY INNOVATIONS

The Big Spring Basin area in northeastern Iowa is characterized by shal-low fertile soils that have been deposited by wind and have weathered from the limestone parent material from what was once a large ocean floor. As water works its way through the creviced limestone in this Karst topogra-phy the rock is rapidly eaten away and caverns are formed underground. At times these collapse naturally due to water infiltration or from the weight of farm equipment, and there are sink holes formed that lead from the surface directly to the groundwater. These holes in fields or farmsteads are a nui-sance to the equipment operator and dangerous to livestock, and the logical solution was to fill them with whatever was available. Unfortunately, some of the garbage for disposal on many farms included livestock carcasses and chemical pesticide containers. When these were thrown in sink holes, and additional rain coursed through the depressions, this moisture went straight to the groundwater, carrying with it the dissolved chemicals and other waste products from a modern agriculture. This was a recipe for water quality disas-ter, and the point sources were quickly identified as a major culprit in the declining quality of water in farm wells for people and livestock, as well as for rural communities.

An early example of a multidiscipline, multiagency project that was in a large part farmer driven, was the Big Spring Basin Demonstration project. In this small watershed of about ten square miles there was a high level of farmer participation and rapid appreciation of the problems caused by con-tinued high rates of nitrogen use for corn production and point sources such as pesticide containers that still had traces of chemicals and other sources of pollutants that were going straight to the groundwater. From the demonstra-tions, farmers learned how to adjust N-rates for optimum yields and prof-itabilty and how to quickly fill these sinkholes in other ways and to cycle containers back though safer channels.

The Southwest Conservation District started another grassroots pro-gram at about the same time in collaboration with agricultural scientists from ISU and environmental and health scientists from University of Iowa (UI). The Iowa Department of Natural Resources (DNR) completed the local and state collaboration in studying impacts of nitrate and atrazine residues in the surface and groundwater. There were measured pulses in concentration of these contaminants in streams in the spring just after application in adjacent fields, and initially the health impacts of pollutants were uncertain. What caused a major change was the discovery that reduced levels of nitrogen applied to corn fields could substantially lower the source of soluble nitrate that left the corn system to enter the water system, and N applications could be reduced by 50 pounds/acre without significantly changing corn yields. This change occurred based on sound research evidence and demonstrations in farmers' fields, and the advance was both profitable for farmers and better for groundwater quality.

Another key first step was the creation in the mid-1980s of the farmer organization, the Practical Farmers of Iowa (PFI, see Chapter 12). The Thompson's on-farm research and the well-attended field days starting with the major activity on Dick and Sharon Thompson's farm east of Boone, Iowa eventually stretched to two days with more than a thousand people attending each year, often from two dozen different states. The field tours, slide shows, ridge-till equipment on display, and cooperation from the Rodale Institute made this a must-see event for farmers considering changes in their own operations.

A major break-through in collaboration between the PFI and Iowa State University occurred in 1987 when Dick and Sharon Thompson were invited by the Plant Pathology Department to present a seminar on campus. This opened doors to a further sharing of information, and soon there were several graduate students doing research together with the Thompsons and other members of PFI. As described in the chapter about on-farm research, there were appropriate experimental designs tested and perfected that allowed full-sized equipment and drive-through plots. These were not only highly visible and credible for farmer audiences but also compared to small plots, the paired comparisons with six or more replications provided statistically valid tests between two treatments (Rzewnicki et al., 1988). Farmers often refer to experiment station trials as *those garden plots the researchers run,* and the large plots designed by farmers immediately brought interest in the research process and a multiplication of experiments on many more farms. The results have been published each year under the title of "Thompson On-Farm Research" (e.g. Thompson, 2003).

Another milestone event was organized in 1987 by the Department of Fisheries and Wildlife Ecology. The *Ecological Farming* half-day seminar was attended by over 150 people, and it featured both farmers and ISU faculty as speakers and panelists at the Scheman Continuing Education Center.

Not all was smooth. Some talked too much and too long. Some did not listen long enough. During one panel discussion, a member of the faculty stated that research was the concern of the university and that sustainable agriculture was not the way of the future. After a heated exchange, cooler heads prevailed and a useful and more cooperative discussion followed. This successful workshop was held again the following year.

It is noteworthy that these first initiatives in bridging the gap between farmers promoting sustainable agriculture and the university community were started by Botany and Plant Pathology and by Fisheries and Wildlife Ecology Departments, and not by traditional specialists in crops, plant protection, and soils. In keeping with the thesis put forth by Thomas Kuhn (*The Structure of Scientific Revolutions,* 1962), major changes most often begin not in the mainstream but around the edges of a given community. Kuhn cites the examples of Protestantism, of women's suffrage, and of civil rights in the United States, that all started as small movements on the periphery of society and eventually permeated the culture. Thus it has been with sustainable agriculture. Thus it was at Iowa State University.

The passage of the groundwater protection act by the Iowa General Assembly in 1987 was another pivotal point in the emergence of sustainable agriculture in the state. With the insistence of key bi-partisan supporters—among them Paul Johnson, Sue Mullins, David Osterberg, and Ralph Rosenburg—this act established funding for four Iowa centers including the Leopold Center for Sustainable Agriculture at ISU, and essentially legitimized this activity with support from the state. Other key people who influenced political decisions in the state were George Hallberg, Jim Gulliford, and Bernie Hoyer. The funding for The Leopold Center came from the Iowa General Assembly to the ISU College of Agriculture and from a tax on nitrogen fertilizer and state pesticide registration fees. Interestingly, this newly imposed tax was a formalization of a self-imposed tax requested by the Iowa agricultural chemical industry several years earlier. The Leopold Center has since grown to be the preeminent sustainable agriculture center in the United States, as described later.

Finally, a natural event occurred in this time frame that underlined the importance of conservation and long-term planning. Large rains in 1993 in northwest Iowa caused a major flooding of the Des Moines River, resulting in a major loss of property and requiring costly repair. It finally became clear to thoughtful observers and analysts of agriculture that the vast spread of monoculture cropping, primarily corn and soybeans, and the loss of buffering capacity in this large watershed had caused a serious decline in ecosystem services, especially the potential to capture and store rainfall and mitigate such a major rainfall event. The stage was set for some important changes throughout Iowa agriculture, and the initiation of a regional SARE-funded program in research and professional development for Extension and NRCS (then SCS) personnel further supported this environment for change (see Chapter 8).

MAJOR INITIATIVES AND PROJECTS
IN SUSTAINABLE AGRICULTURE

With so many precedents both in the farmer organizations and at state level, and with ISU faculty rapidly coming on board with research on farms as well as the experiment stations, it is difficult to describe what was the cause and effect during this early time of change. It is clear that collaboration among state agencies, farmer groups, and university was a prime motivator and driver in the process. The PFI and Leopold Center both played key roles in the change. And the concurrent funding available from the North Central Regional SARE (formerly LISA) Program was another impetus that encouraged innovative researchers and farmers to move into testing new and perhaps more risky farming practices and systems. Here is a description of the major programs and how the SARE funding was a catalyst for change.

Integrated Farm and Management Demonstrations

One specific landmark program that reached many corners of Iowa resulted from the 1987 Groundwater Protection Act. The widespread demonstration of practices and systems of management was a cooperative effort of local farmers, the Iowa Natural Heritage Foundation, the Iowa Geological Survey at the University of Iowa, the PFI, local Soil and Water Conservation Districts, and Iowa State University. At this time the EXXON Energy Overcharge funds were made available to key projects through the state Department of Natural Resources, and this financing helped the state agencies, universities, and nonprofit groups to implement the demonstrations of reduced fertilizer application levels in almost every county in Iowa. Approximately 35 percent of these funds were allocated to the Leopold Center. This successful program was operating in parallel with similar programs in neighboring states, for example the program focused on reducing nitrogen application to corn and sorghum in rotation with soybeans in Nebraska (Franzleubbers et al., 1994). Ideas about practical field designs for research were also tested in these projects (Rzewnicki et al., 1988). This widespread program in Iowa was one of the first to expose a large proportion of the Cooperative Extension staff in the field to the importance and potential economics of reduced fertilizer applications.

Iowa State University

Although many in the farming community remarked about the conservatism and slow changes occurring in the state land grant university, there were some important activities and events that did shape the programs of ISU and the directions of research and Cooperative Extension. Perhaps conditioned by the protests of the 1960s and such events as the Kent State demonstrations and student deaths, a new generation of faculty was making its way through the ranks of academia. Although part of the traditional academic establishment, some of these individuals were poised and ready to help implement change when they reached positions of decision making and power within the university.

The innovators in academia were certainly aided by one landmark study and publication. *Alternative Agriculture* (Pesek, 1989), a publication of the National Academy of Sciences, was developed by a committee of well-known experts in science that was chaired by John Pesek, Head of the ISU Department of Agronomy. Beyond the borders of Iowa, Dr. Pesek was prominent in his professional societies and a former President of both the Soil Science Society of America and the American Society of Agronomy. Such involvement by a high profile soil scientist has an impact that is not lost on faculty, especially junior scientists. At this time it was also obvious that specialists and county agents in Cooperative Extension who were closer to the field and had their fingers on the pulse of rural Iowa were well aware of the need for change. Consolidation in the input industry and in the grain trade were causing an acceleration in the loss of family farms, and these rural families were the lifeblood and core of clients for Extension. Survey data at the national level has shown that sustainable agriculture programs have had more of an impact on Extension than on the other two primary arms of land grant universities, research and teaching (Francis et al., 1995).

Leopold Center Research Initiative Groups

The Leopold Center for Sustainable Agriculture at ISU has played a key leadership role in the programs and events that have shaped the university environment in Iowa. Shortly after the founding of the Center and interim leadership by ISU Economist Dr. Robert Jolly, Dr. Dennis Keeney from Wisconsin was named Director. Similar to John Pesek, the career of Keeney was exemplary in the field of soil science, where he also held the highest national offices of his professional society. Such moves have helped immediately and immensely to legitimize the place of sustainable agriculture in the land grant university system. In one of the most successful university conversion programs in the land grant system, the Leopold Center established a series of multidisciplinary groups and provided financing for faculty release time as well as graduate student support in each of several priority areas. The Leopold Center created some of the first interdisciplinary research teams, also called issue teams, at Iowa State University. From 1989 to 1991, six issue teams were created to investigate various components of sustainable agricultural systems. A seventh issue team and two issue-based initiatives were created from 1995 to 1997. Over a period of ten years, these included the following areas:

- alternative swine production systems (also known as the Hoop Group)
- agroecology
- animal management
- cropping systems
- human systems
- integrated pest management
- manure management
- organic agriculture (also called Long-Term Agroecology Research, LTAR)
- weed management

Although there was modest funding provided for each team (up to $50,000 per year), the majority of faculty time and research support came from the scientist's own programs. Thus the Leopold Center successfully leveraged resources from ongoing mainstream programs and redirected this energy and funding toward priority projects in areas related to sustainable agriculture. This center has grown in prominence for over 15 years, and has a bright future leadership role under the highly capable direction of Dr. Fred Kirschenmann, a large-scale organic farmer and prominent national figure in the field of philosophy and ethics. Fred has published various works on the potentials of organic farming, the value systems of rural United States, and the family farm contributions to agriculture (see Chapter 15).

In cooperation with ISU Cooperative Extension and PFI, and a close collaboration with key individuals at University of Northern Iowa in Cedar Falls such as Dr. Kamyar Ensayan, the Leopold Center has been very active in the development and promotion of local food systems. This has involved attracting business and industry as well as public institutions to use local food and in-season products in their planning of menus and purchase of foods for large groups. One of the impressive results of this project is the growth in farmers markets, especially community supported agriculture (CSA) activities around the state. A short decade ago, there were less than five of these sales activities where farmers contract with consumers for a seasons' worth of produce delivery. In 2004 there were 60 in Iowa. More detail is given on the local foods project in a later section on current projects.

Statewide Sustainable Agriculture Summit

In 1993 there was a seminal meeting of key people, including a wide range of stakeholders, NGO's, and groups, in a summit on sustainable agriculture hosted by ISU Extension. About 30 leaders assembled in Ames to drink organic milk from the local dairy of Francis Thicke, to review research results, to examine the outcomes of a series of listening sessions held around the state, and to make plans for the future. The opinions of clients in Iowa ranged from being highly positive about the changes that were under way to "there's too much money being spent on sustainable agriculture already." The meeting was held to take stock of the current situation and to forge improved collaboration for the future. This initial effort provided the fundamental framework for the development of the Iowa Professional Development Program (PDP) in sustainable agriculture.

Two initiatives emerged from the summit that illustrate the integration of sustainable agriculture into the mainstream of education and Extension. Pesticide applicator training is a key program of Extension that provides recertification training to all personnel, both commercial operators and private farmers (more than 30,000 annually), who apply restricted-use pesticides in agriculture and other sectors. Including some units on organic farming, insect refugia, IPM, and sustainable agriculture was an important way to infuse this new thinking into mainstream programs that reached essentially all Iowa farmers and employees of cooperatives who were engaged in handling and applying pesticides. One of the key approaches to reach this audience was "call it what they want, and give them what they need!"

Another popular program, based on the work of ISU researcher Dr. Mark Honeyman, was centered around the hoop house, a low-cost structure that was highly versatile in application on the farm. In a mid-1990s conference on hoop houses held at ISU by Iowa Extension, The Iowa Pork Producers Association, The Pork Industry Center, and the Leopold Center, over 220 people showed up to learn about this alternative technology. This was followed two years later with a conference attracting more than 300 interested participants in alternative pork production strategies. Used for farrowing and finishing hogs, for raising replacement heifers, for broiler or egg production, and for storage of hay or equipment, these structures quickly captured the imaginations of farmers across the state with nearly 3,000 hoops on the landscape by 2004. Hoops proved to be a highly successful innovation. The early efforts in Iowa have resulted in an International Hoops Conference at Ames in 2004 and the formation and active work of a Pork Niche Group of producers and industry representatives supported by the Leopold Center, Iowa Cooperative Extension, PFI, and other partners.

A third and often unheralded outcome of this meeting was the concerted and coordinated efforts by those in attendance to place a sustainable agriculture representative on the state NRCS *State Technical Committee* in order to broaden its membership and thinking. Letters were written and calls were made, and soon the NRCS State Conservationist LeRoy Brown appointed Francis Thicke and Ann Robinson to the *State Technical Committee.* Their impacts were felt immediately, as they provided within one year the inclusion of an approved program in Iowa for support to growers for transition to organic agriculture.

Regional Professional Development Program
Training Sessions

When the NC Regional SARE Administrative Council awarded a training grant to University of Nebraska and Ohio State University (see Chapter 8 by Carter, Francis, and Olson), the organizers naturally looked to Iowa for examples of projects that were already under way. The first regional planning conference was held in Cedar Rapids in early 1995, with representatives of farmer groups, university Cooperative Extension, NRCS, and private industry from all 12 states in the region. From this first organizing session of 66 regional participants, the program grew to include 10 workshops for educators and five planning and evaluation workshops that were held in ten different states over a period of four years. In the last year of the program, one of the workshops was held in Ames with a field tour that included the Bear Creek Watershed management project. It was fitting that this program began in Cedar Rapids and ended in Ames, just some 100 miles apart, and that it built on much of the interest and research data that had been generated in Iowa (see Carter and Francis, 1995, 1996; Carter et al., 1997, 1998).

CURRENT AND ONGOING PROJECTS

Many of the initiatives described above have been the key to establishing sustainable agriculture in Iowa, and to creating public awareness among consumers and policy makers about the importance of how and where food is produced. In addition to the widely publicized problems and solutions around the major issue of water quality, there are other environmental as well as economic and social implications of this current transformation in the heartland. Several of the most important current projects are described here.

Sustainable Agriculture Education
at Iowa State University

Early visions and a critical mass of interested people led to the formation of a university-wide graduate program in sustainable agriculture serving the diverse needs of students in Iowa and the nation. Under the initial leadership of Drs. Michael Bell, Tom Richards, Ricardo Salvador, Matt Liebman, and others, this program cuts across colleges and serves 15-20 new students annually with a graduate team of more than 80 ISU faculty.

Development of PFI Farmer Network for Research

On-farm research was first high-lighted in the mid 1980s with the initial work of PFI and Dr. Rick Exner (see Chapter 12). This network of interested farmers with challenging ideas cooperates with ISU Extension and research to meet the needs of local farmers. More than 20-25 farmers participate annually in this work, and host PFI field days statewide. Added to this effort was the formation of a campus group in 2004 of faculty who serve as departmental *On-Farm Research Coordinators* for the College of Agriculture under the combined leadership of Jerry DeWitt and PFI.

Direct Marketing Initiatives of PFI
and Leopold Center

Direct marketing and food systems have been vigorously addressed through special projects from the Leopold Center and PFI, and often with Extension field staff participation. The Extension value added program at the Ag Marketing Resource Center (AMRC) at ISU has helped coordinate and support these efforts. Programs include a number of Leopold Center projects and work, including The Pork Niche Group, Value Chain Opportunities, Food Miles study, Vitculture report, Ecolabels, and A Geography of Taste. PFI efforts have centered around Buy Fresh, Buy Local; All Iowa Meals; Marketing to Iowa Institutions; and the Iowa Café workshops.

Development of Organic Production Systems
and Markets

An organic Extension field specialist was hired in 1993 to work in organic agriculture in eastern Iowa. By then, Extension ANR Program Leader

Jerry DeWitt was implementing a field-wide restructuring for Cooperative Extension. Its success and great visibilty led to the proposal for a tenure-track faculty position on campus. With the hiring of Dr. Kathleen Delate in 1996, ISU continued its journey towards a more sustainable agriculture through research and Extension. Delate's field work is conducted statewide with direct support from Extension, the Research Station, and the Leopold Center plus others. Areas of interest include soybean production, alternative crops, economics of organic systems, long-term rotation research, and vegetable production. She works closely with field staff, farmers, and the Extension PDP program, and sponsors an annual Iowa Organic Conference now drawing more than 300 people each year.

Wallace Chair of Sustainable Agriculture at ISU

First established in 1996, the Henry A. Wallace Endowed Chair in Sustainable Agriculture attracted internationally-known social scientist Dr. Lorna Butler. Funded primarily by an endowment from Jean Wallace Douglas and the W. K. Kellogg Foundation, this chair position serves the broad interests of faculty, students, and citizens of Iowa. This office has been instrumental in the development and implementation of the ISU graduate program in Sustainable Agriculture, and also the development of the John Pesek Colloquium in Sustainable Agriculture in cooperation with the Leopold Center and ISU Extension.

North Central Regional Community Development Center

The mission of the North Central Regional Community Development Center (NCRCDC) which is located at Ames is to initiate and facilitate rural development research and education programs to improve the social and economic well-being of rural people in the region. The NCRCDC also provides leadership in rural development regionally and nationally by identifying, developing, and supporting programs on the vanguard of emerging issues. It works closely with sustainable agriculture interests on campus and is directed by Dr. Cornelia Flora. There are five major program areas where the NCRCDC invests resources:

- Increasing economic competitiveness, diversity and adaptability of small and/or rural communities
- Linking natural resource industries, particularly agriculture, with community and environmental resources
- Increasing community capacity to deal with change
- Enhancing self-reliance of families and communities

• Facilitating development of policies that enhance the well-being of rural people and small towns.

Summer Agroecosystems Analysis Course

Dr. Mary Wiedenhoeft, an agronomist at ISU, coordinates a field course for new entrants to the graduate sustainable agriculture program each summer. Enrollment is also open to students from cooperating universities in neighboring states and beyond. The class and faculty travel for one week across Iowa, Minnesota, and Nebraska for farm site visits, study and interaction with farm families, and student groups work to prepare them well for a future course of study in sustainable agriculture at ISU or elsewhere (Wiedenhoeft et al., 2003).

LESSONS LEARNED IN THE IOWA EXPERIENCE

Perhaps most valuable to readers of this chapter are the lessons that have been learned over the past 15 years about what to do and how to do it in a state that is generally conservative and slow to change to new systems or ideas. By the way, this resistance to change is not all negative, and farmers around the world have survived for millenia because they were willing to improve current systems and take only the risks that seemed necessary to increase production or profits without risking their farms or their families' livelihoods. But given the financial crises in agriculture and the serious nature of groundwater pollution in Iowa, it has been obvious that some things would have to change to maintain some level of equity in communities and keep people on the farm, assuming that the family farming system is somehow more desirable than the industrialized alternatives. Here are some of the lessons learned in Iowa from this process.

• Add and integrate sustainable agriculture components or modules to *current* Extension programs for widest reach to clients; integrate, do not compete.
• Maintain a dynamic, practical, and topical training program for newly-hired local educators and specialists in Cooperative Extension and NRCS; reach out to young professionals.
• Establish and maintain dialogue with intended clients through surveys, listening sessions, and focus groups to learn what should be included in sustainable agriculture training; listen more and talk less.

- Build trusting and lasting partnerships for planning and delivery of training to mobilize needed resources; learn from the past, and look to the future.
- Recognize and emulate progress in sustainable agriculture from other Extension programs, and take people beyond their current comfort levels; challenge and reward staff for risk-taking.
- Recognize the importance and uniqueness of situation and place, and design flexible programs that are appropriate to each audience with local planning and implementation; involve local stakeholders and clients.
- Approach sustainable agriculture programs within the appropriate context of complex social, economic, and environmental factors; seek a practical balance, and not a formula.
- Plan and implement on-going training, keeping in mind that the process should be inclusive and involve all potential stakeholders and experts such as farmers and people in NGOs; be sure to include all those who will be impacted by programs.

CONCLUSIONS AND POTENTIALS
FOR FUTURE IOWA AGRICULTURE

At the same time this major transformation is taking place on the family farms and in the marketing of agricultural products in Iowa, there is a continuing trend toward the industrialization of much of the agricultural sector. Thus there are emerging two parallel tracks in production and the food industry—one toward a larger and industrialized commodity activity where the ownership and control is in fewer hands, and another toward a decentralized, farmer- and community-driven, and diverse production system where the benefits are more evenly distributed across the rural Iowa population. There is no doubt that the SARE funding for research, education, and farmer experiences have made a major contribution to the credibility and profitability of the decentralized system. Many thoughtful observers conclude that this system will be more sustainable in the long term. Marty Strange (1988) concludes in his book *Family Farming* that with consolidation in agriculture and business we are slowly re-creating a financial and social stratification in the Midwest that most of our ancestors left northern Europe to escape. In Iowa the programs are designed to reverse this trend.

Although the overall trends are toward an industrial agriculture and global food system, the initiatives taken in Iowa and elsewhere in the Midwest are providing a viable and profitable alternative for some farmers who are willing to think beyond standard feed grade number 2 yellow corn, conventional soybeans, and other commodity crops. The organic food demand has grown by more than 20 percent per year for the past two decades, and now represents over one percent of the U.S. food system (see Chapter 15). It is obvious that a farm family cannot make a decent living today on 80 acres of even the best and most fertile Iowa farmland, at least if they are trying to "farm the government" and only raise commodity crops. SARE programs and the Leopold Center have financed the search for viable alternatives, and these are being implemented by thoughtful and innovative farmers in the region—many of whom do not come from farm backgrounds. Although there is a trend toward industrialization of commodities, reliance on the global markets, and consolidation of farms and businesses, we are working with government and nonprofit groups in finding reasonable alternatives for family farmers in Iowa. As educators at land grant universities our purpose always should be to provide choices, options, support, and hope to farmers, and not to confirm, select, or promote only our personal choices or preconceived dogmas. As educators we also need to encourage our students and other clients to do better than accept the status quo, and to envision a positive future. We agree with the words of Nobel laureate, René du Bos: "Trend is not destiny."

REFERENCES

Carson, R. 1962. *Silent spring.* Houghton Mifflin. New York.

Carter, H. and C.A. Francis. 1995. *Everyone a teacher, everyone a learner.* Extension and Education Materials for Sustainable Agriculture, Vol. 4. Cooperative Extension Division, University of Nebraska, Lincoln, NE. May.

————. 1996. *Shared leadership, shared responsibility.* Extension and Education Materials for Sustainable Agriculture, Vol. 5. Cooperative Extension Division, University of Nebraska, Lincoln, NE. December.

Carter, H., C. Francis, and R. Olson. 1997. *Linking people, purpose, and place: An ecological approach to agriculture.* Extension and Education Materials for Sustainable Agriculture, Vol. 7. Cooperative Extension Division, University of Nebraska, Lincoln, NE.

Carter, H., R. Olson, and C. Francis. 1998. *Facing a watershed: Profitable and sustainable landscapes for the 21st century.* Extension and Education Materials for Sustainable Agriculture, Vol. 9. Cooperative Extension Division, University of Nebraska, Lincoln, NE.

Francis, C., C. Edwards, J. Gerber, R. Harwood, D. Keeney, W. Liebhardt, and M. Liebman. 1995. Impact of sustainable agriculture programs on U.S. landgrant universities. *J. Sustain. Agric.* 5(4):19-33.

Franzluebbers, A.J., C.A. Francis, and D.T. Walters. 1994. Nitrogen fertilizer response potential of corn and sorghum in continuous and rotated crop sequences. *J. Prod. Agric.* 7:193-194,277-284.

Kuhn, T.S. 1962. *The structure of scientific revolution.* University of Chicago Press, Chicago, IL.

Pesek, J. 1989. *Alternative agriculture.* National Research Council, Board on Agriculture, National Academy Press, Washington, DC.

Rzewnicki, P.E., R. Thompson, G.W. Lesoing, R.W. Elmore, C.A. Francis, A.M. Parkhurst, and R.S. Moomaw. 1988. On-farm experiment designs and implications for locating research sites. *Am. J. Altern. Agric.* 3:168-173.

Smiley, J. 1991. *A thousand acres.* Random House, New York.

Strange, M. 1988. *Family farming: A new economic vision.* University of Nebraska Press, Lincoln.

Thompson, R. 2003. *Thompson on-farm research.* Practical Farmers of Iowa, Boone.

Wiedenhoeft, M., S. Simmons, R. Salvador, G. McAndrews, C. Francis, J. King, and D. Hole. 2003. Agroecosystems analysis from the grass roots: A multidimensional experiential learning course. *J. Nat. Res. Life Sci. Educ.* 32:73-79.

Chapter 8

Regional Training Workshops for Sustainable Agriculture

Heidi Carter
Charles A. Francis
Richard Olson

INTRODUCTION

Conventional agricultural extension programs have been slow to encompass the integrated production, economic, environmental, and social concerns of farmers interested in developing long-term, sustainable production systems. For this reason, the national Sustainable Agriculture Research and Education (SARE) Program was initiated in the late 1980s to fill a recognized need for more targeted training. This chapter describes a series of sustainable agriculture training workshops held during a four-year period for Extension educators and specialists, Natural Resource Conservation Service (NRCS) personnel, and others interested in answering farmers' questions related to production systems. We describe the background and need for these workshops, comprehensive planning approach, themes, topics, evaluation, lessons learned, and recommendations for the future (expanded from Carter et al., 1999).

Extension education programs have traditionally focused on new advances in component technologies, such as choosing the right hybrid or fertilizer rate, or on learning special skills, such as pesticide application safety or farm record-keeping. Sustainable agriculture Extension programs were not part of the mainstream training activities before the 1990s. The national SARE program and four regional programs brought both new attention and substantial funding for sustainable agriculture research and education. These programs were funded through three types of grants: research and

Developing and Extending Sustainable Agriculture
© 2006 by The Haworth Press, Inc. All rights reserved.
doi:10.1300/5709_08

education, producer research, and professional development (Bauer, 1996, 1997). An overview evaluation of the first ten years of SARE programs across the United States was provided by Berton (1998).

Early confusion among Extension educators about how to define sustainable agriculture was complicated by the first name of the national program, "Low Input Sustainable Agriculture." Many educators had difficulty moving beyond the initial misconceptions surrounding the term sustainable agriculture, defined as low-input, and including implications of low productivity, low technology, and low profits (Francis, 1990). One impact of the regional training program was to broaden an appreciation of sustainable agriculture as incorporating sustainable productivity and profits, minimal negative environmental impact, and long-term social viability (Agunga, 1995; Dunlap et al., 1992; Paulson, 1995). Another perspective was that current farming and food systems should not limit the ability of future generations to meet their needs for food, fiber, and a healthy environment (Harwood, 1990). There was an emerging awareness that sustainable agriculture is a process toward long-term goals and not a specific set of defined practices.

Within the framework of conventional agriculture and Extension education, we created a new type of learning environment in a series of training workshops in sustainable agriculture in the North Central Region from 1995 to 1998. With funding from the regional SARE Professional Development Program (PDP), a group led by the University of Nebraska through its Center for Sustainable Agricultural Systems began to develop and implement a comprehensive training program. The goal was to prepare a cadre of teachers to conduct innovative training in their own states. Two overall guiding principles emerged from the first planning meeting in 1995, well articulated by Jerry DeWitt of Iowa State University (Carter and Francis, 1995):

- Sustainable agriculture must be viewed in a complex framework of social, economic, and environmental factors, and
- Training must be inclusive, both in terms of trainers who plan and implement the sessions and the audiences.

To avoid institutional stereotypes and to help participants move beyond the conventional mindset, workshops were not held on university campuses. The sites chosen provided easy access to farms for tours, and included educational conference centers, a state park in Indiana, and research stations in North Dakota and Ohio. During the four-year period, ten workshops were conducted in ten different states, plus five planning and evaluation sessions.

FOCUS ON PLANNING

Central to the planning process was recognizing and valuing diverse types and sources of knowledge. Although our primary contact persons in each state were in Extension administration and those designated as state-wide leaders in sustainable agriculture, we intentionally brought together people with different backgrounds to serve on state planning committees. Key representatives from university Cooperative Extension, NRCS, farmer groups, and nonprofit research and education centers were involved in planning the program's focus and specific activities. Members of the planning team were later asked to conduct learning sessions, lead field tours, and recruit participants. This process led to a broad-based representation from the agricultural sector, including private industry, at each training workshop. From the beginning, we believed that by involving these various client groups in planning, we were more likely to address real issues for them, and there would be a higher probability of their adopting the concepts and learning methods addressed in the workshops.

As part of the planning process, a review meeting was held at the beginning of each year to assess the results from the workshops and explore training needs across the region. Recommendations from these reviews were used to set program goals and objectives, and to guide the selection of workshop themes and content. The people that we recruited for state planning teams represented the intended client audience. Their input was crucial to setting the meeting agendas, suggesting local speakers, and choosing tour sites. Although many arrangements were accomplished from Lincoln, Nebraska, state teams were essential in organizing most local logistics.

Decentralized planning was a key to success. Frequent communication with state teams in the months leading up to each workshop increased the local sense of ownership in the process and learning activities, as well as affirmed their expertise and local knowledge. Each team took great pride in planning a workshop that would show off their unique resources, people, and accomplishments in sustainable agriculture. We regularly mailed information items to the state teams and to other leaders in that state to be sure that all were well informed. This communication helped in recruiting participants within the host state.

During the four years, ten sustainable agriculture training workshops were co-sponsored by the North Central Region PDP and the state planning teams in Illinois, Indiana, Iowa, Kansas, Michigan, Nebraska, North Dakota, Ohio, Minnesota, and Wisconsin. The total number of attendees was more than 850. Participants and presenters were Extension educators and specialists (55 percent), state and federal agency employees (14 percent), farmers and ranchers (13 percent), nonprofit representatives (11 percent), college and graduate students (6 percent), and private consultants or industry people (1 percent). The target audience according to the grant included Extension educators, NRCS specialists, and other government agency employees; thus the predominance (69 percent) of attendees from these groups was consistent with the grant guidelines. Farmers and ranchers were most often presenters in the field or were field-tour leaders, although some attended sessions as members of state delegations. Other attendees added richness to the discussion and credibility to the workshops, fulfilling our intent to be inclusive in both trainers and audiences.

An explicit goal during the planning sessions was to use and later evaluate a range of learning methods. We had observed from years of experience that the majority of formal Extension events resembled the classroom lecture, with primarily one-way communication of information and limited time for discussion or hands-on practice. An exception was the field tour, especially where farmers were the leaders. We included at least one multifarm tour in each workshop. To enable a careful comparison of learning methods, we included classroom presentations; participatory exercises; facilitated discussions; outdoor tours of farms, research stations, and field trials; and evening fellowship gatherings or artistic presentations. Written evaluations surveyed attendees about each learning exercise and the entire workshop.

CHOICE OF WORKSHOP THEMES

The themes for the yearly workshops were chosen by the regional organizers after the planning workshop, and following in-depth discussions with Extension administrators and state leaders. There was a progression in scale from farm level to community to watershed over the four years (Figure 8.1), and the theme of building leadership was integrated through all the workshops. Before the first workshops were held, Jerry DeWitt of Iowa State University summarized the region's concerns about sustainable agriculture training (DeWitt, 1995). These points became guidelines for program design:

- Maintain integrity of sustainable agriculture programs and avoid duplication of traditional Extension programs.
- Update the program on a continuing basis to meet the needs of new staff in Extension and other information providers.

- Listen to the targeted audience, including farmers and ranchers, to learn what they think Extension and NRCS specialists should know about sustainable agriculture.
- Build partnerships among the appropriate organizations for program planning and delivery of education and include the private sector and nonprofit organizations.
- Seek methods for direct as well as indirect delivery of information, using a variety of methods; not all planning and education need to be "face to face."
- Make sure the program is different and not perceived as just another Extension training or in-service event; if everyone is comfortable, maybe you have not gone far enough.
- Make the training appropriate to audience and location—not all training should be multi-state; provide options that require less travel for staff.
- Involve farmers and ranchers, as mandated by Congress, which is also a good idea; farmers will be some of our teachers, and we are not the only holders of knowledge.

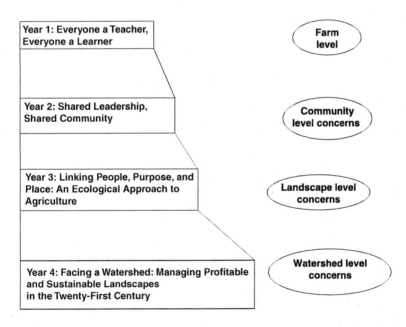

FIGURE 8.1. Progression in a spatial hierarchy in successive workshop years (not to scale).

ANNUAL THEMES FOR WORKSHOPS

Year 1: *Everyone a Teacher, Everyone a Learner* (Carter and Francis, 1995)

In the first year's workshops it was important to discuss definitions of sustainable agriculture. Some confusion was obvious almost every time the theme came up, and we needed to assure people that this concept was one that described a goal rather than a prescribed set of practices that would fit all sites and situations. In a presentation by John Ikerd (1995), we explored the questions of sustaining what? (agriculture), for the benefit of whom? (humanity), and for how long? (forever). Ikerd pointed out that "agriculture, by its very nature, is an effort to shift the ecological balance so as to favor humans relative to other species in production of food and physical protection." Although this same challenge of definitions still faces research and Extension specialists, there is a broader acceptance today of the fact that sustainable agriculture is a process that leads to positive economic, environmental, and social impacts in food systems. The details of the first year workshop schedule are shown in Table 8.1.

Everyone a Teacher, Everyone a Learner included topics on integrated crop and animal systems, social issues related to agriculture and communities, environmental impacts of practices and systems, and economics of alternative production systems. Integrated resource management was presented as a holistic approach to planning. Methods of education were described and demonstrated, including case studies, focus groups, decision cases, on-farm research options, demonstrations, and tours. We discussed the integration of cultural resources, as well as the latest information available from around the region. Evaluation was included as an important topic. Workshops at the Arbor Day Farm Lied Conference Center in Nebraska and the Turkey Run State Park near Marshall, Indiana, were well attended.

Year 2: *Shared Leadership, Shared Responsibility* (Carter and Francis, 1996)

Transformational Leadership, including "full-range leadership" as described by Bass and Avolio (1994), was the major workshop topic and involved a full day of participatory exercises on different leadership styles. These leadership styles are especially important in working with volunteer groups and with university faculty or agency employees, where each individual has a high degree of autonomy in choosing work priorities and program directions. We explored models for whole-farm planning and again

TABLE 8.1. Workshop program from the first year, "Everyone a Teacher, Everyone a Learner," including speakers or discussion leaders (Indiana, Nebraska); speaker names are examples, as these differed between workshops.

Day	Topics	Topics
1	Orientation, Concepts, and Training in Sustainable Agriculture	• Welcome and introductions: mandate from 1990 Farm Bill (D. Petritz, C. Edwards, G. Bird) • What is sustainable agriculture? (J. Ikerd) • State strategic plans for training in sustainable agriculture (state coordinators) • Planning learning activities for Extension and NRCS (C. Francis) • Integrated sustainable systems of crop and animal production (C. Edwards) • Social issues related to agriculture, communities, and new technologies (J. Flora)
2	Integrative Systems and Environmental and Economic Dimension of Sustainability	• Integrated resource management in sustainable agriculture (B. Vining) • Role of system science in the design of pest management systems (G. Bird) • Evaluating local challenges: the opinion/editorial discussion method (C. Francis) • Economics of sustainable agricultural systems (J. Ikerd) • Environmental impacts of conventional practices (F. Hitzhausen and panel) • Building soils for better crops (F. Magdoff) • Constraints on promoting and implementing sustainable agriculture (C. Edwards) • Organizing and conducting farm tours and on-farm research (C. Francis & panel) • Open forum to share materials and methods; time for state meetings
3	Learning Methods and Resources for Sustainable Agriculture	• Participatory strategies for learning: the decision case method (S. Simmons) • Resources for sustainable agriculture (G. Hegyes) • Listening session with farmers (panel)
4	Evaluation of Training and Future Planning	• Importance and methods of evaluation (C. Francis) • Future implementation of state strategic plans (state coordinators) • Future activities of North Central Region training program (C. Edwards & C. Francis) • Evaluation of train-the-trainer workshop (H. Carter & C. Francis)

this year presented a workshop on Holistic Resource Management. Entrepreneurship and Marketing were identified as key interests of concern to Extension specialists and educators, and these topics were prominent on the program. We also presented active learning about study circles and networks for information exchange, and updated the latest resources in sustainable agriculture. Fellowship was a key component of all workshops, and we encouraged people to meet specialists and farmers from other states during the meals and tours. The evenings had some organized activities, such as the play in Wisconsin, not a required activity, but one that was popular and everyone attended. Workshops were held at George Williams College on Lake Geneva in Wisconsin, and at the Carrington Research Extension Center near Carrington, North Dakota (Table 8.2).

Year 3: *Linking People, Purpose, and Place:* *An Ecological Approach to Agriculture* (Carter et al., 1997)

Growing awareness of the importance of understanding the ecology of natural systems, as well as agroecosystems, led to a focus on an ecological approach to agriculture. People are an integral part of agricultural systems, and their goals and purposes become important drivers. This case is obvious with a farmer or rancher who makes decisions and lives with the economic outcomes. As the Extension educator or NRCS advisor becomes more familiar with clients' farms, they also become an "internal resource" or important member of the team. The workshop was designed to explore this role of educators and the uniqueness of place, an essential characteristic of ecology and natural ecosystems (Table 8.3).

Broad workshop objectives were to demonstrate how an understanding of ecological principles can help us design farms and ranches that provide commodities and income while retaining some of the beneficial ecosystem services of natural systems, such as clean air, water, and biodiversity. Because people are central components of agroecosystems, another goal was to explore characteristics of local communities that promote or impede sustainability.

The three workshops were tailored to the ecosystem and agricultural systems of each site. In Ohio we focused on the transition zone from forest to prairie. In Kansas and Minnesota we looked at tallgrass and mixed grass prairie regions that are now dedicated to rainfed agroecosystems. We heard about presettlement conditions and how farmers arrived at current agroecosystems. Ecological principles were related to agriculture, such as resource use and cycling in intensive grazing systems. Other activities imbedded in the landscape and culture included dinner with an Amish family in Ohio, a walk at Konza Prairie in Kansas, and a canoe trip down the Pomme

TABLE 8.2. Workshop program from the second year, "Shared Leadership, Shared Responsibility," including speakers or discussion leaders (Wisconsin, North Dakota); speaker names are examples, as these differed among workshops.

Day	Topics	Topics
1	Introduction and Review of Sustainable Agriculture Education Activities	• Welcome from state team • Washington DC Update (J. DeWitt) • Participant Introductions • State and Natural Resources Conservation Service Reporting Sessions
2	Whole-Farm Planning, Tools for While Farm Planning, and Educator Highlights	• Introduction to PLANETOR Version 2.0 (B. Craven and W. Richardson) • Introduction to Holistic Resource Management (A. Arner and D. French) • Participatory Exercises on PLANETOR and HRM • Approaches to Whole-Farm Planning (J. Enlow and S. Bonney) • Finding the Balance between Systems Thinking and Specific Tools (V. Mundy) • Financial Analysis for Sustainable Agriculture (R. Levins) • Developing a State-Wide Sustainable Agriculture Team (M. Hogan and M. Bennett) • Dinner and *Rural Voices* (theatric presentation on rural and farm issues)
3	Regional Update and Leadership Training	• Regional Update (G. Bird) • Full Range Leadership (E. Birnstihl and D. Wheeler) • Tapping into Your State's Resources for Training (J. DeWitt and D. Mayerfield) • Tour of Michael Fields Agricultural Institute and Krusenbaum Family Farm (WI) • Tour of Carrington Research Station (ND)

TABLE 8.2 *(continued)*

Day	Topics	Topics
4	Tour, Discussion, Policy, Entrepreneurship, Marketing	• Entrepreneurship, Marketing Products from Sustainable Ag Systems (J. Baker and H. Carter)
		• Using Study Circles in Sustainable Ag Training (D. Cavanaugh-Grant and M. Coté)
		• Farmer Networks as learning/ Teaching Methodology (A. Hager)
		• Tour of Bison Ranch (ND)

de Terre River in Minnesota. These events broadened our appreciation of the uniqueness of place, the people who work there, and the current condition of their agroecosystems. Workshops were held at the Ohio Agricultural Research and Development Center in Wooster, the Holiday Inn near Manhattan, Kansas, and the Western Minnesota Research and Extension Center near Morris.

Year 4: *Facing a Watershed: Managing Profitable and Sustainable Landscapes in the Twenty-First Century* (Carter et al., 1998)

Watersheds, landscapes, and water quality were the primary issues in the fourth year. We presented an overview of freshwater use, watershed management, and developing management plans. There was a session on group dynamics in designing and implementing a watershed management plan. It included a participatory group exercise to practice the skills needed to accomplish this task as a team. Information resources for watershed management as well as sustainable agriculture and education were provided. There were also sessions on conservation buffers, riparian management, farmland protection, green corridors, and suburban sprawl as it affects agriculture (Table 8.4).

It was particularly important to include NRCS specialists in the planning and implementation of workshops related to their principal area of interest—resource management and protection. Accomplishing a wider participation by NRCS was a challenge, because their planning framework required long-term anticipation of workshop dates, and these events occurred

TABLE 8.3. Workshop program from the third year, "Linking People, Purpose, and Place: An Ecological Approach to Agriculture" (Ohio, Kansas, Minnesota); speaker names are examples, as these differed among workshops.

Day	Topics	Topics
1	Introduction and Linking Agriculture and Ecology	• Update from Washington DC (G. Bird) • Linking Ecology and Agriculture (R. Olson) • Role of Agroecology in Sustainable Agriculture (C. Edwards) • Ecological Principles of Integrating Perennials into Farming Systems (R. Janke)
2	Whole-Farm Planning and Ecological Principles	• Redesigning State Lands Using Ecological Principles (J. Butler and T. Hurford) • Whole-Farm Planning in the Community Context (W. Monson) • Understanding Your Whole Farm (B. Stinner) • While Farm Planning: What it is? Why do it? (J. Lamb) • Soil Quality (R. Olson) • Agroforestry (J. Vimmerstedt, R. Olson) • Grazing Systems (P. Ohlenbusch, R. Jones)
3	Pest Management, Soil Management, Rural Communities, Resources	• Weed and Insect Management (J. Scrimger, L. Dyer) • Farmer Groups and Communities (J. LeCureaux, J. Jost) • Farmland Conversion (R. Olson, S. Taff) • Resources and Information Sources (C. Francis, H. Carter)

TABLE 8.4. Workshop program from the fourth year, "Facing a Watershed: Managing Profitable and Sustainable Landscapes in the twenty-first Century" (Illinois, Michigan, Iowa); speaker names are examples, as these differed among workshops.

Day	Topics	Topics
1	Introduction, Overview of Watersheds and Planning	• Update from Washington DC (G. Bird) • Overview of Freshwater Use and Watershed Planning (J. Church) • Group Dynamics in Designing & Implementing a Watershed Management Plan (J. Rendziak)
2	Watershed Leadership, Team Approaches	• Leading and ommunicating (M. Brown) • Conflict Management (W. T. Schenck) • Information Resources for Watershed Management (R. Olson) • Conservation Buffers and Riparian Management • Buffer Strips for Water Quality (M. Plumer) • The Stream Team (B. DeVore) • Impacts of Intensive Grazing on Stream Ecology & Water Quality (L. Paine) • Grazing Management of Streamside Pastures (D. Cates)
3	Farmland Protection, Green corridors, Suburban Sprawl	• Farmland Protection, American Farmland • Trust (AFT Representative) • The Green Corridor Project (Project Representative) • Information Sources for Sustainable Agriculture and Education (R. Olson & H. Carter) • Study Circles and Case Studies (D. Cavanaugh-Grant)

late in their fiscal year, when funds were limited. Workshops were held at Pheasant Run Resort near St. Charles, Illinois; Northwest Michigan Horticulture Research Station near Traverse City; and the Holiday Inn Gateway Center in Ames, Iowa.

MATERIALS AND LEARNING METHODS

Based on the results of planning sessions and on experiences in workshops, we continued to build information resources and test learning methods for four years. One goal was to bring the newest and most current materials to the Extension audience and make them available for wider use in the region and beyond. Another was to experiment with different learning activities and see which ones were most appropriate to our clients.

Educational Materials

One objective of each workshop was the collection and dissemination of educational materials from the North Central states and other regions. Workshop participants, Extension personnel, and planning teams set up display booths in a type of "information fair." The displays brightened the meeting rooms, added focal points for people to congregate and talk, and provided another method for extended learning. We posted carefully chosen quotations from the literature and from participants that were relevant to the workshop theme and the education process. Participants often referred to these quotes during group reports or presentations. One favorite from the 1998 workshops was, "You can't solve a problem by applying the same consciousness that created it" (Albert Einstein). Another was, "Trend is not destiny" (René du Bos). To draw attention to the North Central Region booth with training materials, we displayed an album with photos from previous workshops and meetings.

Planning included development of a 600-page loose-leaf binder of information on practices and systems in sustainable agriculture, in addition to papers related to the annual theme. State sustainable agriculture leaders, workshop speakers, and members of two sustainable agriculture resource person lists were asked to contribute materials to the binders. The resource notebooks supplemented the workshop sessions and provided educators with background readings, references, and sources for more in-depth information. The contents reflected the latest available information and were as diverse in topics and sources as the participants themselves. The overall guiding principle of inclusiveness was upheld in development of the resource binders. We made the effort to include information from Extension, government agencies, nonprofit organizations, farmers, and presenters at the workshops. According to follow-up surveys, these materials were incorporated into state training programs, copied for cooperators, used in newsletters, and cited in grant proposals. For example, about 70 percent of the 1995 regional materials were combined with local bulletins and Extension materials and used in training programs that reached every Extension educator and specialist in North Dakota and South Dakota.

Besides the 600-page resource notebooks, an abridged collection was edited and published in spiral binding each year. These materials were included in the *Education and Extension Materials for Sustainable Agriculture* series published by the Center for Sustainable Agricultural Systems and Nebraska Cooperative Extension Division. The books were offered to Extension and other information providers at cost or less, supported in part by the regional grant (Carter and Francis, 1995, 1996; Carter et al., 1997, 1998). The books consisted of 250 to 300 pages, and 7,000 copies were distributed using grant funds or sold during a six-year period. Many were ordered by Extension personnel who had not attended the workshops, or by other states outside the region. Although there is no documentation, this series is perhaps the largest distribution of Extension manuals since the start of the SARE program.

To increase the potential for networking, we prepared detailed participant and speaker lists with addresses, phone and fax numbers, and e-mail addresses to help people stay in contact after the workshops. These lists provided the foundation for an informal speaker's bureau and were used by many program organizers to locate appropriate presenters.

PARTICIPATORY LEARNING METHODS

In addition to the innovative content in the workshops, we introduced a variety of teaching methods. From experience we knew that many subject matter specialists in Extension relied solely on lectures (Carter and Francis, 1995; Francis and Carter, 2001). In discussions with presenters, we urged them to use approaches other than lecture format. Our ambitious objective was a balance of 50 percent conventional presentations and 50 percent other participatory learning approaches. We added exercises that required a high level of participation and involved attendees in the interactive learning process. We consulted with educational specialists at universities and in the private sector to ensure that methods were useful and transferable. For example, small groups used a modified version of the USDA/ARS Soil Quality Test Kit to measure differences in soil characteristics between cultivated fields and native prairie, and between conventional tillage and no-till fields. Participants were asked to discuss soil quality indicators, answer questions on sampling, and consider the impacts of different management strategies on soil quality. These learning methods were carefully evaluated, as described later.

Methods used during the four years, besides conventional slide shows and group discussions, included the following:

- *Decision case studies.* Specific questions or decisions faced by farmers were presented using brief video segments or other format, followed by small group discussions on how to handle that situation. Decisions by each group were then shared and compared within the larger workshop audience. The theory and application of decision cases in agriculture were developed at the University of Minnesota, and the method is now widely used in education (Simmons et al., 1992).
- *Study circles.* At several workshops state reports, sustainable agriculture issues, and urbanization challenges were discussed in simultaneous study circles. Small groups met in a circle, which promoted good interaction and discussion that is difficult with a larger group.
- *Jig-saw puzzle exercise.* This method can be used to learn about several topics at once, mix a group or participants, and make everyone responsible for sharing information. Groups subdivided around questions of common interest, brainstormed those questions for 15-20 minutes, and then reported their conclusions back to the whole group. People reported that this encouraged divergent thinking, expanded their horizons, and allowed every voice to be heard.
- *Gallery walk.* Small groups were given topics to brainstorm, and their results were posted on flip chart paper and read by other groups. Comments were added to the charts as people circulated around the meeting room, and the results were summarized by having the entire group tour the room and discuss what was on each flip chart. People found that this promoted participation by everyone in the group, created energy and activity, and caused thinking outside the conventional arena.
- *Watershed planning.* After learning basic planning tools, small groups designed management plans for a watershed-level area and discussed how neighbors could work together toward implementation. Groups then reported their strategies and conclusions to the larger learning community. People found that having a concrete task increased understanding, involvement, and relevance of the activity.
- *Managing conflict.* Small groups practiced resolution of conflict and consensus building skills by role playing in situations that often generate strong differences of opinion in the rural sector.
- *Building group commitment.* Methods of getting stakeholders to the table and finding common ground were demonstrated and then prac-

ticed. Role play exercises in small groups were especially effective to build community.

- *Full-range leadership.* The methods used in consensus building and participatory leadership models were demonstrated and discussed in large and small groups.
- *Farm tours.* Considerable thought went into selecting the farms for tours, and farmers were prepared to stimulate questions during the visits. Handout materials included family goals, maps, and enterprise descriptions, and these provided background to discussion and hands-on exercises during the field or farm visits. The exercises during tours were much more than observations and listening.
- *Paired evaluation interviews.* As an alternative to only filling out evaluation forms for each presentation, pairs of participants were asked on the last day to interview each other and record the results. We still do not know why this evaluation step was given a high rating as an educational activity, but this was a consistent result across several workshops. Apparently the interview process introduced a higher level of responsibility and rigor to the activity, compared to an anonymous individual filling out a form, and the activity was well received.

EVALUATION

Evaluation was a continuing and integral component of the workshops, not merely an isolated activity reserved for the last hour or a follow-up survey after completion. We decided during the planning process that we would continuously evaluate the sessions, give rapid feedback to the presenters, and adjust the program and activities to the extent possible to meet the needs of the group. This process changed after the first year, and we stopped the too-frequent evaluations. Yet the surveys were still done hourly several times each day and proved invaluable to the workshop planners, who met each evening to read through the results and adjust the next day's program wherever possible.

Our first experience at adjusting schedules was in the initial planning workshop in Cedar Rapids, Iowa, in 1995. The original plan to move sequentially from individual reflection and identification of issues, to small groups, to a plenary session proved far too slow for the participants. To meet their expressed need to move quickly to quantify and prioritize issues and set an agenda for action, we completely rewrote the script the first evening and presented a new series of activities the next day. This change proved useful to the group and resulted in valuable guidelines for subsequent workshops. We learned that it was essential to know the audience well and to plan appropriate topics and activities. Another important lesson was that free coffee was essential—no one appreciated our efforts to save money by eliminating the expensive catering fee and having each person buy their own cup!

A strategy for evaluation was established in the first year. After each hour's program, or a longer activity, we handed out a half-page evaluation sheet and asked participants to quickly reflect on the session and write their comments. These were collected immediately and scanned by one of the organizers and provided to the speakers or activity leaders. Although there was some grumbling about the frequency of evaluations, most people took the task seriously in response to our requests for frank comments about both the content and presentation method. We concluded that this intensive evaluation provided a series of windows on learners and learning. It gave everyone an opportunity to evaluate the sessions, to contribute to ongoing modification of the agenda, and to assume more ownership of the learning process. It also provided chances for brief reflections and opportunities to internalize some of the information presented. We did modify this procedure in subsequent years with one evaluation each day or after each major block of topics, and we ended with the paired interviews as the final workshop evaluation.

The most highly ranked sessions in the four years of workshops, 1995 to 1998, are listed in Table 8.5. The lists include both subject matter and learning methods. The items within each of the four categories are listed in approximate order of their ranking by participants, according to numerical scales that were on the evaluation forms. In whole-farm planning, the Holistic Resource Management™ (HRM) model was presented by specialists with experience in the interactive teaching methods that have made HRM training highly successful across the United States. Linking farms to communities and farming practices to watershed management proved to be successful learning areas, and farm tours were always highly ranked. We were pleasantly surprised that ecological principles and their links to agriculture were considered important by participants. However, because this theme was a major advertised focus of the workshop series, it is possible that participants selected themselves in response to the theme and topics.

Economic and social issues were a high priority identified in all the planning sessions. Those most highly rated were financial analysis, marketing, cooperatives, and value-added enterprises. Economics of intensive rotational grazing were of interest because of the growing acceptance of this practice across the North Central Region.

Several surprises came from the evaluation of learning methods. We don't know why the paired interviews were seen as so favorable and useful, but speculate that this type of reflection and interaction introduced a new

TABLE 8.5. Categories of highly ranked sessions at regional workshops, 1995 to 1998.

Whole-Farm Planning

- Holistic Resource Management, introduction, and participatory exercise
- Whole-farm planning in the community context
- Innovative farming practices as potential tools in watershed management
- Four farm tours on diverse topics

Ecological Principles and Land Management Strategies

- Linking ecology and agriculture
- Role of systems science in design of pest management strategies
- Ecological principles and pest management
- Long-term soil management and its effect on crop health
- Ecological principles of integrating perennials
- Water quality exercise, stream ecology, and sampling
- Tour of an alternative waste disposal site

Economic and Social Issues

- Financial analysis for sustainable agriculture
- Marketing and cooperatives
- Sustaining rural communities by adding value to agricultural products
- Economics of grazing systems and alternative enterprises
- Constraints to adoption of sustainable practices and systems
- Extension's role in sustainable agriculture
- Sustaining agricultural communities: the Family Farm Project
- Environmental landscape: culture and agriculture
- Purchase and transfer of development rights

Methodology

- Partner interviews for evaluating a workshop
- Planning learning activities
- Decision case studies
- Study circles
- Jig-saw activities
- Gallery walks
- Putting together a watershed plan
- Managing conflict
- Working with groups to build commitment
- Full-range leadership

degree of accountability to the evaluation process. Planning was also identified as an important form of learning. Specific activities such as the decision case, study circle, jig-saw interaction, and gallery walk were identified by participants as useful for learning, along with more complex processes, such as watershed planning and managing conflict. We have no long-term evaluation of how often participants later used these methods.

The open-ended comments given by participants provided another level of evaluation. "I continue to be invigorated by the synergism which takes place when different people, agencies, and organizations come to the table with a common vision and their individual goals and strategies." One of the most valuable benefits of the workshops was the interaction among in-state trainers, planners, and educators to network and share ideas. Holding a workshop in each state helped identify key players and bring them together to work toward a common goal. Visitors from other states also learned from field projects and found access to new Extension materials.

"Being able to see first-hand what's going on in other parts of the region has inspired me to think outside the box about our programs in Illinois." Workshops instilled a sense of regional community and recharged batteries for those participants who felt isolated in their daily work environment. "The most important part of the workshops was exposing people to different methods with the opportunity to see them in action and learn how to use them in local efforts." Varied teaching methods helped to hold people's attention. Some participants said that education about learning styles was just as important as technical subjects, because sustainable agriculture training needs to reach multiple audiences. A major suggestion for improving workshops was to encourage speakers to use more interactive styles and involve participants. "Interaction is critical. Otherwise, people do not feel valued, and they then lose interest." The best activities were hands-on projects, small group discussions and reporting, critical thinking exercises, and outdoor events that combined concepts with practical experience.

Because we kept the pace rapid and squeezed many activities into each day, some people felt that sessions were rushed, often too short to cover complex subjects. "I wish more time could have been spent on this topic," was a common complaint. Keeping the momentum is a good objective, but it must be balanced with depth and time for reflection. Also, the balance between teaching how to teach a subject and actually teaching with new information was hard to achieve. We chose not to go deeply into most subjects because our audience was trainers, not end users, and we expected these educators to go beyond the workshops to get all the information needed for their own programs.

Evening fellowship gatherings were carefully chosen to fit the workshop topics and local context. We found that socializing was important to building networks and having fun kept a positive attitude toward the workshops. The play *Rural Voices* (Wisconsin in 1996), dining at an Amish farm (Ohio in 1997), canoeing the Pomme de Terre River (Minnesota in 1997), and touring a family-run winery (Michigan in 1998) were all local events that generated much interest. Attendance at the evening events was always optional, and though there were several complaints that the days were too long, almost everyone participated.

LESSONS LEARNED AND RECOMMENDATIONS

From short surveys at the workshops, we learned that participants in their current jobs most frequently used lectures, one-to-one visits with farmer clients, field days, and small group discussions. Based on comments from evaluation forms, we concluded that observing and practicing new methods in the nonthreatening environment of the workshops provided a good opportunity for people to experience new learning approaches and gain confidence in using them (see Table 8.5).

The same surveys revealed that the main sustainable agriculture information resources were written materials, people (especially farmers), demonstration farms, organizations, university research, and videotapes. The large binders packed with new information, the resource displays at each workshop, and the individual interactions with other participants all added to this resource base. The contact information on each person for future communication established a regional network of educators and trainers in sustainable agriculture.

Field tours continue to be a prime learning method, especially in linking ecology with agriculture. Combining principles with practices has long been a central tenet of pasture walks, hybrid demonstration plots, fertilizer and herbicide comparisons, and hands-on calibration of planters, sprayers, and combines. Our tours were on farms, research stations, processing plants, long-term ecological research sites, and wildlife conservation areas. Participants wanted to hear from more farmers doing their own research or in collaboration with universities.

Economics need to be addressed, whatever the practice or system that is being evaluated. Crop diversity, rotations, and alternative weed management options are ecologically interesting, but must also be financially sound. New enterprises, diverse products from the farm, and value-added activities are useful concepts, but must fit into the activity calendar and be achievable by farmers. Whatever the topic presented, education should include an economic dimension.

Some farm visits provided a better look at broader issues than the specific practices or in-field systems that were shown, as well as meeting the goal of learning in context. Farmers often related their systems and input use to health and safety, environmental impacts, relationships with neighbours and community, and their finances. Many farmers discussed family issues, including potential for succession of ownership of the farm, and the quality of life questions they put forth gave a holistic view of why they chose certain enterprises and systems.

CONCLUSIONS AND FUTURE DIMENSIONS

Concepts and applications generated during the period of SARE training grants have increased the awareness of current challenges in our Midwest agriculture. In spite of some successes by larger farmers and corporations, often because of high levels of government program supports, the fragility of the economics of farming for most small- and mid-scale farm owners and managers has come into clearer focus. The impacts of current farming practices on the environment—especially the improper use of chemical fertilizers and pesticides and excessive cultivation—have been identified and quantified in ways that were not possible even two decades ago. Social consequences of farm consolidation and near monopolies in the agricultural input and commercial commodity sectors are being felt by farm families and rural communities. Even those farmers who have survived and those communities that have prospered have experienced the effects of economic changes in agriculture. There is a widening gap between large land owners and those who work for minimum wages; these growing inequities cause changes in social relationships with neighbors and greater economic stratification in communities. The SARE programs in Extension have explored these difficult issues and helped educators and specialists acquire tools to deal with them.

We have identified several recommendations for future directions that will help Cooperative Extension in renewing the social contract with the U.S. rural population.

- Consolidation of farms continues, with a resulting bimodal distribution of ownership of farmland and increasing concentration of federal benefits and rural resources in the hands of a minority. Some have concluded that we are reestablishing an agricultural system that many of our ancestors left Europe to escape (Strange, 1988). We need to seek alternative farming systems and value-added options, plus examine the real limits of economics of scale in farming and communicate this information through effective Extension channels.

- Purchase of seed companies by major chemical corporations and development of integrated production packages that pair hybrid or variety with specific herbicides have reduced opportunities and choices for small, family farms and increased input costs for everyone in agriculture. In spite of claims that some new technologies will reduce pesticide use and improve the environment, there are serious questions about the overall impact of these technologies. Extension must be alert to all research results and continue to present an unbiased evaluation of technologies to farmers across the farm-size spectrum.

- We must recognize the tremendous economic power of today's industrial complex and its impact on the information available to farmers. The promise of cheap food through an industrial and global agriculture and food system may have very negative long-term economic, nutritional, environmental, and social costs. Such costs are often not calculated in the short-term bottom line used to evaluate economic net returns from an individual enterprise or system on the farm each year. Chapter 6 by John Ikerd describes several alternative economic strategies to identify these costs and explore viable options for the future.

- Although the four years of workshops reached 850 educators and specialists, and the *Education and Extension Materials for Sustainable Agriculture* series another 7,000, there is a continuing and compelling need to communicate information on sustainable practices and systems to new people in Cooperative Extension. In spite of cutbacks in budgets and personnel, we have new people in Extension each year, both educators and administrators, who need a broad range of information and experiences to effectively meet their challenges, including a sustainable agriculture perspective.

- Most systems and practices are site-specific, yet many principles of sustainable agriculture, including site-specificity, are widely applicable and most efficiently addressed through regional workshops, tours, and other activities. Workshops are more than regional clearing houses for information and new recommendations. With appropriate preparation, background papers, and reflection on current challenges these events can be valuable "think tanks" for making new connections and exploring new directions that will help Cooperative Extension deal effectively with the future challenges in the agricultural sector.

- Most Extension training should be done at the state and district levels, because the organization and management of Cooperative Extension is somewhat unique to each state. We are learning from ecology that the majority of effective practices and systems are site-specific, a drastic departure from the homogenizing approach offered by uniform, high-technology packages of hybrid variety + fertilizer + pesticides. The parallel in Extension education is that solutions to local problems will be generated in local programs through participatory design and learning activities.
- We have learned that interactive education strategies can be well received and highly effective in an Extension training setting. They provide a stark contrast to the lectures and Power Point© presentations that dominate the current educational landscape. The predominantly one-way communication styles persist in defiance of educational research results that people learn best from practical applications, hands-on activities, and relating a topic to personal experiences. Workshops focused on process and learning about learning may be more important than those focused exclusively on content.
- Multidisciplinary and systems-oriented topics are less likely to receive attention than those related to single issues and practices. Because most educators and specialists have been trained in specific disciplines, their continuing specialization will be a factor in people choosing topics within their personal comfort zones or their most frequently requested areas of expertise as perceived by clients. It is essential that Extension provide expertise and programs in efficient cropping and crop/animal systems, because commercial information providers will continue to emphasize systems components and inputs they have for sale.
- Although the treatment of broad and long-term issues in agriculture falls beyond the individual expertise of most Extension educators and specialists, we need to find an appropriate forum for these concerns to surface if Cooperative Extension is to find a visible and viable role for the future. Budget reductions in the past several years have hit Cooperative Extension disproportionately in comparison to research and teaching, the other two major activities of land grant universities. To renew the contract with U.S. rural people and communities is to provide a forum for discussion of the difficult issues and to be able to respond to crises when they occur. These responses must go beyond issues such as drought, an insect infestation, or current farm prices. We need to ask and seek answers to difficult questions.

It is unrealistic to suggest that a few regional workshops have drastically changed the course of Cooperative Extension toward a new contract with rural U.S. farmers. Yet the issues that were explored and the discussions on key questions of sustainable agriculture have broadened the social consciousness of those who attended. When we meet these participants today, they clearly remember some topics and events from the workshops. The collective memory across the region suggests to us that the regional training workshops were highly successful activities from 1995-1998. Participants have moved into positions of planning and leadership and caused positive impacts in their own state programs. From the recommendations listed above, we conclude that the job is far from done. However, a focus on sustainable agriculture practices and systems can go a long way toward renewing the contract, and to bringing increased emphasis to the economic, environmental, and social challenges facing agriculture in the region.

REFERENCES

Agunga, R. 1995. What Ohio extension agents say about sustainable agriculture. *J. Sustain. Agric.* 5:169-187.

Bass, B.M. and B.J. Avolio. 1994. *Improving organizational effectiveness through transformational leadership.* Sage Publications, Thousand Oaks, CA.

Bauer, L. (ed.). 1996. *Planting new possibilities, harvesting a healthier future.* North Central Region SARE, 1996 Annual Report, University of Nebraska, Lincoln. 181 p.

———. 1997. *Celebrating a decade of sustainable agriculture research and education: 10 years of SARE.* 1997 Annual Report, University of Nebraska, Lincoln. 160 p.

Berton, V. (ed.). 1998. *Ten years of SARE: A decade of programs, partnerships and progress in sustainable agriculture research and education.* SARE National Office, U.S. Department of Agriculture, Washington, DC. 96 p.

Carter, H. and C.A. Francis. 1995. *Everyone a teacher, everyone a learner.* Extension and Education Materials for Sustainable Agriculture, Vol. 4. Cooperative Extension Division, University of Nebraska, Lincoln. May.

———. 1996. *Shared leadership, shared responsibility.* Extension and Education Materials for Sustainable Agriculture, Vol. 5. Cooperative Extension Division, University of Nebraska, Lincoln. December.

Carter, H., C. Francis, and R. Olson. 1997. *Linking people, purpose, and place: An ecological approach to agriculture.* Extension and Education Materials for Sustainable Agriculture, Vol. 7. Cooperative Extension Division, University of Nebraska, Lincoln.

———. 1998. *Facing a watershed: Profitable and sustainable landscapes for the 21st century. Extension and Education Materials for Sustainable Agriculture,* Vol. 9. Cooperative Extension Division, University of Nebraska, Lincoln.

————. 1999. *Blueprint for regional training in sustainable agriculture.* Center for Sustainable Agricultural Systems, University of Nebraska, Lincoln. 5 p.

DeWitt, J. 1995. Concerns and comments on Chapter 3 training. In: *Everyone a teacher, everyone a learner,* H. Carter and C.A. Francis (eds.). Extension and Education Materials for Sustainable Agriculture, Vol. 4. Cooperative Extension Division, University of Nebraska, Lincoln. May. p. 3.

Dunlap, R.E., C.E. Bues, R.E. Howell, and J. Waud. 1992. What is sustainable agriculture? An empirical examination of faculty and farmer definitions. *J. Sustain. Agric.* 3:5-39.

Francis, C.A. 1990. Sustainable agriculture: Myths and realities. *J. Sustain. Agric.* 1:97-106.

Francis, C.A. and H.C. Carter. 2001. Participatory education for sustainable agriculture: Everyone a teacher, everyone a learner. *J. Sustain. Agric.* 18(1):71-83.

Harwood, R.R. 1990. History of sustainable agriculture. In: *Sustainable agricultural systems,* C.A. Edwards, R. Lal, P. Madden, R.H. Miller, and G. House (eds.). Soil & Water Conservation Soc., Ankeny, IA. pp. 1-19.

Ikerd, J. 1995. Economics and quality of life issues in sustainable agriculture. In: *Everyone a teacher, everyone a learner.* Extension and Education Materials for Sustainable Agriculture, Vol. 4. Cooperative Extension Division, University of Nebraska, Lincoln. May. pp. 127-138.

Paulson, D.D. 1995. Minnesota extension agents' knowledge and views of alternative agriculture. *Am. J. Altern. Agric.* 10(3):122-128.

Simmons, S.R., R.K. Crookston, and M.J. Stanford. 1992. A case for case study. *J. Nat. Resour. Life Sci. Educ.* 21:2-3.

Strange, M. 1988. *Family farming: A new economic vision.* University of Nebraska Press, Lincoln.

Chapter 9

Regionalization of a Research and Education Competitive Grants Program

Steven S. Waller
Elbert C. Dickey
Charles A. Francis

RATIONALE FOR A REGIONAL PROGRAM

The U.S. Congress authorized funding in the 1985 Food Security Act for a small national effort in research to promote a more sustainable agriculture. In December 1987 funds were appropriated and the Low-Input Sustainable Agriculture (LISA) program was launched. On January 19, 1988, the Secretary of Agriculture's Memorandum 9600-1 provided the initial United States Department of Agriculture (USDA) policy statement on sustainable agriculture:

> The purpose of this memorandum is to state the department's support for research and education programs and activities concerning "alternative farming systems" which are sometimes referred to as "sustainable farming systems". . . . The department encourages research and education programs and activities that provide farmers with a wide choice of cost effective farming systems including systems that minimize or optimize the use of purchased inputs and minimize the environmental hazards . . . . The Assistant Secretary for Science and Education is responsible for encouraging and guiding the development of **research** and **extension** programs that best meet farmer's needs for facts, information, and guidance concerning alternative farming systems.

Developing and Extending Sustainable Agriculture
© 2006 by The Haworth Press, Inc. All rights reserved.
doi:10.1300/5709_09

Within a few years the program was renamed from LISA to the Sustainable Agriculture Research and Education (SARE) Program, in response to widespread concerns about the implications of the term "low-input" that to some people meant low technology, low output, and low profits. The SARE name has served the program well for over a decade.

The authors acknowledge the important perspectives and contributions of the late Dr. Patrick Madden that were summarized in an unpublished paper from 1998, *The Early Years of the LISA, SARE, and ACE Programs,* and applaud his many years of work at the national level to help establish and lead this program from its infancy to adolescence. His opinions are quoted several times in this chapter. Dr. Madden's understanding of the issues (Madden, 1987) and leadership provided much of the stimulus for others to achieve what has been done in the North Central Region and elsewhere in the SARE programs.

The competitive grants program was developed to expand knowledge and adoption of sustainable agriculture practices that were economically viable, environmentally sound, and socially acceptable. The funded projects were to provide both research and education for the future economic viability of U.S. agriculture.

The national program was administered through a single USDA agency, the Cooperative State Research Service (now CSREES). One of the most innovative and significant administrative decisions in the development of the program was to regionalize the administration and organizational structure of this federal competitive grant program. The concept of regional control builds on the assumption of similar natural resources, cropping and animal systems, and production constraints across a relatively homogeneous ecoregion. The relationship between the federal government and the land grant university system was a fundamental strength in designing a decentralized grant program that could address the unique regional needs of U.S. agriculture. The historical collaboration between Extension and the Experiment Stations and the precedent of regional research projects and extension education programs provided ample evidence for the utility of regional programming. The program was aligned with the existing USDA regions. The four regional programs (northeast, southern, north central, and western) were autonomous and each had leadership, policy, and grant-making responsibility within the region. While the regional approach was not unique, the autonomy and local ownership and accountability signaled a new way to administer a federal grant program of national scope.

National guidelines for program implementation were developed in January, 1988 by a USDA ad hoc committee with representatives of CSRS, experiment stations, cooperative extension service, and the four USDA regions. These guidelines, with minor revision, were codified into Subtitle

B of Title XVI of the Food, Agriculture, Conservation and Trade Act of 1990, Chapter 1. Host institutions were identified in each region, based primarily on evidence of activity in sustainable agriculture. The University of Nebraska-Lincoln was selected for the North Central Region and the late Dr. Warren Sahs became the first Regional Coordinator.

Each region would be administered by a regional Administrative Council. Membership was to include representatives of the Agricultural Research Service; CSRS, the national Extension Service, state cooperative extension services and state agricultural experiment stations; the Soil Conservation Service; state departments engaged in sustainable agriculture programs; nonprofit organizations with demonstrable expertise; agribusiness; the State or United States Geological Survey; and other persons knowledgeable about sustainable agriculture and its impact on the environment and rural communities including professional agriculturists (farmers and ranchers). The diversity of the membership reflected the program's commitment to serve the broad spectrum of people in the agricultural community.

The SARE grant program sought projects that would closely coordinate research and extension activities; indicate the manner in which findings of the project would be made readily usable by farmers; maximize the involvement and cooperation of farmers and ranchers, including projects involving on-farm research and demonstration; involve a multidisciplinary systems approach; and involve cooperation among farmers, nonprofit organizations, colleges and universities, and government agencies. Each of the four regions began with the same mandate, though regional autonomy allowed different approaches.

In the early days of the program, the North Central Region was ahead of its time, leading the other regions in several important innovations while defending the integrity of the national program according to Patrick Madden. The evolution of a unique "social contract" with all North Central Region stakeholders was rooted firmly in the land grant philosophy, and the process fostered a very progressive and productive program, one that offers insights into synergistic organizational structure and policies.

A brief historical review of broader issues is relevant. There have been numerous critiques of the land grant system through the years by those who fear that this large group of educational organizations has strayed from its goal of meeting the needs of family farmers and rural communities (e.g., McDowell, 2001; Strange, 1988; Thompson and Stout, 1991). There is ample evidence that consolidation of farms and businesses has drastically affected rural communities, and the consequent loss of infrastructure is directly impacting family farms and smaller communities (Allen, 1993; Flora, 2002; Flora and Flora, 2004). One response of the federal government to these growing challenges was the founding of the LISA/SARE program, and its successes have helped land grant universities to connect better with family farms. Much of the success has been because of the autonomy and unique programs developed in each region. Summaries of projects and programs of the North Central Region have been described by Bauer (1996, 1997), and these have been firmly imbedded in the national program, whose overall successes and impacts have been summarized by Berton (1998).

NORTH CENTRAL REGION ADMINISTRATION AND PLANNING

The North Central Region SARE program played an innovative and seminal role in developing awareness of sustainable agriculture in the Midwest, and in the process provided a model for several creative grant programs that were initiated and later adopted in other SARE regions. The history of the region is filled with many key leaders in universities, nonprofit organizations, federal agencies, and farmer/rancher groups who were instrumental in making this program successful. They are not mentioned by name in this chapter, but rather by position, and those familiar with the region's development will recognize their influence on the program. The approach to regional administration in the North Central Region is credited with the evolution of innovative programming. Ultimately, the regional successes are products of a highly diverse group of individuals who passionately shared a common vision. It is the maturing of their collective relationships and the persistence of an organizational culture that transcended individuals and permeated the region which deserve attention. The North Central Region illustrates a model of selfless collaboration in the SARE Program with exceptional people who were willing to put the economic success and quality of life of farmers and ranchers before their own state, agency, organization or discipline interests. One indicator of the richness of our human capital is the number of key people from the North Central Region who have served in roles as leaders in the national SARE program. There are useful lessons to be learned from this region's experience in utilizing organizational structure and policy to catalyze significant change.

The North Central Region includes 12 states: Michigan, Ohio, Wisconsin, Indiana, Illinois, Iowa, Missouri, Minnesota, Kansas, Nebraska, South Dakota, and North Dakota. The region has been identified as the nation's breadbasket and for many it is the *Heartland.* The early success of the North Central Region LISA/SARE program was not independent of the history, tradition, abundant resources, and people of the region.

Corn and soybeans are dominant crops in the eastern and central areas with higher rainfall, or where irrigation is available in the western reaches of the Dakotas, Nebraska, and Kansas. Dryland cropping in the western area is dominated by wheat, and there are many important forage and specialty crops including alfalfa, sunflower, sugar beets, dry beans, and minor small grains and vegetables. Livestock is important to the economies of all states, and increasingly is found in confined animal operations. Increasing farm size through consolidation threatens the future of small and mid-sized family farms and communities across the region (Flora, 2002; Flora and Flora, 2004).

Although there are some commonalities in crops and livestock systems, the topography and climate in farming and ranching areas vary across the region. Elevation ranges from near sea level at the Great Lakes to more than 5,000 feet at the western edge. Rainfall varies from as little as 10 inches in the western areas to nearly 40 inches at the eastern borders. Such climatic variation suggests that few projects could successfully focus on the entire region, but multistate activities would be appropriate in subregions with similar systems and conditions.

The region also benefited from a long history of collaboration among the major research universities. Public research is concentrated in the 13 land grant universities (12 of them founded through the 1862 Morrill Act and one founded by the 1890 Morrill Act). The region also had a viable private, nonprofit community engaged in sustainable agriculture research and education. Their role in the establishment of the federal program was significant. Nonprofit research centers and collaborative work with state-wide farmer groups was commonplace. On-farm research was prevalent and there was a regional environment of interagency, interinstitutional, private for-profit and not-for-profit organizations, and multistate collaborations.

The host institution [University of Nebraska, Lincoln (UNL)] had a long history of close collaboration between research and Extension, with most faculty holding joint appointments to conduct research and deliver extension education. There was also a strong record of on-farm research and farmer/rancher involvement in the testing of new technologies, for example, the testing of crop hybrids and varieties, developing fertilizer application strategies, and exploring options in pest management. Cooperation of UNL with private industry had been evident since the early development of no-till planters and cultivators, as well as self-propelled irrigation systems.

These historical considerations provided a foundation for trust as well as for fostering and administering collaborative research in SARE programs. The land grant philosophy of openness, accessibility, and service to the people was embedded in UNL's Institute of Agriculture and Natural Resources (IANR). As the program has evolved, it has become more and more apparent that the spirit of collaboration among people in research, extension, and teaching with their stakeholders and their attitude of service to clientele had to begin at home. Nebraska's IANR served as a successful collaborative model that encouraged and supported programmatic partnerships which were consistent with the goals of the federal program. Within this framework, the North Central Region LISA (now SARE) program was initiated.

UNL volunteered its facilities for housing and fostering development of the management group for the North Central regional program, and used funds from the research division to set up offices and infrastructure. From the outset of the program until 2002, UNL identified a faculty member to serve as regional coordinator who was approved by the regional Administrative Council. The Administrative Council also annually evaluated the Regional Coordinator and the host institution staff.

ORGANIZATIONAL STRUCTURE FOR REGIONAL RESEARCH AND EDUCATION

Administrative Council

The regional program began in an environment of mutual distrust. The private organizations supporting sustainable agriculture perceived the land grant universities to be historically opposed to organic or low-input farming technologies. The North Central Region's private organizations had played an important role in getting Congress to appropriate the funds to start the grant program according to Patrick Madden in a personal communication. The first approach in the North Central Region for appointing the Administrative Council members was assigned to the land grant universities, alienating the private organizations and placing the entire program in jeopardy. Some of the most politically powerful support for the LISA program originated in the region, and any reduction in support would threaten the program. To resolve the impasse of membership on the administrative councils, the national guidelines were revised to embrace regional self-determination to assure acceptable representation.

While all regions were conforming to the revised guidelines, the North Central Region went much further in defining the process. A rotation schedule was developed that was based on a four-year term and provided representation for every category of groups listed in the enabling legislation. The terms also allowed for continuity from year to year, providing a stability in programming and a policy that served the stakeholders well. Unlike the other parts of the country, the North Central Region provided a membership slot for each state to be filled by either research, extension, or private category (including professional agriculturists—farmers and ranchers). This provision honored the intent of the original legislation and developed program ownership region-wide. It was a key step, and resulted in a relatively large Administrative Council; through time the rotation schedule and membership slots dramatically increased the base of knowledgeable advocates. The process also allowed the region to embrace its diversity of agriculture and provide a voice for the wide array of stakeholders. All groups and affiliations were offered the opportunity to nominate individuals for any vacancy. This openness in process alleviated the concerns of many that the program would develop a narrow base of leadership.

Regional Coordinator

Unlike the other regions, the Regional Coordinator did not chair the Administrative Council. The Regional Coordinator and staff have provided leadership and direction for the program, but it has been a subtle and highly supportive type of leadership that invested major decisions in two stakeholder groups: the Administrative Council and the Technical Committee (described later). The Administrative Council set the policy and priority for programs and established procedures for requesting and reviewing proposals. The Regional Coordinator assumed a staff role, allowing the Council to select its own leadership from within. This self-determination was a visible sign of a coalition of ownership, rather than a vested and autonomous leadership arising from the land grant universities. It promoted inclusiveness and accessibility, and valued the contributions of all. The council fostered a broad and sincere respect among the members and defused the antagonism between the sustainable agriculture community and the land grant universities. Thus, it became everyone's program and a beacon for program-wide cooperation.

During the early years, the North Central Region had farmers, ranchers, agency personnel, and representatives of private organizations, agribusiness and universities serve as Chair of the Administrative Council. This rotation substantially changed the culture of the region compared to others and ultimately led to credible, innovative, and successful programming. The access to a leadership role for farmers, ranchers, and nonprofit group representatives was unexpected within the agricultural community. The program became a *people's program* rather than a federal program. Every Administrative Council member became a bridge builder. This step was crucial in establishing and maintaining a practical orientation in the project review and evaluation process. Unlike other competitive grant programs, the end user had meaningful and instrumental input into the direction and priorities of the program. Consistent with the history of the region, farmers and ranchers were fully integrated into the program, enriching the dialogue, focusing the program, and ensuring responsiveness to a changing agriculture.

Here we must comment on the important roles and recognition given to farmers and ranchers in the regional leadership, as mentioned in Chapter 1. Both conceptually and in practice, these agricultural professionals were considered as equals in the administrative and technical review process. By recognizing and validating the experience and ideas of farmers and ranchers, and placing equal weight on their opinions along with those of the federal and state agency people, we explicitly legitimized their status and role in the regional program as that of professionals, and provided a clear statement that we were not dealing with "producers" on an assembly line of some type in agriculture. This strategy has been applauded by farmer organizations in the region.

Operating Guidelines

The North Central Region Administrative Council employed operational procedures to a degree not found in the other regions, in the opinion of Patrick Madden. The importance of the program and the responsibility of the first administrative councils to reaffirm the wisdom of the regional approach warranted a business-like strategy. Formal parliamentary procedures were used. Votes were taken and conflict of interest guidelines were developed and utilized. The Administrative Council developed a policy book to ensure consistency year-to-year, formalized procedures, and archived their records to maintain complete accountability. While the process was formal, it guaranteed fairness and ensured integrity of the program. It provided the foundation for sound management independent of the Chair affiliation, or the members of the Administrative Council. The formality of the guidelines also empowered every member with an equal voice. This was critical in the meaningful engagement of farmers and ranchers in the process. One relatively minor policy with significant symbolism was funding the time of the self-employed members to participate in meetings and review. Without that decision, a two-tiered class of membership would have been an unfortunate reality. Other regions subsequently developed similar operating guidelines.

Technical Committee

The Technical Committee was appointed by the Administrative Council and given responsibility for evaluating the technical merit of proposals. This highly representative group included farmers and ranchers, nonprofit and agency representatives, and university faculty. The chair has traditionally been a farmer or rancher, or other nonuniversity person. Representing the same major constituencies as the Administrative Council, this group evaluates and scores projects based on technical merit, proposed budget, and capabilities of the investigators to accomplish the research. Once their recommendations are completed, the process returns to the Administrative Council.

Unlike in other regions, the North Central Region Technical Committee was structured with three-year terms. The primary criterion for membership was expertise; however, the Administrative Council also sought to create geographical distribution of members. The term ensured that each year, any resubmitted proposals would benefit from having two-thirds of the Technical Committee familiar with the previous review. This continuity and consistency was critical to keep authors of the proposals engaged in the process. The Technical Committee was also empowered with more program responsibility than the committees in other regions. The Chair of the Technical Committee was a voting member of the Administrative Council. In many respects the Technical Committee served as a formal partner in the regional management team. Joint meetings and workshops with the Administrative Council and Technical Committee were often a part of the process. Technical Committee input was critical in developing regional priorities. Technical Committee members often became Administrative Council members. As with the Administrative Council, the Technical Committee process formally engaged more stakeholders in the regional program, providing each the opportunity to become a knowledgeable advocate.

STRATEGIC PLANNING

The North Central Region was the first region to prepare and publish a strategic plan, as described by Patrick Madden. The planning process was effective in defining a common vision that transcended affiliation, membership, position, agency, or organization. It diminished bias and assured that past history became a foundation for the future of the program, but did not hinder current decisions. The common vision of a shared future lifted the Administrative Council's discussion above the historical rhetoric. Completing the strategic plan was a benchmark activity that solidified a culture of trust and respect. Programmatically, it moved the Administrative Council from a reactive board funding the best of what was submitted to a proactive board that targeted priorities based on the strategic plan. The regional program was maturing in philosophy, policy, and practice.

Regional Mission Statement: The mission of the North Central Region's Sustainable Agriculture Research and Education Program is to create and manage a system designed to encourage the involvement of farm and non-farm citizens in the process of discovery and learning that leads to achieving a more sustainable, environmentally-benign agriculture.

Although there have been many debates through the years in the Administrative Council about program priorities, selection of projects, and balance in the overall program, a high degree of personal respect among the members has assured that each debate was conducted on a professional level. The entire group was focused on regional goals, practical applications, and making a difference in people's lives. With this orientation, most of the differences in personal opinions were resolved and the program was able to move ahead. With this administrative and planning process in place, the regional programs grew and matured in both project scope and reach of the results. Several unique positions were pioneered in the North Central Region SARE office, and these became models for other regions.

GRANT MAKING PROCESS

One of the regional goals was to personalize the grant making process, to make it accessible to all applicants with or without previous grant experience, constructive, nurturing, and ultimately accommodating for the grantee. The region sponsored grant writing workshops for prospective applicants. Because the Regional Coordinator and staff did not vote on the Administrative Council, they could provide assistance to the authors of preproposals and proposals. A second goal was to ensure that the Administrative Council and Technical Committee viewed the grant making process as a joint activity.

The region used a preproposal process early in its development. This first step was to recognize the large efforts needed by grant seekers in preparing full proposals, to let the Administrative Council direct the program toward priority questions, and to provide the Technical Committee with a moderate sized portfolio of proposals to review. Preproposals were evaluated by the Administrative Council through a structured protocol that ensured that each reviewer considered the same general criteria, provided quantitative feedback to authors, and allowed preliminary ranking before the meeting. These scores were used to give guidance to the discussion by the Administrative Council; however, the final decision on soliciting full proposals was based on discussions at the meeting. All voices were heard, all points of view considered, and the collective wisdom of the group often yielded a different set of proposals compared to the preproposal ranking. The discussion also allowed the balancing of a research portfolio across the region, and a set of projects that would address regional priorities.

The proposal review included the Technical Committee and selected anonymous external reviewers. Each proposal had at least one member of the Administrative Council as a reviewer. During the deliberations of the Technical Committee, the Administrative Council had a representative to serve as a resource person. Each proposal author received a detailed feedback. The Administrative Council also facilitated collaborations among complementary proposals to develop a single proposal.

The Administrative Council made final decisions on which research projects to fund based on technical merit, available budget, and on how well the proposals met the guidelines and priorities set for each year. The Administrative Council also sought balance among states in the region, while not sacrificing project quality, and a broad portfolio of projects including crops and animal enterprises. Animal projects have been much less prevalent since the inception of the program and they were encouraged. There was also an intent to distribute the grants across the wide array of public and private, profit and nonprofit organizations and agencies.

COMMUNICATIONS SPECIALIST

The North Central Region created the position of Communications Specialist and hired the first one in the nation, as described by Patrick Madden. It was apparent that the region needed a communications plan to fulfill its mandate to make relevant information accessible to farmers and ranchers. Each project was required to have an educational-outreach component, but the quality and accessibility of these reports varied greatly. The first regional publications were modeled after extension circulars and became efficient and effective ways to distribute information. The outreach program for sustainable agriculture quickly evolved to other media and continues to be a critical component of the agriculture literature. A Communications Specialist continues to assist researchers and farmers interpret and package their results for a general agricultural public.

SPEAKERS BUREAU

The region developed a speakers bureau that provided funds to organizations to bring speakers to meetings, conferences, or workshops. Organizations could select from a list of speakers provided by the region, or could recommend other speakers for priority topics.

PRODUCER GRANT PROGRAM

The North Central Region was the first to establish a competitive grant program for farmers and ranchers. With a strong history of on-farm research in cooperation with Extension, Natural Resources Conservation Service (NRCS), and nonprofit farmer groups, many agricultural entrepreneurs in the region recognized the importance of site-specific testing of new technologies and systems under conditions of their own farms. A grant program was established to allow individuals or small groups of farmers to apply for modest funding to try new, often high-risk practices that may not have been tried without encouragement and support from beyond their own resources. A separate review panel was created that included farmers or ranchers from the Technical Committee as well as from outside. This unique approach to technical review was well grounded in the practical world of agriculture.

There were two stipulations in the initial calls for proposals. One, farmers and ranchers needed to work with an Extension educator or specialist in preparing the proposal and monitoring field results. Two, they were required to organize a field day or some other educational event, plus provide written documentation of the results. Large-scale plots appropriate to field-sized equipment were often used to make these trials practical for farmers and relevant to visitors observing the results. Long, drive-through plots such as those pioneered by groups in Iowa and Nebraska were used for paired statistical comparisons on many farms (Franzluebbers and Francis, 1991).

A wide range of different practices was tested in this producer grant program. Alternative weed management strategies were compared. Cover crops, compost and manure applications, and late spring soil sampling were tested as creative ways to reduce purchased fertilizers and thus reduce both farmer costs and potential loss of nutrients from fields. Rotational grazing was a new practice that was featured prominently in the program, and grazing standing maize was another innovation that was tried. As with the research grant program, there were fewer on-farm projects in animal production than in the crops and vegetables. Direct marketing options were an important component in many projects. People who participated in this grant program were often presenters in regional workshops, and their farms were used for tours. The program was highly successful and was soon adopted by the other three regions. This success was the basis for strong pressure by farmers and nonprofit representatives on the Administrative Council and Technical Committee to increase funding available for producer grants. Consequently, the funding for the regional SARE programs has increased over the past 15 years. The producer grant program has been evaluated by den Biggelar and Suvedi (2000).

PRODUCER GRANT COORDINATOR

To no small extent, the success of the producer grant program was because of the appointment of a Producer Grant Coordinator who works out of the regional office. Once again, this appointment was a first for the national program. This person has traveled widely across the region and helped producers in shaping ideas into successful projects, and he has visited all producers with grants in the field. This model has assured that practical results were adequately evaluated and summarized for presentation to farmer and rancher audiences.

DIVERSITY ENHANCEMENT GRANTS

In 1994 the North Central Region Program addressed the specific needs of native Americans, low-income, and other under-represented groups through a pioneering, special call for proposals targeted to these clients. It was developed within the Producer Grant program and the allocation of funds ensured that the region would have activity in the under-represented areas of agriculture. These proposals were mentored by the Producer Grant Coordinator and the evaluations were culture sensitive.

TARGETED AREAS OF RESEARCH AND EDUCATION

Youth and FFA groups were specified as priority clients for programs in some years. Such a program was unique in the nation, and provided another example of innovation by the Administrative Council and regional administrators. The North Central Region was the first to include socioeconomic projects in the annual call for proposals according to Patrick Madden. In 1997, the Administrative Council initiated a special call for proposals on innovative marketing strategies to encourage the development of viable markets for regional farm and ranch products. This call was the result of several marketing roundtables hosted by the region.

GRADUATE STUDENT SUPPORT

Since the initiation of the SARE regional program, graduate students have been involved in many research projects financed by the grants. This funding has always been a popular and practical way to tap into youthful energy and innovative ideas, and to extend the capabilities of university researchers and Extension specialists to implement projects in the field. The North Central Region expanded the support available to graduate students by initiating a call for dissertation and thesis proposals in early 2002. In most cases, these funds were used to expand the research already planned by students who were receiving graduate research and teaching assistantships from their universities. The special grant funds were used to add new dimensions to research and to direct our projects more precisely toward meeting the goals of the SARE program.

TRAINING AND EDUCATION IN SUSTAINABLE AGRICULTURE

In 1992 the Professional Development Program (PDP; Chapter 3 of Subtitle B of Title XVI of the Food, Agriculture, Conservation and Trade Act of 1990) received an appropriation. The North Central Region Administrative Council immediately approved a new organizational structure that included a Regional Extension Coordinator. To the extent possible, the Chapter 1 and Chapter 3 activities would be programmatically integrated and jointly administered. The previous history of collaboration at the host institution made this a successful reality.

There was pressure to allocate professional development funds on a formula basis to ensure that all states were equally involved in providing education. The Administrative Council recognized the need to provide financial support for every state to orient and educate their Extension specialists and field educators. The council did allocate an initial $10,000 per state per year for the development of a state-specific, multiagency, multi-organizational strategic plan for sustainable agriculture education, and has continued this support for states. The Administrative Council demonstrated great courage in resisting pressure to allocate all of the funds on a formula basis. History has shown this decision to be a landmark in the program.

The majority of funds was distributed through a competitive grant process. The North Central Region was the only one to establish a broad program for sustainable agriculture education through regional workshops. The vision of the Administrative Council in establishing a regional program in Extension training and bringing together the diverse groups potentially involved in moving sustainable agriculture to farmers proved to be a fundamental factor in making the training programs available to everyone. One important result was to bring state agricultural program leaders from Extension and designated sustainable agriculture coordinators together for discussion and joint planning and programming.

A consortium that included the University of Nebraska, Lincoln and The Ohio State University, later joined by Michigan State University, established workshops during a four-year period that eventually included among the attendees more than 800 Extension educators, specialists, and administrators; nonprofit and government agency people; farmers and ranchers; and agricultural business people. The workshop binders provided to each participant included more than 600 pages of new material each year. These binders were summarized and reduced to 250-page publications that were provided at cost to a wider audience. This *green book series* edited by the Center for Sustainable Agricultural Systems and Cooperative Extension at UNL included previous volumes on education and training programs in sustainable agriculture and on-farm research, also supported by separate SARE grant funds. By the end of the four-year cycle, more than 7,000 copies of these books had been distributed nationwide and internationally. The *green book* series perhaps reached one of the widest regional audiences of any such training materials in the short history of sustainable agriculture Extension (Carter and Francis, 1995, 1996; Carter et al., 1997, 1998).

Workshops, two or three each year, were held in 10 of the 12 states in the region. They are described in detail by Heidi Carter and colleagues (see Chapter 8). What made them successful in large part was the close cooperation between administrative leaders in Nebraska in Extension and research. Based on the successful joint appointment model for faculty, and the long-term working relationship in Nebraska district centers between research and Extension, this spirit of cooperation in educational programming was infused into the orientation and decision making of the Technical Committee and Administrative Council. Moreover, a long history of collaboration with farmers and nonprofit groups in Nebraska and other states provided the foundation for trust and close cooperation in setting priorities for practical research and training as well as decisions on which projects to fund.

There was an initial reluctance in some states to get Extension involved, and the regional workshops provided a comfortable and nonthreatening environment in which to meet, to grow professionally, and to plan for future training. For a few states where initially there was no interest in sustainable agriculture, the regional workshops provided a forum where committed educators and specialists could come together with farmers and ranchers for learning and sharing visions. The strategy of providing training funds for vans to travel to the workshops proved a valuable time for state teams to ride together and discuss potential programming. The van ride was a 10-hour rolling team meeting for some states, and farmers and Extension people often sat together.

In the regional training workshops, a strong effort was made to provide both new materials and innovative learning methods (see Chapter 8). The latest national publications in sustainable agriculture were accessed and provided to participants. A communication specialist from the SARE program at Beltsville presented an overview of what was available and under development across the country. Workshops also provided information displays with resources from the host state and others. To expand the appreciation of different learning methods, the workshops moved beyond traditional slide shows and lectures to include small group discussions, study circles, case studies, field tours, and other participatory activities. It was considered important to demonstrate and evaluate these methods in a safe environment where it was possible to assess the impacts and appropriateness of relatively nonconventional approaches. As usual, the field tours led by farmers were ranked high by the participants, and paired interviews used with evaluation were also valued as a learning experience.

Beyond the information and teaching methods presented in these regional workshops, the networking opportunities during four years became an invaluable experience for participants. Administrators, Extension specialists and educators, nonprofit and government agency people, and farmers and ranchers were able to build a regional sustainable agriculture community. The detailed contact information that was always provided at meetings promoted further communication and sharing of Extension materials among people with common interests. Program planners in each state used the participant and speaker lists to identify people who could cross state lines and bring new perspectives to their local programs. Specialists and educators were able to meet and gain inspiration from leaders in sustainable agriculture, and learn about successful approaches used by the agricultural program leaders in other states. Perhaps, most important of all was a chance for participants to step out of their daily routine, meet and discuss sustainable agriculture challenges and opportunities with like-minded educators, and see practices and systems in other states.

Success in the training program was also due in no small measure to the early recognition by the Administrative Council and Regional Coordinator that education was extremely important in the transition to a more sustainable agriculture. Close cooperation between the Administrative Council and the state leaders and incorporation of sustainable agriculture projects into state Extension priorities were critical in the process.

STATE SUSTAINABLE AGRICULTURE CONFERENCES

In 1996 funds were designated to support state or multistate meetings on sustainable agriculture. States were encouraged to integrate past and current grant recipients into the programming. As an example Illinois, Indiana, Ohio, and Michigan pooled their funds to host a four-state conference entitled "Profitable Farming in a Changing Environment."

EVALUATION OF REGIONAL SARE PROGRAMS

In the research, education, producer, and student grants, evaluation has been an integral and ongoing activity used to identify the needs for in-course adjustments or modifications in methods or measurements. In research grants the process has included addition of field treatments using different practices, as well as alternative approaches to analysis and evaluation of results. All authors of proposals were contacted after the cycle of grant making for input on the grant process. The producer grant program was formally evaluated under a contract with evaluation consultants (den Biggelar and Suvedi, 2000). The research and education program was also formally evaluated. In the PDP workshops, hourly evaluations have been used to assess learning and to adjust program content and delivery methods to better achieve workshop goals (see Chapter 8). In the very first PDP regional planning workshop, the entire program for days two and three was reorganized, based on evaluations from day one. The Extension administrator with regional oversight wisely recognized that a major change was needed, and the workshop coordinators responded quickly.

Among the comprehensive evaluation methods used were several ones unique to each program and type of grants. Research project leaders have been required to submit annual reports to the regional office, to describe the specific roles of farmers and educators in getting their results out to clientele, and the potential economic consequences. In the PDP workshops, there was an initial reluctance to hourly evaluations that assessed each activity and provided immediate information to speakers and to workshop coordinators. When these results were quickly summarized and results given back to the group, the value of such imbedded evaluations became more accepted. When the regional workshop coordinators later summarized the evaluation results, they became an invaluable resource for planning future workshops, identifying the most important content areas, identifying which teaching methods were most effective, and writing project reports.

Producer grant activities were monitored and evaluated in a collegial way by the Producer Grant Coordinator, who happened to be a farmer with many years of experience and a capacity to relate well to his peers. Thus, evaluation is an important and integrated component of all programs that helps regional administrators and Administrative Council members to explore cost/benefit results of projects and guide future directions. The long-term, cascade effects of SARE grant support to researchers, educators, and producers are described by Shirley Trout and colleagues (Chapter 14).

CONCLUSIONS

The North Central Regional SARE program has been innovative in a number of ways and has initiated several new programs, all focused on improving openness, accessibility, and service to people. Many methods and programs were adopted by other regions. The successes can be attributed to a coalition of leadership that was fully participatory, and where everyone was a peer, respect was mutual, and farmers and ranchers were considered professionals. The philosophical foundation of the NCR that was so instrumental in creating an organizational environment that fostered innovation also reaffirmed the wisdom of the Morrill Act of 1862. Passage of the first Morrill Act reflected a growing demand for agricultural and technical education in the United States and created the land grant universities. The Morrill Act was subsequently expanded by the Hatch Act of 1887 which created the state agricultural experiment stations:

> Section 2: It is further the policy of the congress to promote the efficient production, marketing, distribution, and utilization of products of the farm as essential to the health and welfare of our peoples and to promote a sound and prosperous agriculture and rural life as indispensable to the maintenance of maximum employment and national prosperity and security;

and the Smith-Lever Act of 1914 that created the Cooperative Extension Service:

> Section 2. Cooperative agricultural extension work shall consist of the development of practical applications of research knowledge and giving of instruction and practical demonstration of existing or improved practices or technologies in agriculture.

The second Morrill Act of 1890 and the Improving America's Schools Act of 1994 expanded the land grant responsibility to embrace the diversity of predominantly black and Native American institutions. The enabling legislation for the SARE program, as administered at the regional level, is the modern integration of the founding legislation to meet the needs of an evolving agriculture. The inherent values and ethics, openness, accessibility, and service to people have remained the same, while the challenges have dramatically changed.

The history of cooperation among states in the region and between researchers and farmers in practical, on-farm research provided a healthy foundation for a successful program through SARE. Many of these prior friendships and working relationships were the basis for new on-farm research projects, cooperative education efforts, and producer grants designed by farmers in collaboration with Extension educators. Especially valuable was the organizational and administrative model used in Nebraska with joint appointments and close working relationships between research and extension divisions. This spirit of collaboration was modeled during early negotiations with the Administrative Council and organization of the regional structure and decision- making procedures.

An emphasis from the outset on collegial decision making, recognition of the value of farmers' and ranchers' opinions, and leadership in the region have been important to the program. A high profile for farmers, ranchers, and nonprofit organizations has assured a practical direction for all programs and given the SARE activities credibility with the agricultural community. The service and support roles of the Regional Coordinator and coordinators for education and producer grants have provided programmatic and administrative resources for the Administrative Council. Farmers and ranchers were given a key role in decision making and overall leadership. The grassroots nature of the regional administration and decisions, the practical focus of research and education, and the vital participation of farmers and ranchers have made the national and regional SARE programs a viable model for future federal competitive grant programs.

REFERENCES

Allen, P. (ed.). 1993. *Food for the future: Conditions and contradictions of sustainability.* John Wiley & Sons, New York.

Bauer, L. (ed.). 1996. *Planting new possibilities, harvesting a healthier future.* North Central Region SARE, 1996 Annual Report, University of Nebraska, Lincoln. 181 p.

―――. (ed.). 1997. *Celebrating a decade of sustainable agriculture research and education: 10 years of SARE.* 1997 Annual Report, University of Nebraska, Lincoln. 160 p.

Berton, V. (ed.). 1998. *Ten years of SARE: a decade of programs, partnerships and progress in sustainable agriculture research and education.* SARE National Office, U.S. Dept. Agric., Washington, DC. 96 p.

Carter, H. and C.A. Francis. 1995. *Everyone a teacher, everyone a learner.* Extension and Education Materials for Sustainable Agriculture, Vol. 4. Cooperative Extension Division, University of Nebraska, Lincoln. May.

Carter, H. and C.A. Francis. 1996. *Shared leadership, shared responsibility.* Extension and Education Materials for Sustainable Agriculture, Vol. 5. Cooperative Extension Division, University of Nebraska, Lincoln. December.

Carter, H., C. Francis, and R. Olson. 1997. *Linking people, purpose, and place: An ecological approach to agriculture.* Extension and Education Materials for Sustainable Agriculture, Vol. 7. Cooperative Extension Division, University of Nebraska, Lincoln.

Carter, H., R. Olson, and C. Francis. 1998. *Facing a watershed: Profitable and sustainable landscapes for the 21st century.* Extension and Education Materials for Sustainable Agriculture, Vol. 9. Cooperative Extension Division, University of Nebraska, Lincoln.

den Biggelar, C. and M. Suvedi. 2000. Farmer's definitions, goals, and bottlenecks of sustainable agriculture in the North Central Region. *Agric. Human Values,* 17:347-358.

Flora, C.B. 2002. *Interactions between agroecosystems and rural communities.* CRC Press, Boca Raton, FL.

Flora, C.B. and J.L. Flora. 2004. *Rural communities: Legacy and change* (2nd ed.). Westview Press, Boulder, CO.

Franzluebbers, A.J. and C.A. Francis. 1991. Farmer participation in research and extension: N fertilizer response in crop rotation. *J. Sustain Agric.* 2(2):9-30.

Madden, J.P. and S.G. Chaplowe. 1997. *For all generations—making world agriculture more sustainable.* OM Publishing, Glendale, CA. 642 p.

Madden, P. 1987. Can sustainable agriculture be profitable? *Environment* 29(4):19 34.

McDowell, G.R. 2001. *Land-grant universities and extension: Into the 21st century.* Iowa State University Press, Ames.

Strange, M. 1988. *Family farming: A new economic vision.* University of Nebraska Press, Lincoln.

Thompson, P.B. and B.A. Stout. 1991. *Beyond the large farm: Ethics and research goals for agriculture.* Westview Press, Boulder, CO.

Chapter 10

Expanding Visions
of Sustainable Agriculture

Lorna Michael Butler
Cornelia Butler Flora

INTRODUCTION

Although agriculture has been an integral part of our lives for generations, it conveys different meanings to each of us. It may provide the satisfaction of working in the soil, hiking through a newly mowed pasture, purchasing vegetables at the local farmers' market, or appreciating a wildlife
habitat. On the other hand, if one has little direct experience with farms or
ranches, agriculture may trigger images of foul-smelling livestock feedlots
or contaminated streams. Agriculture may even induce images of decaying
rural towns and deserted farmsteads.

AGRICULTURE AS A MULTIFUNCTIONAL ACTIVITY

In the past decade we have begun to recognize the multiple contributions
of agriculture beyond food and fiber production. The view of agriculture
as a food factory is too narrow, and has been a barrier to our taking full
advantage of agricultural resources that benefit society. Agriculture has a
profound impact on our landscapes, local, national and global economies,
community vitality, and ecosystem diversity. It can be a valuable tool for
replenishing and sustaining ecosystems, and subsequent agroecosystem
services can benefit human populations.

As Jules Pretty (2002) has stated, agriculture is *fundamentally multifunctional* because it impacts the very assets, or forms of capital—natural, so-

Developing and Extending Sustainable Agriculture
© 2006 by The Haworth Press, Inc. All rights reserved.
doi:10.1300/5709_10

203

cial, human, physical, financial—on which it depends for success. Multi-functional Agriculture is a perspective that charges agriculture with creating strong rural economies, independent farmers, rural employment, and vibrant rural cultures. Environmental contributions include conservation of biodiversity, water quality and quantity, clean air, bio-energy and healthy soil, as well as other amenities such as food quality and safety, food security, animal welfare, scenic landscapes, and farmland preservation (DeVries, 2000; Jervell and Jolly, 2003).

The European countryside's movement away from the core production activities of agriculture, and toward greater *pluriactivity* is giving rise to new rural development processes and a rejuvenated rural economy (Ploeg, 2000). European farming strategies have become more diversified with respect to technology, marketing, and policies. Where this has occurred, the shift has promoted rural income and employment opportunities, increased quality of life, and social cohesion (Flora, 1995). The decreased dependency on external inputs has fostered diversification of on-farm income earning opportunities, increased environmental sustainability, resulted in healthier landscapes, and promoted the integration of new farm-based rural development activities (Ploeg, 2000).

For generations, innovative farmers in developing countries have resorted to multiple strategies to enhance their agricultural options and the health of their communities. On the small-scale farms of Latin America you will find wide variation in slopes, microclimates, elevations, soil types, cropping patterns and market strategies, all of which are intricately adapted to local conditions and cultures. A good example is the adaptation of various potato cultivars at different elevations in the Andes Mountains of Peru. These systems, which are combined with unique household resource management systems, are sustainably managed with little dependency on mechanization, chemical fertilizers, pesticides, or other modern technologies. A move to this kind of agriculture will require a different mindset and organizational structure than those most useful for industrial agriculture. In the U.S. heartland, we see a new agriculturalist emerging, supported by new social movements and value chains.

THE SUSTAINABLE AGRICULTURALIST

Initial research with sustainable agriculturalists (Chiappe and Flora, 1998; Flora et al., 2001; Meares, 1997) and analysis of agricultural shifts suggest that those moving toward a more sustainable agriculture are creating the jobs and the worker/manager/owner of tomorrow. Farms that are environmentally sound and economically competitive are an emergent part of a new economy.

Fordism is a term that represents the old economy, originating from Henry Ford and his automobile. The black Model T, mass-produced and made on the assembly line, was the only car available to buyers. Post-Fordism, the new economy moves from commodities that are undifferentiated—oriented toward a mass market—to differentiated products that are oriented to very specific markets and produced to meet those markets' needs. In a post-Fordist industry, and in sustainable agriculture, there are tightly integrated supply chains, as the product is produced with a specific user in mind who has committed to purchasing it prior to its production. Sustainable agriculturalists must prepare and organize to understand, develop, nurture, and control those networks, rather than merely supply the products. Flexibility, which is a key, means responsiveness for sustainable agriculturalists, a new generation that is aware of the conditions under which they produce, open to opportunities, and shifting comfortably in response.

In Fordist industry the production process was designed in a manner that the individual did the same thing repeatedly. Manual work was separated from mental work, and workers were easily replaced with little requirement needed for training. Post-Fordist economy demands highly skilled, team-oriented, and versatile workers, with continued training and skill enhancement. Sustainable agriculturalists form positive relationships and flexible networks, have the ability and willingness to learn from others to improve their knowledge and skill level, and can separate problems from solutions. They determine which problems are the most important for immediate solution, then identify alternative solutions. Sustainable agriculturalists attend to detail and are aware of all activities across the operation and other value chain components. There is continual attention to advancing their skill level and to knowing the land better, consistently shifting to become more sustainable—economically, environmentally, and socially. Sustainable agriculturalists have the initiative and are willing to take risks, defining unfulfilled expectations not as failures, but learning opportunities. Sustainable agriculturalists are continually challenged to identify niche markets and link to them, so that qualities that are intrinsic in sustainable products can be transferred and supported by well-identified and well-connected markets. This perhaps is where today's sustainable agriculturalists have the furthest to go to become successful.

Sustainable agriculture is part of the new economy because it involves the ability to constantly innovate, monitor, assess, and react to new situations. Farms or related organizations are run democratically, with workers who share in the vision and have a stake in the decision making and outcomes. Reflection is a key for sustainable agriculturalists, which means that good record keeping is important so that past performance can be evaluated.

Clearly, sustainable agriculturalists have more to contend with than conventional agriculturalists, since there is less stratification and differentiation in what is done on the farm and off the farm.

Because there is more certainty of a base level of income due to the close links between producer and consumer, sustainable agriculturalists will also learn to budget with flexibility because of shifting markets as well as changing environmental conditions. Sustainable agriculturalists plan for the worst year rather than the best year, and this is what allows them to both protect the environment and enhance rural communities. The concept of sustainable agriculturists is not limited to farmers or ranchers alone; it applies to each member of the team that works together to create an innovative, responsive, and sustainable food system that brings long-term benefits to society.

A NEW SOCIAL MOVEMENT IN THE HEARTLAND

During the past decade scholars and others have speculated on the growing U.S. social movement that is seeking to transform the conventional food and agriculture system into a multifunctional system that "balances concerns of environmental soundness, economic vitality, and social justice among all sectors of society" (Hassanein, 1999). Many participants in sustainable agriculture share a common belief about the need for a new kind of agriculture that provides a wider spectrum of benefits to farmers and ranchers, to the environment, and to society as a whole.

Hassanein (1999) contends that this emerging consensus about the need for change is the outcome of three principle concerns: (1) disenchantment with the structure of agriculture and its intersection with wider industrial interests, all of which have contributed to a reduction in the number of family farms and the decline of rural communities; (2) concern about the ecological damage that conventional agricultural systems have imposed on our landscapes, wildlife, and natural resources; and (3) more recently, the increasing dissatisfaction with the insensitivity of agriculture to social justice and equity issues. Unequal distribution of power is a pervasive theme. This is aptly illustrated by the widespread problems of hunger and poverty, the disregard of farm workers' concerns about the dangers of working with excessive chemicals, and gender discrimination in agriculture (Sachs, 1996; Van Esterik, 1999).

In some communities, agriculture in its reconstituted form is contributing to the health of communities by enhancing human and organizational relationships. This is achieved by strengthening problem solving capacities, enhancing regional economic prosperity, and improving individual and family health. Children are participating in gardening and cooking classes and showing a greater interest in eating a variety of healthy vegetables.

Farmers are becoming more knowledgeable about the biological processes to improve their soil health, retain soil moisture, and control potentially damaging insects and plant diseases. Memberships in community supported agriculture are growing. Family members are asking more questions about how the quality of their food and diet will impact their health. These and other factors signal a growing awareness among consumers as to the importance of buying locally to support local farmers, accessing higher quality food, and decreasing fossil fuel reliance. If these hopeful signs continue to spread, this could provide the momentum for sweeping changes in our agricultural and food system, and for our communities and landscapes.

The following section highlights promising illustrations that support our contention that sustainable agriculture, including emergence of *sustainable agriculturists,* is taking hold in many forms and patterns, and as such is energizing positive changes at institutional, farm, community, and environmental levels.

NEW INSTITUTIONAL MODELS

The Land Institute

The Land Institute (TLI), now over 25 years old, is located near Salina, Kansas. Founder Wes Jackson contends that soil erosion and soil degradation associated with agriculture have had a significant impact on our ecological system over the past 10,000 years. Thus, research, education, and public policy programs at TLI are focused on the integration of soil conservation with agricultural production. Research in *Natural Systems Agriculture* attempts to "understand and mimic" the natural ecosystem that was once characteristic of the Midwest prairies. Jackson proposes that agriculture be designed in the image of nature, rather than follow the single-crop approach found in most fields today (Jackson, 1980). TLI has focused on the development of perennial multi-species plant systems, their fertility and recycling mechanisms, soil building capacity, and resilience in the face of pests, pathogens, and droughts. Knowledge of mixtures of perennial crop plants is important because of their abilities to hold the soil, retain fertility, take advantage of the sun, manage moisture efficiently, and control weed, insect and disease problems. The sustainable agriculturists who lead TLI are constantly advancing knowledge of the land. As they learn more about nature-based agriculture, they have taken risks and seized upon opportunities to learn more through creative research design, training others, and advancing public awareness of ecology and sustainability issues.

In 2001 and 2002, the breeding program generated hybrid offspring cool season grasses (wheat, rye, other species) and warm season species (sorghum, sunflower, perennial legumes) for future evaluation. The potentials of Maximilian sunflower, Illinois bundle flower, and chickpea are being investigated, as well as wheat and alfalfa intercropping, perennial and annual crops with different root systems, the ecology of beneficial soil fungi, opportunities for perennializing grain crops, and long-term comparisons of prairie and agriculture. Long-term fieldwork has produced data on the energetics of farming in order to draw comparisons with current farming practices. A Prairie Writers Circle helps to communicate to the public about the intersection of agriculture, community, and environment (Bontz et al., 2002; The Land Institute, 2003).

Leopold Center for Sustainable Agriculture

The localization movement that is reshaping food relationships between producers and consumers may be a reaction to a relatively recent acknowledgement that the distance we move our food has a direct influence on local economies, product freshness, fossil fuel consumption, and the subsequent release of carbon dioxide to the atmosphere. Research conducted by the Leopold Center for Sustainable Agriculture has documented the distance that 30 different types of fresh produce travel before reaching the Chicago Terminal Market from points in the United States and Mexico. Six fruits and vegetables traveled more than 2000 miles to their destination. Mexico was the source for 21 of the 30 items studied. In local food system projects, conventionally sourced meat and produce traveled 34 times as far as foods sourced locally. The distance that produce travels by truck within the United States has increased by 22 percent between 1981 and 1998 (Pirog and Van Pelt, 2002; Pirog et al., 2001).

University of Northern Iowa Local Food Project

Dr. Kamyar Enshayan, University of Northern Iowa (UNI) Center for Energy and Environmental Education, leads the Local Food Project (LFP). The project, funded totally by grants, works with institutional food buyers to identify ways they might purchase their food from local or regional farmers and food processors. Through these efforts, farmers and processors are linked with nursing homes, hospitals, schools, and restaurants. Between 1998 and 2002, eight institutions purchased $783,000 worth of meat, fruits, and vegetables from farmers in Black Hawk and neighboring counties.

Through a customer awareness program in cooperation with Practical Farmers of Iowa, LFP has published and distributed 23,000 copies of a *Buy*

Fresh, Buy Local directory to the six-county area to encourage everyone to learn about local food. Other mechanisms are used to make it easy to procure local food, for example, postcards listing farmers' market days, farmers' price lists faxed to local food buyers, farmer delivery of retailers' orders, farm tours for chefs, radio announcements, newspaper advertisements, and direct mailings. Partner food retailers also promote local food. For example, one restaurant uses a *table tent* to list farmers' markets; a large wall poster to introduce customers to *Rudy's Farmers*; and communities display farmers' market signs. A special label and several point-of-purchase materials have been developed to assist farmers with marketing. Kamyar believes that keeping local food dollars in the immediate area makes good community economic development sense. Six years of data has shown that for every grant dollar, the program generates $6.5 local food dollars (Enshayan, 2003).

In 2001, Iowa spent in excess of $8 billion dollars on food; of this, residents of the Waterloo/Cedar Falls area spent almost $300 million on food, groceries, and restaurants. The LFP has taken advantage of local buying power to encourage a more diverse agriculture and to strengthen the regional economy. In 2001, one collaborator, Rudy's Tacos, spent 71 percent of its food budget on fresh, locally-grown ingredients. This amounted to $143,000 (Enshayan, 2001; L. Miller, 2003).

However, there is still work to do at the institutional level. Kamyar facilitates all of this activity as a volunteer who is housed at UNI. The program operates primarily on grant funds. Unfortunately, the program is not yet integrated in any way with UNI's academic or outreach efforts. While it is apparent that the UNI community is aware of the LFP through the attention it attracts, the university has so far shown little inclination to take advantage of the program's success, for example, by integrating student courses on *food, agriculture, and business in Iowa* (90 percent of UNI's students are from Iowa), or through the adoption of a local food procurement policy in the campus dining service. With more institutional support a program like the LFP could benefit the university and the region by expanding the knowledge of young people and the public about the opportunities that exist to improve the local economy through local food production and use. They can also educate them about the ways in which high-quality food can contribute to better individual health. Perhaps it will take a coalition of institutions, working together, to capitalize on the innovations and risk-taking efforts of the LFP.

Food Policy Councils

In April 2000, Iowa Governor Thomas Vilsack appointed Neil Hamilton, Director of the Agricultural Law Center at Drake University Law School, as chair of the newly formed Iowa Food Policy Council. The Council is composed of 22 food system stakeholders from different food-related sectors (farming organizations, professional associations, industry), six ex-officio members, and a policy advisor from the Governor's office. The council provides a voice for the concerns and interests of stakeholders who may be under-served by agricultural institutions. Some of the policy issues that have been examined include food security for at-risk citizens; public awareness of the relationships among food security, agriculture, health, and economic issues; promotion of local food systems; diet, health and obesity; institutional purchasing of locally-grown foods; value-added agriculture through direct marketing; and sustainable agriculture production practices.

In December 2003, the Iowa Food Policy Council submitted a set of recommendations to the governor of Iowa for improving the state's food system. Recommendations targeted opportunities for institutional purchasing, promoting a state-wide buy fresh and buy local campaign, expanding food processing as a form of rural economic development, and improving the state's food assistance and food security programs. The Iowa Food Policy Council partners with food councils in Connecticut, New Mexico, North Carolina, Oklahoma, and Utah (Iowa Food Policy Council, 2004).

Interest in food policy councils is on the rise. In addition to state level councils, they exist at community, county, and city levels, for example, in Berkeley (California), Hartford (Connecticut), Knoxville (Tennessee), Lane County (Oregon), Portland (Oregon), Tacoma (Washington), and Toronto (Ontario, Canada). There are many issues that food policy councils can pursue, such as improving citizens' access to quality food produced by local farmers, or promoting a deeper understanding of the current food system and how it might be improved. Some councils work to protect farmland or to create a state food security report card. Because food policy councils serve as a forum among diverse interest groups, they can be an effective tool to engage consumers, farmers, food processors, retailers, anti-hunger activists, and policy makers in constructive policy discussion, planning, and action (Hamilton, 2002; Iowa Food Policy Council, 2004).

Multi-Level Alliances Between Farmers and Institutions

Declining profit margins and concerns about vertical integration in the food chain are compelling sustainable agriculturists to take greater control of their futures. Farmers are joining with each other to form new alliances or new business entities, with joint goals, in order to capture additional value from a particular product. The underlying assumption is that by retaining ownership of the product through the entire value chain, farmers will be in a better position to capture added value from subsequent value chain steps.

Various terms are used to describe the organizational structures formed to create alliances, for example *value chain relationship, consortium, strategic alliance, cooperative, limited liability company, partnership, joint venture,* or *corporation.* Partnerships with value chain participants are just beginning to tap the opportunities of the agricultural sector. The ability to cooperate to create win-win business alliances seems to be the key (Fulton, 2000; Government of Alberta, 2002).

Iowa's Value Chain Partnership

In many parts of the world, farmers are forming alliances with other farmers, as well as with processors, distributors, and retailers in order to gain a more equitable share of the profit. However, the distinguishing factor is the group's commitment to higher standards of community and environmental stewardship in addition to profitability. The *Value Chain Partnership for Sustainable Agriculture* (VCPSA) facilitates farmers, commodity groups, nonprofit agencies, and university and community partners in activities that focus on the challenges found across food and fiber value chains, including issues at the production, processing, distribution, and retail levels. VCPSA is funded partly by Kellogg Foundation, with additional support from the Leopold Center for Sustainable Agriculture, the ISU College of Agriculture, the SYSCO Corporation, as well as other organizations.

The goal of VCPSA is to make the entire value chain—production, processing, and distribution—more collaborative and rewarding to mid-sized farmers who use more sustainable production practices. It is seeking ways to encourage practices that are environment friendly, profitable, and equitable in sharing information, risk, and rewards among all partners. VCPSA is encouraging Iowa State University and its partners to provide the farmers, processors, distributors, and retailers of these value chains with the tools they need to improve the existing value chains or develop new ones rooted in sustainable agriculture. Using a shared leadership team made up of three key cooperators and a broad-based advisory group, VCPSA stresses ways to improve business structures across the value chain. This is being done through three working groups, each of which addresses a particular challenge found within the chain. The groups are the: (1) Pork Niche Market Working Group (PNMWG), (2) BioEconomy Working Group, and (3) Regional Food Systems Working Group. Each is driven by the priorities of its diverse membership. The 30-member PNMWG is targeting the need for highly differentiated and profitable pork value chains that incorporate farmers' ownership and control, and that contribute to environmental stewardship and rural vitality. Two strategies are employed to achieve these goals: funding projects that address challenges faced by alternative pork producers, and facilitating quarterly meetings for participants to share information, discuss progress, and to respond to opportunities. As of spring 2004, there were 18 projects in various stages of operation. Project topics include the examination of the costs of alternative pork production, analysis of business planning and feasibility, and facilitating study tours of consumer-driven niche pork marketing and certification systems. A major project is developing a USDA-AMS Process Verified program for family-sized pork producers, and advancing a meat quality assurance program for pork niche market opportunities. The U.S. Department of Agriculture's Agriculture Marketing Service (AMS) and Grain Inspection, Packers and Stockyards Administration (GIPSA) have developed voluntary testing and process verification programs to facilitate agricultural product marketing. The emergence of value- enhanced commodities and a niche market for non-biotechnology-derived (non-MO) commodities has created an increased need to differentiate products in the handling system. Information on GIPSA's programs and services is available from: http://www.usda.gov/gipsa/programsfgis/inspwgh/processver/.htm. For livestock and seed, USDA Process Verified suppliers are able to make marketing claims such as: breed, feeding practices, or other raising and processing claims verified by the USDA and marketed as "USDA Process Verified." Further information is available from: http://www.ams.usda.gov/lsg/arc/prover.htm.

The working group also has been engaging SYSCO Corporation in discussions about the potential of carrying more highly differentiated non-SYSCO brands of pork that do not carry the SYSCO name. There is still much to be done to ensure that mid-size swine producers achieve economic sustainability and fair contracts (Pirog, 2004).

The two other working groups are in early stages of development. As with sustainable agriculturists, their participants work with a diverse membership; they give continual attention to the challenges of bio-based business development, and to regional food systems business development—two areas of future opportunity. The Bio-economy group will leverage funds for research and development projects, and will test farmer-community models for equity investment in bio-based businesses and bio-refineries. Early projects to receive support include a feasibility study of growing and processing kenaf as an alternative crop, analysis of kenaf business strategies for farmers to organize as a *feedstock link* in supply chains for bio-based businesses, development of a quality testing system for natural fibers, and evaluation of the sustainability of a one million ton stover collection and delivery system along the Missouri River.

More than 30 people representing 20 institutions and businesses have joined the Regional Food System Working Group (RFSWG). This group is focused on the development of a food system that can bridge the gap between small direct marketing systems and the global systems that predominate in most supermarkets. Objectives of RFSWG will guide research to target the investigation of the multifaceted impacts of community-based food and fiber enterprises; development of farmer and community-based projects and forums to advance the capacity of higher education to support the needs of farmers, entrepreneurs and businesses attempting to develop regional and community-based value chains; and communicate results to people and organizations that need this information. The first round of funded projects addresses such topics as Iowans' understanding of sustainable agriculture issues with respect to regional food systems, the extent to which northeast Iowa wins or loses financially from prevailing food systems, the economic impacts of farmers' markets on Iowa's economy, and the documentation of local food purchase data and related economic benefits in the Black Hawk county region (Pirog, 2004).

Participants in these working groups are clearly a new breed of sustainable agriculturists who, with diverse collaborators, are analyzing the problems faced by mid-sized farmers, entrepreneurs, and businesses. They are attempting to shift their production and marketing strategies in response to new consumer demands. The VCPSA working groups provide an ideal mechanism for reflection, learning, and shared risk-taking. While each has a slightly different interest, they are continually focused on what it takes to become more sustainable economically, environmentally, and socially.

CHANGES AT THE FARM LEVEL

New Forest Farm, Viola, Wisconsin

Hybrid hazelnuts have been found in the American Midwest since the 1930s. Some feel that these hybrids may contribute to a promising regional industry. For example, in southwest Wisconsin, Mark Shepherd has been managing a permanent ecosystem of hazelnut and chestnut trees, both native to the area, for over seven years. The plants, which represent varieties that have adapted to their environment over many years, grow in contoured hedgerows somewhat like a lilac. These varieties are different than traditional hazels in that they are bushes, not trees, and are cold-hardy, disease resistant, and require no pruning. Other crops such as asparagus, raspberries, grapes, and prairie flowers are planted between the rows. The nuts, which are expected to reach their peak production in several more years, are marketed in the United States (Rutter and Shepherd, 2002; Tanzilo, 2002).

Being true to the model of the sustainable agriculturist, Mark Shepherd is continually monitoring his system and innovating. He takes advantage of a wide range of information networks, and finds support among other farmers who are also committed to the establishment of permanent ecosystems.

North River Produce Limited, Cresco, Iowa

Northeast Iowa farmers Mike Natvig and Amy Miller produce hogs, corn, soybeans, and beef cattle on 400 acres. Their rotation takes advantage of species diversity to get the most out of local conditions. One of the biological highlights is the 10-acre pasture seeded to native prairie, and home to at least 45 different plant species. The farming system includes small grains and forage plants, organic hogs raised in a low-cost pasture-farrowing system, and an intensive pasture rotational grazing system for cattle. Livestock are moved in rhythm with the grass growth cycle, taking advantage of the perennial pastures' natural ability to hold moisture and soil (see Chapter 4). An innovative small grains-hay-row crop rotation includes 10 different species: oats, wheat, barley, field peas, three species of clover, alfalfa, timothy, and orchard grass. The small grains serve as climbing stalks for field peas, and when harvested together, the diverse mixture produces a nutritious hog feed. Livestock manure is added to fields according to crop needs. Because Natvig wants to learn what the impact of his grazing system is on the stream that runs through his farm, he monitors the stream for insects in the aquatic food chain. In doing so, he has discovered that his grazing system has not damaged the stream banks. In fact, over 80 percent of the insect larvae that he has identified are only found in high-quality habitat. Grazing the stream banks for two days, twice a year, seems to be improving stream habitat (DeVore, 2002; Jackson, 2002; A. Miller, 2003).

A conservation mindset has been evident on this farm for at least three generations. As for the sustainable agriculturist that we have described, the system is an example of continual innovation, gathering new knowledge, and constant problem analysis. These farmers evaluate their system for soil loss, have a long-term plan for crop rotation to ensure healthy soil, and take advantage of grant funds to experiment with prairie restoration. They are now beginning to grow open-pollinated corn. With assistance from the Federal Government's Conservation Reserve Program (CRP), they are working to provide native habitat for wildlife and collecting prairies seeds. For these sustainable agriculturists, this represents a promising alternative to crop subsidy payments (A. Miller, 2003).

Radiance Dairy, Fairfield, Iowa

Near Fairfield, Iowa, Francis Thicke runs an organic dairy operation. He notes a growing interest among farmers and consumers who want to work together to develop a *food* circle, in which all possible food is produced and consumed locally in order to close the circle. This means consumers don't have to purchase their food from far away. Playing his part in the circle, Francis raises Jersey cows and processes organic milk and milk products, then sells the products locally. The 176 acre farm has 40 acres in the CRP, with the remainder planted in forages for a controlled grazing system. Through close observation, Francis is able to add to the plant diversity, ensure that species do well, and meet his cattle's preferences. Thus, he is able to meet several goals at once: maintain a healthy natural environment, generate a reasonable profit, manage his time so he can participate in community activities such as mentoring beginning organic farmers, and contribute to the local food circle (Vagnetti and DeWitt, 2002).

Francis Thicke's organic dairy operation is another example of how the sustainable agriculturist is redefining the community and the visions of other farmers. His commitment to mentoring new, young farmers, and his many contributions to sustainable agriculture communities of interest will influence the next generation of farmers and consumers. The environmental values exhibited on his farm are well known among other farmers, academics, and regional consumers.

What is it that fosters innovative behavior in a state like Iowa that epitomizes the agricultural heartland? Iowa has long been dominated by corn, soybeans and hogs, one of the most concentrated, vertically integrated, hightech systems of farming in the United States, and possibly in the world. Yet, within Iowa there is an amazing collection of initiatives that counter this industrialized model, providing evidence of a movement that is thriving on mutually rewarding relationships among farmers, consumers, local processors and retail outlets. This growing trend is commonly referred to as a *local food system, community food systems, community supported agriculture, field-to-family,* or *field to fork.* While clearly an economic move for some farmers, there are other reasons behind the drive to procure and market food from local sources to local consumers. A growing number of customers want more information about the sources of their food, methods of production and processing, and they value the relationship they have with their farmer. These customers are placing a higher value on *fresh* and *healthy,* the opportunity to retain money within the community, and lowering the use of fossil fuels for long-distance transportation. The new sustainable agriculturists are rising to the challenge.

NEW MODELS AT THE COMMUNITY LEVEL

In the best of all worlds, food, agriculture, and nature would be an integral part of all community and economic planning. This vision would include the engagement of citizens in many different walks of life in crafting ethical plans for their desired future living and working environments. While this sounds almost too idealistic, communities have attempted, or are attempting, to move in this direction. Where this momentum exists, the process has built community decision-making capabilities, public ownership in a vision, and the necessary actions to carry it out, and it has prepared residents to respond to problems as they occur. If well orchestrated, therein lies the potential for a wide cross-section of benefits—the enactment of principles to promote a community land ethic (Leopold, 1949) that includes incentives for farmers and land owners to protect both the ecological and economic values of their properties (Arendt, 1996, 1999); diverse forms of agriculture that contribute to social, environmental and economic capital (Pretty, 2002); and regionally based food and agricultural systems that build relationships among farmers, communities, and regional marketing systems (Green and Hilchey, 2002). A well-integrated food and agricultural system can enhance the relationships between different types of individuals and organizations, and encourage institutional, business, and policy linkages that support a region-wide quality of life.

Growing Greener, Pennsylvania

In 1998, Pennsylvania initiated a statewide collaborative planning effort to address state environmental concerns, and to protect interconnected networks of open space at the local level. It began through the joint efforts of the Department of Conservation and Natural Resources, Natural Lands Trust, and the Pennsylvania State University Cooperative Extension Service, and included multiyear funds from the William Penn Foundation and state government entities. The central *growing greener* message was, "The open space that is conservable in nearly every new residential development can be required to be laid out so that it will ultimately coalesce to form an interconnected system of protected lands running across your community" (Arendt, 1999).

The types of open space that can be conserved following the *growing greener* approach include wetlands, floodplains, and steep slopes, but also upland woodlands, wildlife habitats, scenic meadows, pastures or fields, garden plots, small farms, existing hedgerows, and historic, cultural, or scenic amenities that communities value. The model links natural resource conservation, historic preservation, land development, and real estate development interests through conservation zoning and conservation subdivision design, something like the *clustering* concept (Arendt, 1999).

In Pennsylvania, the growing greener concept was introduced at participatory workshops in which attendees were asked to design a residential development according to an agreed upon set of principles. The plan was to include the conservation and development of a particular piece of property. Some of the recommended elements for inclusion were: wetlands and their buffers, floodways and floodplains, moderate and steep slopes, groundwater resources and their recharge areas, woodlands, productive farmland, wildlife habitats, historic, archaeological and cultural features, and scenic view sheds. Central to the approach is the modification of the density on development parcels to confine house lots to smaller pieces of land. In contrast to conventional developments, half, or less, of the buildable land is taken up by lots and streets (Arendt, 1999).

A number of communities from Maine to California have demonstrated the *growing greener* approach to resource conservation; some of these have incorporated open meadows, agriculture or food production areas, wildlife habitat, and energy and water conservation. Examples include Michigan's Livingston County which has protected over 1,000 acres; Pennsylvania's Garnet Oaks (Delaware County) which has confirmed home buyers' preferences for homes that face onto woodland space or open space; and Illinois' Grayslake (Lake County) which has included horse pastures, nature trails, variable-sized lots overlooking protected lands, home owner gardens, and a 10-acre community-supported organic farm (Arendt, 1999).

Village Homes, Davis, California

Village Homes were among the first to pioneer *green development*. Applying concepts of sustainable design, developers Michael and Judith Corbett wanted to minimize any damage to the environment, limit dependency on nonrenewable resources, and respond to people's need for social interaction, shared neighborhood spaces and activities, and a sense of control over their shared resources (Corbett and Corbett, 2000).

The project, developed between 1975 and 1981, grew from a vision of about 30 people who wanted a community that provided a shared sense of meaning, social support, and resource efficiency. The 60-acre site supported 210 single-family homes on lots of about 4,500 square feet, and 30 attached rental units. Village Homes was designed to enhance community social life and create a more sustainable environment. The following Village Homes design features emerged from these goals:

- Economical use of money, resources (land, water, energy), and a safe environment, for example, earth covered houses, roof/wall/ window insulation, solar technology, shaded south-facing windows, water-saving devices, and natural drainage systems.
- An increased sense of community and social interaction (spaces for walking, biking, gardening, eating, playing), for example, mini-parks, community centers, common space that family clusters develop according to their wishes.
- A nature-oriented lifestyle, for example, walking and biking paths, bridges across creeks, paths along open spaces, no through streets, home orientation to promote neighborhood interaction, private courtyards, rooftop rain water channeled into meandering streams, graded channels, ponds, and greenbelts.
- A diverse resident population, for example, mixed income housing, culture-friendly welcome strategies for ethnic groups.
- Democratic governance in which residents have equal but voluntary votes in community concerns.
- Incorporation of agricultural land, including space for gardening, for example, nut and fruit trees on common lands, vegetable gardens, community gardens to supply local restaurants, opportunities for residents to produce their own food, or to market produce cooperatively. (Corbett and Corbett, 2000)

Over 20 years later, Village Homes has fostered an attractive, close-knit community with dispersed agriculture. It has been possible for people to live in a beautiful, natural setting, in harmony with the environment. Community participation, while less than envisioned, works for residents. They are able to solve their problems, gather for community almond and grape harvesting, and produce vegetables for themselves or for the market. The natural drainage systems work, and save money; the environment is safe for families. Home size and style is affordable and comfortable. However, despite increased public interest in these features, many would be difficult to get approved under existing development policies in the United States (Corbett and Corbett, 2000).

Arendt (1999) recognizes the challenges in implementing conservation development for obtaining funds for land purchase, or arranging the donation of easements. Until there is more public support for land purchasing, Arendt feels the most feasible way to protect large land areas in a coordinated way is to link the conservation subdivision design approach with multioptioned conservation zoning. This model involves no public cost or landowner charity, requires no shift of land rights, and is not dependent on adjacent landowner cooperation.

Community Supported Agriculture

Community supported agriculture (CSA) has become a way for rural, urban, and suburban residents to connect with each other through a mutual interest in learning about the food system, increasing access to locally produced, better quality foods, and supporting local or regional farmers. There is a wide variety of CSA models throughout the United States, and in other countries. Green and Hilchey (2002) estimate there may be 728 CSAs in the United States. In some cases, a CSA is a way for food customers to support local farms by buying a "subscription" to support a farm, and in return to receive a weekly allotment of food or other farm products during the growing season. Products may come from the farm with which there is a contract, or through a cooperative arrangement whereby several farms work together to supply customers with produce, for example, meat, vegetables, cheese, wine, eggs, flowers, and herbs. Customers share some of the risk of farming in exchange for a share of the harvest. In some cases, customers provide farm labor for part of the cost.

There are many benefits to be gained, in addition to the products, through participation in a CSA relationship. For example, customers learn about farmers' growing practices. they develop a personal relationship with farmers, and many feel they receive fresher, higher quality products—and they know how their food is produced. Farmers perceive CSAs as a more profitable way to market their produce since they can avoid intermediaries and are assured a customer base. They also learn more about their customers' preferences. CSAs can serve both social and nutritional needs, or form part of a social movement to save local farms, promote healthy food, protect farmland, and counteract the globalization of the food system. Some CSAs work hard to gain public visibility through participatory activities such as school programs, volunteer opportunities, special celebrations, and donation to charities. The greater the public involvement, the more likelihood of reaching some degree of public consensus around the values associated with the sustainability of food, agriculture, and natural resources. Often food serves as a common denominator around which human relationships are built, and conversations begin.

Kansas City Food Circle

A food circle extends the CSA concept to the community level. It creates a network among food producers, processors, marketing entities (e.g., CSAs), concerned citizens, and other interest groups to facilitate the growth of local food and agriculture systems. The food circle model was developed largely by the Food Circle Networking Project at University of Missouri (Green and Hilchey, 2002).

Kansas City Food Circle provides a link among consumers, farmers, retailers, nutritionists, advocates, and others who are committed to work together for a just and sensible food system for communities and regions. The group promotes sustainable agriculture and land stewardship, and provides a voice for those who are concerned about the power of multinational corporations over the food system. Over ten years ago, the Kansas City Greens had a project called the Organic Connection, which sponsored workshops on safe food, CSAs, sustainable agriculture, and rural-urban partnerships. From this, a voluntary, member-based, organization emerged. Over time the Food Circle has been the catalyst for a number of community food initiatives: a distribution center linking consumer families to fresh, organically grown produce; a hotline and voice mail number to help consumers connect with organic farmers; buying clubs and other food access systems; a published list of growers and farmers' markets; public meetings and events; a speakers bureau; promotion of community-based value-added enterprises; state policy initiatives; and useful research, for example, on making local food available to restaurants.

Midwest Food Alliance, St. Paul, Minnesota

The Midwest Food Alliance (MWFA), a partner of the Food Alliance in Portland (Oregon) and a project of the Land Stewardship Project, is a co-alition of Upper Midwest farmers, processors, distributors, retailers, and consumers that recognizes the value of food that is produced in an environmentally and socially responsible manner. Through its seal of approval, or eco-label, MWFA rewards the stewardship of local farmers, raises consumer awareness about the value of these farms and the food they produce, and calls attention to the benefits of participation in a farm-food-consumer alliance. When consumers see the certification seal, they know that the food originates from a local farm, is environmentally friendly, and is produced under socially responsible conditions. Farm or ranch certification is dependent on commitment to the following principles:

- Protecting and conserving water resources
- Protecting and enhancing soil resources
- Reducing pesticide usage
- Conserving and recycling nutrients
- Providing safe and fair working conditions for employees and their families
- Certifying crops that are not genetically modified organisms (GMOs)
- Providing healthy and humane care for livestock
- Continually improving farming/ranching practices (Midwest Food Alliance, 2001)

The Food Alliance, which began in 1994 with a grant from the W.K. Kellogg Foundation, became an independent nonprofit organization in 1997. In the Midwest, over 65 farms and ranches produce over 100 certified products. Over 49 different retailer partners are involved. One of the more recent developments is the partnership between Sodexho USA and MWFA. Sodexho, one of the leading providers of food and facilities management services in North America, is featuring foods from farms certified by the Food Alliance on its menus at college and university campuses in Minnesota, Wisconsin, North Dakota and Iowa. The Food Alliance is unique in that it promotes enhanced use of sustainable agriculture practices using market-based incentives. Currently it certifies over 200 agricultural products, including fruits, vegetables, wheat, wine, livestock, and dairy (Midwest Food Alliance, 2001). Through a broad coalition of interests, and a strong public outreach and promotion program, the MWFA is proving to be an effective regional catalyst for expanding community awareness and action toward a more sustainable future. The potential impact of this regional effort is multiplied through its partnership with a similar organization in the northwest region—a strategy that sustainable agriculturists are increasingly adopting.

CONCLUSIONS

With a view towards the multifunctionality of agriculture, there are increasing opportunities for new sustainable agriculturists to make an impact at the farm, community and institutional level. While the sustainable agriculturist may be a farmer or rancher, there is also an important place for community activists, processors, retailers, distributors, government leaders, developers, planners, and academics. With a common vision for the future, a flexible network, and a committed team, this diverse social movement will continue to grow and impact local farms, local and regional businesses, communities, and consumers. It will require supportive policy with rewards for environmentally sound practices; flexible research programs both on-farm and in-community with active local and regional participation; and extension programs often in partnership with private voluntary groups to support alliances of sustainable agriculture organizations—the key to high performance collaborative programs. It will also demand a new mindset on the part of each sustainable agriculturist—one that sees the advantage of diverse alliances and partnerships, constant change and innovation, and continual reflection on the impacts of decisions. All of these qualities will strengthen the work organization's knowledge and capability to create an expanded vision for agriculture. The new agriculture can and should play a vital role in enhancing the lives of a broad segment of society.

REFERENCES

Arendt, R.G. 1996. *Conservation design for subdivisions.* Island Press, Washington, DC.
————. 1999. *Growing greener. Putting conservation into local plans and ordinances.* Island Press, Washington, DC.
Bontz, S. et al. (ed). 2002. *The Land Institute Annual Report.* The Land Report. No. 73, Summer. Salina, KS. pp. 22-24.
Chiappe, M.B., and C.B. Flora. 1998. Gendered elements of the sustainable agriculture paradigm. *Rural Sociol.* 63(3):372-393. Available from: http://www.ncrcrd.iastate.edu/pubs/flora/gendered.htm (Accessed 3 April 2006).
Corbett, J. and M. Corbett. 2000. *Designing sustainable communities. Learning from village homes.* Island Press, Washington, DC.
DeVore, B.A. 2002. Stewards of the wild. In: *The farm as natural habitat: Reconnecting food systems with ecosystems,* D.L. Jackson and L.L. Jackson (eds.). Island Press, Washington. pp. 97-106.
DeVries, B. 2000. *Multifunctional agriculture in the international context: A review.* Available from: http://www.landstewardshipproject.org/mba/MFAReview.pdf [Accessed: October 4, 2003].
Enshayan, K. 2001. *University of Northern Iowa local food project.* Available from: http://www.uni.edu/ceee/foodproject/ [Accessed: October 26, 2003].
————. 2003. Buy fresh—buy local: Using the food routes initiative to create a national "buy local" effort. *National Workshop on State and Local Food Policy Councils: Innovative Collaborations in Food and Agricultural Policy.* September 4, 2003.
Flora, C.B. 1995. Social capital and sustainability: Agriculture and communities in the Great Plains and the Corn Belt. *Res. Rural Sociol Dev.: Res. Annu.* 6:227-246. Available from: http://www.ag.iastate.edu/centers/rdev/pubs/flora/soccap.htm (Accessed 3 April 2006).

Flora, C.B., G. McIssac, S. Gasteyer, and M. Kroma. 2001. Farm-community entrepreneurial partnerships in the Midwest. In: Interactions between agroecosystems and rural communities, C. Flora. (ed.). CRC Press, Boca Raton, FL. Available from: http://www.ag.iastate.edu/centers/rdev/pubs/farm-communitychapter9.pdf [Accessed 3 April 2006].

Fulton, J. 2000. *Are producer alliances/networks an alternative for producers?* AAEA. Available from: http://www.ag.iastate.edu/centers/rdev/pubs/flora/ soccap. htm; http://www.uni.edu/ceee/foodproject/ [Accessed April 3, 2006].

Government of Alberta. 2002. Alberta agri-preneur, October 30, 2002. Agriculture, Food and Rural Development. Available from: http://www1.agric.gov.ab.ca/ $department/newslett.nsf/homemain/agpr?opendocument [Accessed: October 27, 2003].

Green, J. and D. Hilchey. 2002. *Community, food and agriculture program.* Cornell University. Ithaca, NY.

Hamilton, N.D. 2002. Putting a face on our food: How state and local food policies can promote the new agriculture. *Drake J. Agric. Law* 7(2):407-453.

Hassanein, N. 1999. *Changing the way America farms: Knowledge and community in the sustainable agriculture movement.* University of Nebraska Press, Lincoln, NE.

Iowa Food Policy Council. 2004. http://www.iowafoodpolicy.org/htm [Accessed: March 21, 2004].

Jackson, L.J. 2002. Restoring prairie processes to farmlands. In: *The farm as natural habitat. Reconnecting food systems with ecosystems,* D.L. Jackson and L.L. Jackson (eds.). Island Press, Washington DC. pp. 137-154.

Jackson, W. 1980. *New roots for agriculture.* University of Nebraska Press, Lincoln, NE.

Jervell, A.M. and D.A. Jolly. 2003. *Beyond food: Towards a multifunctional agriculture.* Working paper 2003-19, Center for Food Policy, Norwegian Agricultural Economics Research Institute.

The Land Institute. 2003. *Annual Report.* Fiscal Year July 1, 2002 through June 30, 2003. Available from: http://www.landinstitute.org [Accessed: October 13, 2003].

Leopold, A. 1949. *A Sand County almanac and sketches here and there.* Oxford University Press, New York.

Meares, A.C. 1997. Making the transition from conventional to sustainable agriculture: Gender, social movement participation, and quality of life on the family farm. *Rural Sociol.* 62:21-47.

Midwest Food Alliance. 2001. Web site home page. Available from: http://www.the foodalliance.org/Midwest.html [Accessed: November 04, 2003].

Miller, A. 2003. Federal and State program Opportunities with Organics. *Third Annual Iowa Organic Conference,* November 17, Iowa State University.

Miller, L. 2003. A tale of two local food projects: Communities as well as producers see benefits. Summer 2003 *Leopold Letter,* Vol. 15(2). Available from: http:// www.leopold.iastate.edu/newsletter/2003-2leoletter/localfood.html [Accessed: October 26, 2003].

Pirog, R. 2004. *Value Chain Partnerships for Sustainable Agriculture Project.* Annual Report February 2003 to January 2004 (P0102464). Leopold Center for Sustainable Agriculture, Ames, IA.

Pirog, R. and T. Van Pelt. 2002. *How far do your fruit and vegetables travel?* Revised Reprint from Spring 2002 Leopold Letter. Leopold Center for Sustainable Agriculture. Available from: www.leopold.iastate.edu [Accessed: October 26, 2003].

Pirog, R., T. Van Pelt, K. Enshayan, and E. Cook. 2001. *Food, fuel and freeways: An Iowa perspective on how far food travels, fuel usage, and greenhouse gas emissions.* Leopold Center for Sustainable Agriculture, Ames, IA.

Ploeg, J.D. van der. 2000. Revitalizing agriculture: Farming economically as starting ground for rural development. *Sociol. Rural.* 40 (4, October): 497-511.

Preconference Workshop: Policy Issues in the Changing Structure of the Food System, July 29, 2000. Available from: http://www.farmfoundation.org/tampa/fulton.pdf [Accessed: October 24, 2003].

Pretty, J. 2002. *Agriculture: Reconnecting people, land and nature.* Earthscan, London.

Rutter, P.A. and M. Shepherd. 2002. *Hybrid hazelnut handbook* (1st ed) (draft). Badgersett Research Corporation, University of Minnesota Experiment Station, and The State of Wisconsin Alternative Crops Project. Available from: http://www.badgersett.com/HazHandbook1.html [Accessed October 13, 2003].

Sachs, C. 1996. *Gendered fields. Rural women, agriculture and environment.* Westfield Press, Boulder, CO.

Tanzilo, B. 2002. Shepherd pioneering Wisconsin's hazelnut, chestnut industry. On Milwaukee.com. September 9, 2002. Available from: http://www.OnMilwaukee.com/visitors/articles/hazelnuts.html [Accessed October 13, 2003].

Vagnetti, C. and J. DeWitt. 2002. *People sustaining the land. A vision of good science and art.* Iowa. Acme Printing, Des Moines, IA.

Van Esterik, P. 1999. Right to food; right to feed; right to be fed. The intersection of women's rights and the right to food. *Agric. Human Values* 16:225-232.

Chapter 11

Creating Viable Living Linkages Between Farms and Communities

Michele Schoeneberger
Gary Bentrup
Charles A. Francis
Richard Straight

UNIQUE RURAL/URBAN INTERFACES

We may have become an urban nation, but we remain an agricultural land.

America's Private Lands: The Geography of Hope, 1996

One of the goals of sustainable agriculture is helping farmers with small- and medium-sized family operations to become more efficient in production and marketing so that they can continue to work on the land, produce crops and livestock, and make viable contributions to the food supply as well as to their local communities. Within this goal is the recognition of the role farmers also play in determining overall landscape health and producing the many nonmarket environmental goods society demands from these working lands. "Through their care and stewardship of the land, farmers and ranchers produce safe drinking water, clear-flowing streams, lakes full of fish, skies full of ducks and geese, and scenic landscapes" (NRCS, 1996). However, the connection or linkage between farm and community well-being is becoming more tenuous as fewer people in the total population have roots in the land and understand where and how these goods are pro-

---

Developing and Extending Sustainable Agriculture
Published by The Haworth Press, Inc., 2006. All rights reserved.
doi:10.1300/5709_11

duced. This physical and sociopolitical disconnection between communities and agriculture is most keenly evident at the rural/urban interface, where conflict emerges when people from the city move onto acreages or to the edge of communities where their closest neighbor may be a working farm with crops and livestock.

For several years, we have been studying the problems that arise at the rural/urban interface and have summarized these challenges in a book chapter (Schoeneberger et al., 2001) and in presentations at national conferences (e.g., Bentrup et al., 2001; Francis et al., 2003). As part of the solution, we propose the use of *ecobelts*—linear arrangements of perennial vegetation that when properly planned can ecologically and socially reconnect these fragmented landscapes. Ecobelts can be integrated into working landscapes to create or enhance the multi-functionality of these lands, thereby creating a mutually beneficial situation for both the rural and urban sectors. In this chapter we summarize the problems that can occur at the interface between rural and urban activities and the potentials of ecobelts to reconnect the people with farms and with nature to enhance sustainability of our working lands. We conclude by focusing on the key to implementing successful ecobelts: *creating people linkages to create working landscape linkages.* Several brief case studies illustrate how people have worked together to create viable working landscape linkages, transforming a zone of conflict into one of cooperation and learning.

PROBLEMS FACING OUR WORKING LANDS

The conflict between rural neighbors who are farming and those who are essentially urban people living on acreages arises from a series of very different expectations and activities. Some of the differences in perspectives between these two groups are listed in Table 11.1 (from Francis et al., 2003). Problems that can be harmful or unpleasant to urban homeowners or people on acreages are the physical damages to the immediate environment that result from farming: herbicide or insecticide drift, odors or insects from livestock, or dust from large feedlots or fields being cultivated or harvested. Others are more of an inconvenience, such as noise from equipment at odd hours or slow-moving equipment on roadways that inhibit people commuting to work. Most of these problems are seen by farmers and their families as part of the rural scene, some of the costs of doing business on the farm, and just things that go along with being a farm family. It may be difficult to understand why others do not accept them, and in fact the farm was often there first and others moved into the area. "Surely these urban people knew they were moving into farming country, or next to farms, when they pur-

TABLE 11.1. Sources of conflict and different perspectives of urban and rural residents.

Source of conflict	Urban perspective	Rural perspective
Agriculture-Induced		
Livestock odors	Unnatural and disgusting	Natural part of the farm environment
Herbicide drift	Serious danger for lawn and yard plants	Hard to eliminate or control
Insecticide drift	Serious danger for pets and children outside	Hard to eliminate or control
Dust from fields	Causes health problems and hazards for motorists	Common result of tillage activities
Insects from livestock	General nuisance	Accepted part of farm environment
Noise from equipment	Disturbs outdoor activities	Normal part of farming operation
Pollutants from agricultural runoff	Water quality problems pose health risks and are expensive to correct	Attempt to minimize but hard to control
Slow-moving equipment	Road hazard, slows traffic	Essential to reach fields
Urban-Induced		
High-speed traffic	Need to commute to work	Dangerous to farm operations
Dogs in fields	Normal for dogs to explore	Harmful to livestock
Garbage in/near fields	Over-the-fence, out-of-mind	Interferes with farm operations
Increased runoff volumes	Increase in impervious cover part of urbanization	Creates excessive erosion and loss of farmland along streams
Equipment security	Kids need to explore and learn	Danger of damage to expensive equipment and facilities
People crossing fields	Desirable open space for hikes, skiing and snowmobiles	Invasion of private property, danger to crops and livestock
Gates left open	Kids will be kids and need to learn responsibility	Danger of losing livestock on roads, liability issues
Complaints to authorities	Normal approach to solving problems	Interrupts farm operations

Source: Adapted from Stokes et al., 1997, and Schoeneberger et al., 2001.

chased lots or acreages—they should have no complaints. We have been here for several generations," a farmer was heard to say.

From the farmer's perspective there are some real problems with urban neighbors. Dogs or people crossing fields can cause harm to crops, or open gates can result in livestock running loose or even getting lost. Trash thrown over the fence onto what is *just farm land anyway* can be a nuisance when it plugs up a planter or cultivator, and when blowing plastic gets into crop fields, combines, or feedlots. Curious neighborhood children can often compromise the security of equipment or buildings, which can be difficult to control on farms and ranches. Even moving equipment from one field to another using county roads or crossing the blacktop can be a problem for traffic safety—many roads are barely wide enough to accommodate two lanes of traffic, much less a wide planter or combine header. These are challenges faced by farmers who have an increasing number of urban neighbors, and they become real constraints to those farming as conflict arises, and resources need to be invested in solving the problems. Unfortunately, litigation is often seen as the only way to deal with conflict arising from these challenges. Litigation is rampant in the United States, as we become increasingly unwilling to settle our differences in personal and reasonable ways, and the resulting lawsuits and increased insurance premiums create a lose-lose situation for all parties (Schoeneberger et al., 2001; Stokes et al., 1997). We believe that ecobelts planted between these two contrasting types of activities can help bridge the physical and conceptual gap between people with different perspectives and expectations, minimizing conflict while providing tangible and mutual benefits.

ECOBELTS: CONNECTING THE WORKING LANDS

Vegetation-based buffers or corridors are one approach to reconnecting working landscapes. This basic concept has been used for many centuries from the ancient hedgerows in Europe to the shelterbelts in the Great Plains during the 1900s. More recent examples include the development of linear parkways or greenways in urban communities (Smith and Hellmund, 1993). Our concept builds on this foundation of vegetation-based buffers and greenways to create a more holistic system of green infrastructure that transforms the zone of conflict into one of shared ownership and use. We define this as the concept of *ecobelts* (Schoeneberger et al., 2001). As Figure 11.1 suggests, carefully planned and designed ecobelts can address a wide range of issues from education to visual quality, while creating a sense of place and community. Due to the diversity of potential issues ecobelts can address, they can

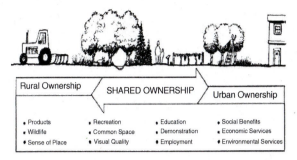

Rural Ownership	SHARED OWNERSHIP		Urban Ownership
• Products	• Recreation	• Education	• Social Benefits
• Wildlife	• Common Space	• Demonstration	• Economic Services
• Sense of Place	• Visual Quality	• Employment	• Environmental Services

FIGURE 11.1. The rural-urban interface: A zone of shared ownership. *Source: modified from Schoeneberger et al., 2001.*

take many forms, such as community shelterbelts, living snow fences, riparian buffers, and revitalized railroad trails, to name a few examples (Bentrup et al., 2001). Exhibits 11.1 and 11.2 provide specific applications.

By adding structural and functional diversity to the landscape, these tree-based plantings can perform ecological functions that have environmental and socioeconomic significance far greater than the relatively small amount of land they occupy. Table 11.2 provides examples of specific benefits. However, realizing this potential is a complex task of determining what opportunities, limitations, and trade-offs exist in each situation, and of designing an ecobelt system that achieves the best balance among them. A planning and design strategy that is flexible but comprehensive should be used for ecobelts, such as the process described by Schoeneberger et al. (2001).

ECOBELT PRINCIPLES

Ultimately, for ecobelts to be successful they must be culturally sustainable. That is, the ecobelts must elicit sustained human attention over time or else the benefits may be compromised as land ownership changes, as development pressure increases, or as different political viewpoints arise (Nassauer et al., 2001). Only with long-term agreements will ecobelts make a substantial and lasting contribution to the management of real landscapes. To promote successful reconnection of working landscapes, several key principles should be considered when designing and implementing ecobelts (Bentrup et al., 2001).

EXHIBIT 11.1. *Working Trees*—Creating Win-Win Practices

Animal operations may have the "smell of money," but to many people, especially at the rural/urban interface, this "smell" can be a major source of conflict. Not only is it the odor, but also dust, noise, and unpleasant views created by animal operations that add to the tension between these rural/urban neighbors. *Working tree* plantings can buffer these problems, creating not only an opportunity for animal producers to demonstrate their commitment to being a good neighbor and an environmental steward but also an opportunity to increase their production efficiency. By modifying the microclimate around the operations, these plantings create better living conditions for the animals while reducing heating and cooling costs for the producers.

This win-win situation created by planting trees around animal operations was recently highlighted in Delaware's *News Journal.* In response to a complaint about the odor coming from a nearby poultry farm, Bud Malone, a University of Delaware poultry extension specialist, promoted the idea of planting trees around the chicken house. This was done not only to address the odor but also to limit the dust and feathers. By serving as wind screens, these same plantings can also help in limiting the spread of airborne animal diseases within and between farms. Like many other places across the United States, Delaware is experiencing rapid acreage or ranchette development out into farm lands, and farmers are having to become "more sensitive to complaints from urbanites who appreciate country living, but not necessarily the smell, noise or sights associated with it."

Modified from: Tadesse, L., "Chickens Grow with Trees." *The News Journal,* July 8, 2003.

Shared Ownership

A primary tenet of ecobelts is shared ownership of the ecobelt between urban and rural residents. Shared ownership is often a necessary component to build a sense of community and responsibility for planning, implementing and maintaining an ecobelt system. If the rural or urban residents do not have a stake in the ecobelt system, the potential to replace the zone of tension with a neighborhood of cooperation is greatly diminished. Shared ownership can take many forms and does not necessarily have to imply traditional deed ownership. A sense of shared ownership can be created simply through the planning and design process that carefully incorporates rural and urban concerns. Part of the ecobelt planning pro-

EXHIBIT 11.2. Community Shelterbelts—
Western Minnesota

Community shelterbelts are plantings of single or multiple rows of trees or shrubs in a farm field, but adjacent to a community. Community shelterbelts are commonly established to minimize the negative impacts from excessive wind, and reduce blowing snow, dust, agricultural pesticides and debris to the local community. They also provide recreational opportunities, create wildlife habitat, and produce useful products for small towns and neighborhoods, reduce home heating costs for residents, enhance the aesthetic diversity of otherwise somewhat monotonous expanses, while at the same time reduce conflicts between agricultural producers and residents of the community.

Since 1990, at least ten rural communities have established community shelterbelts on the agriculturally dominated plains of western Minnesota, usually on the north and west sides of town (the most common prevailing wind direction in this area). A diversity of coniferous and deciduous trees and shrubs are used to enhance aesthetics, and provide fruit, nuts and other products that are valuable to both wildlife and people. The plantings have generally met expectations, particularly for protection against wind and blowing snow. Because these are community-based and community-driven initiatives, they have built community cohesion, cooperation, and pride. Many of these communities have in the past been literally buried by drifting snow creating dangerous conditions for residents, and creating huge snow removal costs for counties and communities. The older plantings have effectively reduced snow deposition within the communities, significantly reducing the burden of snow removal costs. Indeed, a recent study on the benefits of living snow fences in Minnesota shows a benefit/cost ratio of 17/1 to 29/1 for plantings established on private lands. This analysis only considered the reduced costs of snow removal and not the reduced commerce, accidents and casualties due to blowing and drifting snow, nor environmental benefits these plantings also provide.

Source: Josiah S.J., L.J. Gordon, E. Streed, and J. Joannides.1999. *Agroforestry in Minnesota: A Guide to Resources and Demonstration Sites.* University of Minnesota Extension Service, St. Paul, Minnesota. www. extension.umn.edu/environment.

cess will be educational, where stakeholders learn to consider the issues from each other's point of view. This face-to-face dialogue allows a common definition of the issues to be created and addressed, instilling ownership in the ecobelt proposal.

TABLE 11.2. Examples of specific benefits of ecobelts at the rural/urban interface.

Function	Benefit/effect	Reference
Provide corridors for wildlife movement.	Bears in Louisiana used wooded corridors to move between habitat patches in an area heavily modified by agriculture.	Anderson, 1997
Filter pollutants from agricultural runoff.	In a modeling study of a 19,132 acre watershed in Missouri, riparian buffers had the potential to reduce atrazine from 44.44 ppb to 24 ppb for a one-time economic savings of $654,779.	Qui and Prato, 1998
Improve aesthetics of the landscape.	Studies have shown that shelterbelts add positively to the scenic beauty of the Great Plains landscape.	Cook and Cable, 1995
Provide short-term flood storage.	Riparian buffers along the St. Charles river in MA were calculated to provide a value of $79,655/acre for short-term flood protection.	Thibodeau and Ostro, 1981
Improve stormwater management.	Existing vegetation in Salt Lake City, UT reduced stormwater runoff by 17 percent (43.2 million liters) during a 6-hour storm event.	USDA, 1985
Improve air quality.	Assuming 1990 air pollutant concentrations, model simulations for Sacramento's urban forest estimated that approximately 1,457 metric tons of air pollutants are absorbed annually, at an implied value of $28.7 million.	Scott et al., 1998
Provide recreational opportunities.	A study of the 26-mile Heritage Trail near Dubuque, Iowa found a total economic impact of estimated $1.3 million.	Moore et al., 1994
Provide opportunities for environmental education.	Researchers in Massachusetts determined a value of $3,980/acre for nature study along riparian buffers of the St. Charles river in MA.	Thibodeau and Ostro, 1981
Manage drifting snow.	For every $1 spent on living snow fences, $17 dollars are saved annually on snow removal costs, greatly improving road safety.	MNDOT, 2003

Function	Benefit/effect	Reference
Increase property values.	A study of greenbelts in Boulder, CO, determined that property values adjacent to the buffer had an average property value 32 percent higher than property located 3,200 feet away.	Correll et al., 1978
Provide employment.	The community-owned forest in Weston, MA, employs students to harvest organic produce and sap for maple syrup production.	Donahue, 2000
Reduce wind erosion.	Velocity reductions for average tree shelterbelts range from 60 to 80 percent near and to 10 times the height of the shelterbelt on the leeward side.	Tibke, 1988
Promote bio-control of insect pests.	Studies have shown that trees increase the abundance of natural enemies of insect pests.	Dix et al., 1995
Increase crop production from micro-climate modification.	Windbreaks can provide net yield increases in crops (i.e.,15 to 25 percent for winter wheat and 6 to 28 percent for soybeans).	Kort, 1988
Filter dust and other particulates.	Trees have reduced dust particulates and odors from poultry houses by 50 percent.	Tadesse, 2003
Source of decorative, medicinal and edible products.	Over 103 products from 73 plant species are collected by a wide variety of ethnic and socio-economic groups in Baltimore, MD, parks and greenways.	Community Resources, 2000
Source of products for commercial markets.	Woody florals grown in ecobelts can yield annual net returns ranging from $400-$3500 per 1,000 linear feet for one row of plants.	Josiah and Skelton, 2003
Provide energy savings.	Buffer plantings have been shown to reduce annual heating of nearby residences by 7 to 15 percent and summer shade can reduce cooling needs by 50 percent or more.	McPherson, 1988
Reduce noise levels.	A 5-m buffer of evergreens resulted in a 7.5 decibel decrease at 10m behind the screen.	Cook, 1978

Problems As Opportunities

The urban-rural interface zone and associated issues are often viewed as problems rather than as opportunities to create amenities for the community. For instance, dust originating from agricultural fields is considered a negative issue for nearby homeowners trying to keep their houses clean. However, it can be seen as an opportunity to mobilize residents into creating an ecobelt that can filter dust while also providing other environmental, social, and economic services. By reformulating the problem into a positive framework, residents can use the issue to bring resources together to benefit the larger community.

Agroforestry Products

Ecobelts can often incorporate agroforestry, which is the combination of agriculture and forestry technologies to create integrated, diverse, and productive land use systems (Garrett et al., 2000). Through careful management, products can be sustainably harvested from agroforestry systems. Such products include edible foods like berries and nuts, medicinal products such as ginseng and goldenseal, and horticultural materials such as evergreens for floral wreaths or colorful woody stems for the floral industry. An example of an agroforestry system is a riparian buffer planting that can attenuate flooding effects and protect water quality, while providing wildlife habitat and harvestable products, such as edible berries, medicinal herbs, and decorative willows. The integration of ecobelts and agroforestry systems can be a perfect combination to reconnect agriculture and urban communities. In addition to providing inexpensive and tangible goods for residents, the process of managing an agroforestry system can foster a sense of community (Corbett and Corbett, 2000). For instance, annual harvest parties can bring urban and rural residents together for a common purpose.

Landscape Linkages

Ecobelts should not be created as isolated elements in the landscape but should instead be designed as part of the green infrastructure: a network of connected corridors, working lands, and natural preserves that function as a system. Based on concepts from landscape ecology, connected ecobelts will offer more benefits than fragmented ones. Environmental services such as wildlife movement, reduced flooding, and improved water quality all benefit from connectivity (Forman, 1995). Pedestrians, joggers, and cyclists also benefit from ecobelt connectivity when the corridors are designed with pathways, especially when there are minimal road crossings.

A system of ecobelts offers the flexibility to meet the desired objectives of rural and urban residents. To accommodate various objectives, ecobelts will vary in width and size, much like a road system designed to carry different traffic flows. For instance, an ecobelt in one location may be a narrow corridor primarily designed to address noise and dust issues while producing community Christmas trees. In another location, a wide corridor may be required to provide opportunities for wildlife movement and recreational benefits.

Economic, Social, and Ecological Integration

Many successful community-based projects blend together economic, ecological, and social issues into a well-balanced system that addresses residents' goals for their area. Projects that emphasize one set of issues at the expense of other issues will rarely have the community support necessary to implement the plan. Community support is especially critical for ecobelts, which must satisfy a wide range of rural and urban objectives. Traditional greenway projects have succeeded in integrating ecological and social issues such as water quality, wildlife habitat, environmental education, and recreation (Smith and Hellmund, 1993). Economics is sometimes overlooked in this equation and yet may be a particularly powerful issue to reconnect urban communities and agriculture. Ecobelt economics can include employment opportunities for youth in maintenance of the ecobelts, increased property values, agroforestry products, and environmental services such as reduced costs for snow removal and water quality improvement, minimizing the need for expensive treatment. By exploring the range of economic, ecological, and social issues, the glue required to hold together divergent rural and urban interests may be discovered.

CREATING PEOPLE LINKAGES TO CREATE WORKING LANDSCAPE LINKAGES

Only in the success of our abilities to work together coupled with our skills in assessing the land, will we realize our public as well as individual conservation objectives.

America's Private Lands—The Geography of Hope, 1996

The discussion of ecobelt principles reveals that the success of ecobelts depends on bringing together a multitude and diversity of interests in order to reconcile society's needs with private land rights. Options can then be developed that not only reconnect the lands ecologically but also provide benefits that reconnect the neighbors. In essence, the planning and implementation of ecobelts are as much about creating functional relationships between urban and rural residents as creating physical features in the landscape.

Partnerships among landowners and local, state, and federal agencies are critical in the delivery of successful conservation strategies including ecobelts. Innovative partnering entities, such as the Resource Conservation and Development Councils (RC&Ds) and Regional Councils of Government can serve as a bridge between federal, state, and local resource management agencies and local land managers, and provide the facilitation and leveraging of resources needed in ecobelt implementation. The payoff from these partnerships is not just getting projects implemented on the ground, it is also the synergy created—the teamwork, relationships, and buy-in established among the partners. To illustrate these types of partnerships, several brief case studies are discussed, highlighting the linkages among different stakeholder groups.

USDA National Agroforestry Center

Because of its breadth of issues and projects, agroforestry is a natural activity for involving many people. Agroforestry, the basis of ecobelts, crosses many boundaries: scientific disciplines, such as agriculture and forestry, landowners and land users both rural and urban, and the agencies that develop and deliver the agroforestry technology. It must necessarily involve a large suite of participants if it is to be successfully deployed on our working landscapes to meet both landowner and societal objectives.

With its origins in the 1990 Farm Bill, the USDA National Agroforestry Center (NAC) was created in 1992 with a mission to accelerate the development and application of agroforestry technologies. The Center's model is to serve as a catalyst and facilitator in this effort, relying on partnerships rather than solely on internal infrastructure to deliver its programs. NAC began as a partnership between the research arm of the USDA Forest Service (Rocky Mountain Research Station) and the delivery arm—State and Private Forestry. In 1995, NAC expanded into a formal partnership with the USDA Natural Resources Conservation Service.

The tight coupling between research and development and with technology transfer, the leveraging of resources, and the reduction in duplication of efforts to produce agroforestry technical support by both agencies have resulted in accelerating the development and delivery of improved technologies. However, to accomplish its mission more fully, NAC further relies on a national network of partners and cooperators to conduct research, develop technologies and tools, establish demonstrations, and provide technical training. It is only with the multiplier effect obtained through these partnerships that NAC, with its small staff, has been able to develop and deliver its program nationally (see Exhibit 11.3).

EXHIBIT 11.3. Partnering to Train Today's and Tomorrow's *Working Tree* Professionals

Agroforestry, or *working tree* technologies, has been recognized as a highly versatile land use option that can assist landowners in balancing production objectives with environmental stewardship. This is especially true to the small-to-mid resource farmers who are having a difficult time surviving today's agricultural environment. One of the primary teaching/extension delivery systems to this group, especially minority farmers, is the 1890 Land Grant Colleges and Universities. In order to increase agroforestry awareness among the 1890 universities and to encourage the faculty to incorporate agroforestry into their courses and extension efforts, the USDA National Agroforestry Center (NAC), designed and coordinated an annual workshop that provides the background, field experience, and curriculum materials for specific agroforestry practices.

Hosted by Alabama A&M University, the annual workshop is produced through a partnership with USDA Forest Service (Washington Office of Civil Rights, Southern Research Station, and Region 8), USDA Cooperative States Research Education and Extension Service (SARE), and NAC (USDA FS - State & Private Forestry and Rocky Mountain Research Station, and USDA Natural Resources Conservation Service). To date, there have been 4 workshops delivered: *Agroforestry: Blending Agriculture & Forestry* (2000), *Riparian Forest Buffers* (2001), *Silvopasture Systems* (2002), and *Agroforestry Solutions for Communities: Green Infrastructure/Stormwater Management* (2003).

The workshops already have had a tremendous multiplier effect. Several faculty members have incorporated one or more of the agroforestry sections into their existing courses. Another member has included a chapter on agroforestry in a textbook prepared for Prentice Hall. As a result of the workshop, the 1890 Agroforestry Consortium was formed by the participants. This consortium can play a key role in helping to deliver agroforestry technology to 1890 faculty and students, in particular the underserved landowners throughout the United States, greatly extending NAC's capabilities to help teach today's and tomorrow's professionals and to reach today's and tomorrow's users.

Recently, NAC has developed planning and design tools for ecobelts that foster dialogue and linkages among the stakeholder groups while providing information on the complex and dynamic interactions and potential trade-offs of ecobelts (http://www.unl.edu/nac/conservation/). These tools incorporate multiple issues because stakeholders by necessity consider a variety of economic, biophysical, and social issues in their decision-making process. End-users have been directly involved in the tool development process to prevent making ineffective tools that do not respond to users' problems and needs, creating a waste of project funds and bitter feelings between developers and users (Hoag et al., 2000; Turner and Church, 1995). Further, to encourage participation in the ecobelt planning process, tools need to allow the exchange of ideas among stakeholders without the risk of anyone being ostracized (Buchecker et al., 2003). One tool that promotes this type of collaborative environment is the computer-based visual simulation that can depict photo-realistic future scenarios. NAC is currently developing a low-cost simulation software application *(CanVis)* that will enable stakeholders to conceptualize the ecobelt proposal and make valuable contributions to the design, influencing acceptance and adoption. In essence, the right tools can facilitate the necessary people linkages to create sustainable landscapes.

SARE Agroforestry Producer Grants

As with any new or changing agricultural technology, the success of technology transfer efforts, and ultimately the adoption of the technology, depends on gaining local interest and support. Early in its existence, NAC targeted part of its technology transfer and applications (TT&A) funding for establishing agroforestry demonstrations throughout the United States. These demonstration projects were recognized as an excellent teaching tool benefiting a variety of cooperators, stakeholders, and natural resource professionals. Because demonstration projects are continuously present in their locale, they are available even when a NAC agroforester is not. Consequently, they are a cost-effective way to increase public awareness of agroforestry, promote adoption by landowners, and obtain support from stakeholders. Further, agroforestry demonstrations themselves can assist in developing and furthering the technology and our understanding of its application. A demonstration of planting establishes a working example of an agroforestry technology under local conditions, showing *what* it is, *why* it is used, and *how* it functions (Irwin, 1997).

Declining budgets coupled with increasing workloads in the conservation sector prompted NAC to heavily rely on partnerships to create successful demonstrations. Since it was funded in 1988, the USDA's Sustainable Agriculture Research and Education (SARE) program has sponsored hundreds of projects to explore and apply economically profitable, environmentally sound, and socially supporting farming systems. SARE's primary landowner/producer audiences are commonly early adopters of new ideas who are willing to take risks to explore new technology; early adopters are a prime audience for NAC's TT&A efforts. Beginning in 1999, NAC began partnering with SARE to provide special opportunities to advance *working tree* technologies via the SARE Farmer Producer Grants Program. Building on SARE's national program, innovative clientele, and matching funds, NAC was able to get a wide variety of agroforestry demonstrations established throughout the four SARE Regions (Table 11.3). With over 60 agroforestry demonstration projects in place nation-wide, it was decided in 2004 to shift the partnered funding over to the Professional Development Grants Program to advance the working tree concepts. This program targets another client base, the natural resource professionals who assist producers in

TABLE 11.3. Examples of SARE Agroforestry Producer Grants (1999-2003) throughout the SARE regions.

SARE region	State	Title of project
North Central Region www.sare.org/ncrsare/	Minnesota	Growing various species of Angelica as a forest crop in the Midwest
	Michigan	Building a thermal blast peeler to prepare chestnut for on-farm, value-added processing
	Missouri	Increasing farm production by converting fenceline brush to agroforestry
	Kansas	Using agroforestry to winter cattle
Northeast Region www.uvm.edu/~ nesare/index.html	Massachusetts	Multi-purpose windbreaks for protection of vegetable crops and production of fruit and/or nut crops
	Maine	Improving financial returns in an orchard's life through agroforestry
	New York	Enhancing meat goat production through controlled woodland browsing
	Connecticut	Increasing small farm profits with American Chestnut production and silvopasture

TABLE 11.3 *(continued)*

SARE region	State	Title of project
Southern Region www.griffin.peachnet .edu/sare/	Puerto Rico	Demonstrating the benefits of agro-forestry practices on family farms
	North Carolina	Oriental persimmons and paw-paw: two sustainable crops for the south
	Kentucky	Marketing timber after adding value through the use of one-person sawmills and solar kilns
	Florida	Performance of various forage combinations under thinned pine canopies
Western Region http://wsare.usu.edu	Guam	Evaluation and implementation of nitrogen-fixing species in hedgerow intercropping
	Washington	Tilth-agroforestry niche demonstration project
	Hawaii	Grow your own sustainable barn
	Northern Marianas	Luta windbreak/agroforesty project

Source: USDA, 2003.

applying agricultural technology. By increasing the professional's agro-forestry competence, NAC can begin to reach another producer audience, the nonsustainable agricultural producer. In addition, the newly trained resource professionals will be encouraged to utilize the existing demonstrations and practices, thus increasing the value of the demonstration. Readers are referred to *Inside Agroforestry, Summer/ Fall 2003* (http://www. unl.edu/nac/ia.html) for additional overview of this partnership.

Stormwater Management to Reconnect Land and People

Agricultural activities are a primary contributor to many of the water quality problems that we face today, and many programs have been initiated to try and deal with them. Cities and towns throughout the USA are now facing similar Environmental Protection Agency (EPA) requirements to address their contributions to water quality decline, more specifically stormwater contributions. Though many communities occupy only a small portion of the watershed, they can greatly affect their watershed and are, in turn, affected by the activities of others in their watershed. The water quality goals of both rural and urban governments are often quite similar: reduce soil erosion, reduce soluble contaminants, and reduce flooding. But rarely do the same governmental agencies work together with coordinated policies and cross-boundary efforts. Water quality is an all-lands issue and can only be truly addressed when all residents, urban and rural, learn how their land-use decisions affect one another and how they might work together to achieve common goals.

Ecobelts can be very effective in protecting and enhancing water quality, while providing additional amenities being sought from these rural and urban working lands, such as wildlife habitat and travel corridors, and providing recreational opportunities and alternative sources of income. In 2002, the City of Topeka, Kansas, population 125,000, initiated GREEN TOPEKA to address water quality and quantity concerns (Yoko, 2002). GREEN TOPEKA, a partnership initiated with NAC along with state agencies, Kansas State University, local government, nonprofit organizations, and private stakeholders, is developing stormwater management alternatives that incorporate green (vegetative) technologies with conventional engineering approaches. More important, the approach is being applied in a more holistic manner where multiple concerns and broader landscape impacts are included in the upfront planning in order to avoid the costly retrofitting that generally results after urban growth has taken place. In the development of new ordinances and new green stormwater management projects, the City of Topeka has sought input from residential developers, local industry, and neighborhood organizations. In the process GREEN TOPEKA has developed new linkages among the city, its citizens, businesses, and their water resources, as well as fostering support for ongoing and future stormwater management activities. Besides flood and erosion control and water quality protection, GREEN TOPEKA's efforts will also provide additional walking/biking trails, interpretative paths, environmental education, wetland systems, increased wildlife habitat, and more aesthetically pleasing settings.

The second case study is a look at how this model of linking a community to its watershed through stormwater management issues and ecobelts can be scaled up to link multiple governments, organizations, and businesses in multiple watersheds. The Mid-America Regional Council (MARC) is the metropolitan planning organization for the 116 city and eight county governments in the bi-state Kansas City region of which 70 percent of the planning area is agricultural (Figure 11.2) (Yoko, 2002). Using water quality and storm-

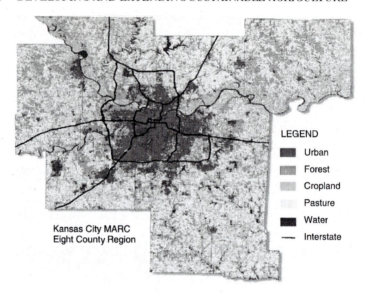

LEGEND

- ■ Urban
- ▨ Forest
- ▨ Cropland
- ▨ Pasture
- ■ Water
- --- Interstate

Kansas City MARC
Eight County Region

FIGURE 11.2. The Kansas City metro planning region depicting the urban/rural interface and mixture of land uses that need to be brought into the planning exercise when addressing environmental issues like water quality.

water issues as a catalyst, MARC is working in close partnership with GREEN TOPEKA, the EPA, the American Public Works Association (APWA), the University of Missouri Center for Agroforestry, and NAC to develop best management practices (BMPs) and engineering guidelines for applying green infrastructure technologies to manage stormwater. MARC had just celebrated a successful Metro Green initiative of developing a network of linear corridors along streams, railroad right of ways, and other greenways, where a number of groups took advantage of an opportunity to develop this new partnership. The public's new-found appreciation of the Metro Green trail system made an easy transition into promoting the use of these same greenways as a link in the larger green infrastructure for stormwater conveyance and other watershed functions. Metro Green became the foundation of linking the management of urban parks, school grounds, waterways, and other public open space to the management of adjacent rural lands and to the rural farm programs.

Through the new green infrastructure partnership MARC is coordinating a strategy to build a capacity within stakeholders for implementing vegetative stormwater management strategies. Three primary components of this strategy are (1) the development of a GIS data set of existing green spaces,

(2) a grants program to establish stormwater demonstration projects that utilize vegetated solutions, and (3) hosting design charettes to build capacity within the local engineering community for green stormwater solutions.

Through a GIS-based planning framework, natural systems and processes considered at the landscape level can help guide the location of ecobelts and other vegetated plantings that can better connect the landscape ecologically. MARC's history of effectively working across multiple government boundaries, including rural and urban entities, will position communities within the Kansas City Metro area to talk with rural watershed partners about shared goals and priorities which can hopefully lead to greater targeting of federal cost-share investments, such as the Conservation Reserve Program (CRP), Wildlife Habitat Improvement Program (WHIP), Environmental Quality Improvement Program (EQIP), as well as EPA programs for clean water.

The establishment of stormwater management demonstrations, such as the SARE funded agroforestry demonstrations, will build a foundation of experience and educational opportunities to expand the utilization of green infrastructure for water quality. EPA funding through MARC initially supported the green stormwater solution projects. A call for proposals went out to county, city, and nonprofit groups to find opportunities to modify traditional gray or hard infrastructure stormwater projects with ecobelts or other technology utilizing vegetation. As a part of the call, applicants were informed that engineering redesign assistance would be provided through MARC. At the same time MARC sent out a request for qualifications to the local engineering consulting community to enlist a firm or firms to provide the expertise necessary for redesigning the selected stormwater project proposals.

The third aspect of the capacity building strategy focused on the engineering community involved with residential and industrial development and consulting for local governments. MARC will host two or three design charettes in the first two years of the initiative and invite engineers and consultants working in the Metro area to participate. Each charette will educate the participants as they work through the development of the design specifications for an actual stormwater project or new development. MARC's demonstrations, training, and common GIS data are encouraging people and governments throughout the metro region to utilize similar technologies in addressing shared goals of water quality throughout their common watershed.

CONCLUSIONS

Based on the practical community applications of woody buffers and multifunctional areas that grew from the initial concepts of ecobelts, we are convinced that this type of initiative is a viable option in establishing a positive interface between different types of human activities. The challenge becomes more real every day, as cities expand and both city-edge subdivisions and acreages create ever more interfaces between people with different and often conflicting activities and basic values. There is a need to provide some logical and acceptable boundary between these people and their different lifestyles. It is also desirable to provide some tangible limits to physical spread of cities, and a need to explore creative ways to infill and increase urban density (Olson and Lyson, 1999).

The case studies in Topeka and Kansas City metro areas provide tangible evidence that innovative partnering and design of green space can effectively reduce conflict at the rural-urban interface as well as serve multiple functions for the local community. By further expanding and testing the idea of ecobelts, we can find additional ways of helping our complex society cope with the complexities of spatial organization of human activities for the future. By recognizing that the most important linkages are those among people who share limited space, we can focus on the process necessary to bring people together to seek common understanding and work in unison to solve the community's problems. Ecobelts can be an important part of these solutions.

REFERENCES

Anderson, D.R. 1997. Corridor use, feeding ecology, and habitat relationships of black bears in a fragmented landscape in Louisiana. *Masters thesis.* University of Tennessee, Knoxville.

Bentrup, G., M.M. Schoeneberger, S. Josiah, and C. Francis. 2001. Ecobelts: Reconnecting agriculture and communities—case studies. In: *Proceedings of the Ecospheres Conference,* June 10-12, W.C. Steward and A. Lisec (eds.). University of Nebraska, Lincoln, NE. pp. 1-12.

Buchecker, M., M. Hunziker, and F. Kienast. 2003. Participatory landscape development: Overcoming social barriers to public involvement. *Landsc. Urban Plan.* 64:29-46.

Community Resources. 2000. *Exploring the value of urban non-timber forest products.* Community Resources, Baltimore, MD.

Cook, D.I. 1978. Trees, solid barriers, and combinations: Alternatives for noise control. In: *Proceedings of the National Urban Forestry Conference,* Washington, DC. pp. 330-339.

Cook, P.S. and T.T. Cable. 1995. The scenic beauty of shelterbelts on the Great Plains. *Landsc. Urban Plan* 32:63-69.

Corbett, J. and M. Corbett. 2000. *Designing sustainable communities: Learning from Village Homes.* Island Press, Washington, DC.

Correll, M.R., J.H. Lillydahl, and L.D. Singell. 1978. The effects of greenbelts on residential property values: Some findings on the political economy of open space. *Land Econ.* 54(2):207-217.

Dix, M.E., R. Johnson, M. Harrell, R. Case, R. Wright, L. Hodges, J. Brandle, M. et al. 1995. Influences of trees on abundance of natural enemies of insect pests: A review. *Agroforest. Syst.* 29:303-311.

Donahue, B. 2000. History, work, and the nature of beauty. *J. Forest.* 98(11):36-41.

Forman, R.T. 1995. *Land mosaics: The ecology of landscapes and regions.* Cambridge University Press, Cambridge, MD.

Francis, C., G. Bentrup, M. Schoeneberger, and M. DeKalb. 2003. Integration of woody buffers at three levels of spatial scale in the urban/rural interface in Lincoln—Lancaster County, Nebraska. In: *Proceedings of the North American Agroforestry Conference,* S.H. Sharrow (ed.), 23-25 June 2003, Corvallis, Oregon, pp. 116-127. Available at: http://www.aftaweb.org/8thNAACproc/Francis.pdf [Verified October 20, 2004].

Garrett, H.E., W.J. Rietveld, and R.F. Fisher (eds.). 2000. *North American agroforestry: An integrated science and practice.* American Society of Agronomy, Inc. Madison, WI (abstract on disk).

Hoag, D.L., J.C. Ascough II, and W.M. Frasier. 2000. Will farmers use computers for resource and environmental management? *J. Soil Water Conserv.* 55:456-462.

Irwin, K. 1997. Guide to a successful agroforestry demonstration project. *Agroforestry Notes* (AF Note-6), USDA National Agroforestry Center, Lincoln, NE. Available at: http://www.unl.edu/nac/afnotes/gen-2/gen-2.pdf [Accessed August 29, 2003].

Johnson, R.J. and M.M. Beck. 1988. Influences of shelterbelts on wildlife management and biology. *Agric. Ecosyst. Environ.* 22/23:301-335.

Josiah, S.J. and P. Skelton. 2003. Productive conservation: Diversifying farm enterprises by producing woody decorative florals in agroforestry systems In: *Proceedings of North American Agroforestry Conference,* S.H. Sharrow (ed.), June 23-25, 2003, Corvallis, OR. pp. 142-143. Available at: http:// www.aftaweb.org/8thNAACproc/Josiah.pdf [Accessed October 20, 2004].

Kort, J. 1988. Benefits of windbreaks to field and forage crops. *Agric. Ecosyst. Environ.* 22/23:165-190.

McPherson, E.G. 1988. Functions of buffer plantings in urban environments. *Agric. Ecosyst. Environ.* 22/23:281-298.

MNDOT. 2003. Living snow fences program. *Minnesota Dept.* Transportation. Available at: http://www.livingsnowfence.dot.state.mn.us/ [Accessed October 6, 2003].

Moore, R.L., R.J. Gitelson, and A.R. Graefe. 1994. The economic impact of rail-trails. *J. Park Recreation Admin.* 12(2):63-72.

Nassauer J.I., S.E. Kosek, and R.C. Corry. 2001. Meeting public expectations with ecological innovation in riparian landscapes. *J. Am. Water Resourc. Assoc.* 37(6): 1439-1443.

NRCS. 1996. America's private lands, a geography of hope. Program Aid 1548. 80p.

Olson, R.K., and T.A. Lyson (eds). 1999. *Under the blade: The conversion of agricultural landscapes.* Westview Press, Boulder, CO.

Qui, Z. and T. Prato. 1998. Economic evaluation of riparian buffers in an agricultural watershed. *J. Am. Water Resourc. Assoc.* 34(4):877-890.

Schoeneberger, M.M., G. Bentrup, and C.A. Francis. 2001. Ecobelts: Reconnecting agriculture and communities. In: *Interactions between agroecosystems & rural communities,* C.B. Flora (ed.). CRC Press, Boca Raton, FL. Chapter 16, pp. 239-260.

Scott, K.I., E.G. McPherson, and J.R. Simpson. 1998. Air pollutant uptake by Sacramento's urban forest. *J. Arboric.* 24(4): 224-234.

Smith, D.S. and P.C. Hellmund. 1993. *Ecology of greenways.* University of Minnesota Press, Minneapolis.

Stokes, S.N., A.E. Watson, and S.S. Mastran. 1997. *Saving America's countryside* (2nd ed.). Johns Hopkins University Press, Baltimore, MD.

Tadesse, L.B. 2003. Chickens grow with trees. *Delaware Online News Journal.* Available at: http://www.delawareonline.com [Accessed August 30, 2003].

Thibodeau, F.R. and B.D. Ostro. 1981. An economic analysis of wetland protection. *J. Environ. Manage.* 12:19-30.

Tibke, G. 1988. Basic principles of wind erosion control. *Agric. Ecosyst. Environ.* 22/23:103-122.

Turner, B.J. and R. Church. 1995. *Review of the use of the FORPLAN (FORest PLANning) model.* Department of Conservation and Natural Resources, Victoria, Australia. 25 p.

USDA. 1985. Salt Lake County named urban forestry research area. *Tree Leaves* 9(4):6.

———. 2003. Inside Agroforestry. Summer/Fall 2003. 12p. Available at: http://www. unl.edu/nac/ia/summerfall03/summerfall03.pdf [Accessed February 15, 2004].

Yoko, G. 2002. Addressing stormwater management the green way. *Land Develop Today.* 1(6) December. Available at: http://www.landdevelopmenttoday.com/ 2002/12/civil/civil_1.htm [Accessed 28 May 2006].

Chapter 12

Testing Ideas and Transferring Capacity Through Farmer Research: The Iowa Model

Rick Exner
Richard Thompson

INTRODUCTION

It might have been easier a few years ago than today to show how on-farm research can lead to sustainable systems, because our definitions were simpler. In the 1980s, the term *sustainability* was new to serious discussions of agriculture. Some people understood that it meant more than simply environmental sustainability, but many of us did not think beyond the apt definition of sustainable agriculture attributed to Wendell Berry—an agriculture that sustains the soil, the water, and the people.

Practical Farmers of Iowa (PFI) was formed in 1985 to be a vehicle for farmer-to-farmer information sharing. About that time Charles Francis of the University of Nebraska observed in a PFI meeting that the problem often faced by sustainable farmers was not one of a shortage of information but of an information glut. He pointed out that the challenge is to extract from this information flood the subset of information that is valid, or useful to farmers in their own fields and conditions. To understand the point, consider the multiple information sources with which farmers are presented, including everything from in-laws and the local coffee shop to public and private sector representatives within the community, to farm publications, and now the Internet. The choices are many, and the stakes of making the right decision are high.

Practical Farmers of Iowa recognized early the need to help farmers validate information about farming practices, and working with university scientists PFI developed simple protocols that farmers can use to design and

Developing and Extending Sustainable Agriculture
doi:10.1300/5709_12

carry out their own research trials (Exner and Thompson, 1988; Rzewnicki et al., 1989). These replicated and randomized large-plot trials can be quite powerful from a statistical standpoint. Rzewnicki et al. (1989) found the typical, six-rep PFI trial equally capable of distinguishing small significant differences as the best experiment station trials.

The experimental designs employed by farmers are generally simpler than those found on experiment farms. An experiment design often reflects a narrow experimental question, typically a reductionist approach, and the question for many on-farm research trials is a straightforward A-versus-B. For example, "What's better, the practice I've been using or this new alternative that someone wants me to buy?" Armed with such simple research protocols, farmers can test ideas and innovations, gaining confidence in their ability to incrementally evolve their farming system. That evolution has in many cases contributed to the gradual transformation of farms.

As suggested earlier, the term sustainable agriculture as many originally understood it emphasized farming practices and the environment. Farmers and scientists pursued a *win-win* in which agriculture benefited the environment by reducing applications of chemical fertilizers and pesticides, and the farmer benefited financially by reduced input costs. As such, a major thrust of PFI on-farm research was the efficient use of inputs and the utilization of management, the farm's biological resources, and the structure of the farming system itself to substitute on-farm alternatives for purchased, off-farm inputs. Good progress has been made in areas related to input use efficiency. PFI research cooperators reduced nitrogen fertilizer use in corn by one-third, and they saved six dollars per acre substituting ridge tillage for herbicides. This philosophy continues to find expression around the world in research on agriculture's relation to the environment.

As early as 1990, however, comments were heard in PFI: "So what, if the whole state is farmed sustainably if every county has only a single farmer left?" The overall trend in U.S. agriculture is indeed toward production systems that are large scale and less diversified. However, Guy and Duffy in an unpublished report, *Labor Constraints on Farm Size and Alternative Farming Systems* (ISU Dept of Economics, 1988), have shown that the two-operator farm captures most economies of scale. Stender and Stevermer (1994) have shown that the scale of swine production systems is only weakly correlated with the cost of production. This suggests that creating a sustainable farm is more than a matter of *efficiency,* as commonly described by agricultural economists. Causes cited for difficulties of small farmers range from restricted access to markets to land ownership arrangements, but that discussion is beyond the scope of this chapter. There is a common opinion in coffee shops that the optimum size of a Midwest farm should be *bigger!*

Suffice it to say that the PFI on-farm research network has documented numerous practices that increase returns per acre or per livestock unit, but few of them have taken mainstream farming by storm. A key reason may be that most of these practices require additional management and labor, tipping the farming balance from extensive to intensive. And although labor inputs are part of the economic calculation for farm practices, the calculus of many farmers is such that if a practice requires time that could be used to farm an additional acre or raise additional livestock, that practice will not be adopted.

Not all producers want to increase the scale of their operations, but there are business and social forces that often make that irresistible. However, other small- and medium-sized producers have responded to this pressure with a change of another sort. They have found or created niche markets that reward them for being who they are (community-based, for example) and farming as they do, for example, with a farm plan, without antibiotics, or organically. These new markets are often associated with very specific production systems. And those production systems bring unique challenges and new questions.

And so we see a new wave of on-farm research, as producers with specific motivation from the marketplace seek to perfect new crops and overcome challenges of new production systems. In some cases these production systems involve crops new to the producer: summer squash, for example. In many instances the crops or livestock are familiar, but the new market requires special practices such as weed management without herbicides or animal disease control without antibiotics.

The research that is required may often be more challenging than on-farm research of a decade ago because it involves understanding biology and manipulating the farming system, and not merely manipulating known production inputs. One of the strengths of research by the farmer is its relevance to the unique circumstances of the farm. Farming systems that are simplified and standardized are likely to find "off the shelf" answers for the production challenges they confront. Farms that are structurally complex or farmers who are exploring new markets are likely to encounter fewer prepackaged answers. Producers in these operations need to know how to get their own answers. As mentioned above, most on-farm comparisons are narrowly defined: a comparison of varieties, weed management alternatives, soil fertility options, or tillage strategies. The advantage is that these components are tested within the context of each farm's unique conditions and systems.

So is there a necessary connection between farmer research and agricultural sustainability? Not to duck the question, but that may depend on how sustainability is defined. It is not controversial to state that sustainable farming is management-intensive farming. And in practice, it is often labor-intensive as well. It is not that sustainable farmers are just too busy to buy production inputs. They are looking for ways to increase the scope of their control over costs and the entire farming system. In this, sustainable farmers are no different from any others. The exercise of this control may start with the testing of purchased inputs. More often control evolves to test utilization of on-farm resources, substitution of management for inputs, and managing the interplay of the elements of a farming system of growing complexity. On-farm research is a catalyst in that progression, as elaborated below.

MATCHING THE DESIGN TO THE QUESTION: PROFILES OF FARMER-RESEARCHERS

Having provided some context for on-farm research in Iowa, we turn to examples. On-farm research can take many forms and can be useful in a variety of different farming operations. To illustrate the point, we asked several producers who have worked with the PFI program over the years to help us understand how they have used research. A review of the archives shows that individual research interests can evolve significantly over the years. In some cases research has helped transform the farm. In other cases, research showed what did *not* work, leading to elimination or reduction of an input, discontinuation of a practice, or change in an entire system. There are few safety nets in farming, but farmer research can help producers make informed choices.

A Farm Evolution, A Personal Journey

Doug Alert, Hampton, Iowa, did not have much handed to him, as Midwesterners refer to careers in farming. Doug did not inherit land or equipment, but he parlayed an ability with equipment (his father ran an earth-moving business) and neighbors taught him the basics to get a start in farming. When Doug began research with PFI in 1991, he was trying to justify ridge-tillage to landlords and perhaps to himself. His trials dealt with rates and placement of fertilizers in the ridge-till system. Soon, he picked up on the PFI interest in strip intercropping, and for several years demonstrated some of the most positive results with that challenging system. However, Alert decided strip intercropping was not for him.

By then, he was renting land from a retired couple who wanted to see the young farmer succeed, and he was married to Margaret Smith, a PhD agronomist with Iowa State University Extension. The fences went in, the farm acquired a beef herd, and the operation became a more diversified, self-sufficient system. Doug writes, "Ruminants are key to the functioning of the farm." Their more recent research has dealt with use of flame weeding, as the fields and crops transitioned to certified organic. Looking back, Alert comments, "Experience with small grains, reducing herbicide use in ridge-tillage, and better understanding of the fertility/crop response on our farm built (my) confidence to forge ahead and make the major changes in production methods." Involvement with the research network and with visiting classes and field day attendees have played a part in a personal evolution for Doug Alert as well. He writes, "The transformation of our farm and the research involved have definitely changed how a shy, reluctantly public farmer interacts with others. From the many impromptu, one-on-one questions all the way to international travel."

The Farmer's Most Useful Tool: Production Questions on a CSA Farm

Angela Tedesco is the operator of Turtle Farm, a fruit and vegetable operation that supplies her 110-member CSA (Community Supported Agriculture) marketing program. In 1998, Angela began a series of trials on two questions that confronted her and similar vegetable farmers. "The number of customers was growing in size. I was looking for more efficiency in greenhouse space and labor in the field," writes Angela. In pursuit of labor efficiency and organic-acceptable practices, Angela examined production practices in onions and then in summer squash.

Onions are a crop that demands considerable labor, particularly in organic production systems. The seedlings are tiny, and the immature root system and spindly leaves offer little competition to weeds. Angela devised what statisticians call a factorial trial in order to evaluate two different practices singly and in combination. She read that Maine vegetable producer Elliot Coleman plants not individual onions but soil blocks containing multiple plants. In this way he maintains the same effective crop population with less labor—both for planting and for weeding. Angela did not have the greenhouse capacity to use soil blocks as large as Coleman's, but she wanted to try planting into a smaller block. She also wanted to see if mulching onions would reduce overall labor requirements by suppressing weeds.

Angela Tedesco carried through this factorial trial for two years. The first growing season was typical of Iowa, with a dry, hot summer. Mulching saved labor and resulted in bigger onions; planting in soil blocks saved planting labor and increased transplanting survival. Year two, however, provided a relatively wet summer. Whereas mulching had conserved soil moisture in the dry season, in a wet summer mulch was a liability. Planting in soil blocks still saved labor, but in the wet year it provided no advantage for plant survival.

This trial and the subsequent one covering summer squash management were both two-year studies and both used factorial designs to evaluate practices that could be implemented by other organic vegetable growers. A university researcher could hardly have done more rigorous work. Tedesco continues to receive inquiries from other growers about the research.

Her own implementation of the methods she researched is flexible, depending on the weather and the availability of mulch and labor. "On-farm research of my own and others continues to contribute to the evolution of the farm as I seek to grow in size and efficiency." Her tips for others interested in on-farm research:

> Plenty of lead time in order to consider what research would be most helpful to her or his operation. Consultation among several farmers about what would be helpful to their operations could help determine that it would be useful to more people. Many people might not have statistical expertise, and PFI's help with this has been essential, from planning the plots to final analysis. Other consultants are also helpful, such as entomologists and plant pathologists at Iowa State. . . . Farmers never stop seeking more help in dealing with these issues, whether it's in farmer networking or in observation of what works and what doesn't on one's own farm. . . . The use of the research is constantly evolving because there are no two years alike in farming. The power of observation is the farmer's most useful tool.

Efficiency in the Conventional Economic Environment

While it is true that farmer research in PFI has increasingly reflected the demands of specialty markets, PFI is a diverse group, and many members simply strive to do the best job of producing crops and meat for commodity markets. Wayne and Ruth Fredericks, of Osage, fall into this category. The Fredericks' farm is a tightly run operation raising hogs, corn, soybeans, and some small grains. Wayne by nature monitors every penny and every minute. His reflections on research:

> I operate a basic conventional farm, yet use many practices which people consider to be sustainable. My internal drive is to find the economics behind these practices and put a dollars and cents value to them. . . . A lot of my research has surrounded the topic of hoop housing for finishing pigs. I put up my first hoophouse basically from gut instinct that it was the right thing to do for our farm operation. It complemented what we were already doing and fit with my mental philos-

ophy. In years following much of my research was directed towards evaluating and verifying that decision.

Our first research with PFI was to track labor requirements in hoop housing. There was a bulletin out using labor estimates with no real substantiation from any type of research. Other PFI on-farm researchers and I went to work and came up with real numbers. These showed that hoops required considerably *less* labor than originally thought. For myself, it gave me personal information to use in comparing my hoops to my other outside finishing systems.

Our next major research project with PFI came out of a joint meeting with university personnel and hoop users and revolved around the topic of what is the best way to handle hoop manure. Should we spread it fresh or is it more profitable to compost it first? I really got into this one. Being a detail person, I got out the stopwatch, notebook and pen, and really put forth a grand effort in getting as much information as I could to analyze this topic. I now know with confidence what to do with the hoop manure when it is removed from the hoops, whether to compost or not, whether to use on corn or beans, and at what hauling distance these economics change.

As I look at my trials over the years, I see a common thread. That is the desire to get information to help me make better management decisions on my farm, and if possible information to help other farmers and university researchers at the same time. Not all trials conclude with the results that we may have hoped for, but that in itself may be the very reason we need to conduct them. We need that information. As I move forward, our farm will be molded around what I learn from my trials. Not all of my trials will interest others, but my research experience with PFI has taught me how to set up trials, replications and all, and will help solidify my management decisions in the future.

Evolving an Integrated Farming System

Tom and Irene Frantzen operate a diversified farm near New Hampton, in northeast Iowa. A past president of Practical Farmers of Iowa, Tom has spoken to hundreds of other farmers at field days and at farm meetings around the country. In the late 1990s, he petitioned the USDA to allow organic certification of livestock, and he now spends much of his time helping other farmers produce for the organic pork market.

Until you visit with Tom Frantzen, it is hard to imagine how his farming has changed. "I *knew* that you couldn't grow corn on less than 180 pounds of nitrogen per acre." He also "knew" that row crops could not be raised without herbicides to control weeds. But beginning in 1989, and encouraged by the examples of other farmers in the research network, Tom began putting new ideas to the test. He found that he had the skills to make alternative practices work. "The weed control without herbicides changed my attitude about organic farming. We needed the confidence that we found in the weed control plots to inspire us to adapt the new idea of organic crop production."

This is not to say that every idea that came under his scrutiny was successful. A previously successful strategy is sometimes reconfigured as priorities change. Frantzen worked with ISU agronomist Richard Cruse to develop a three-crop system of strip intercropping. But before long, the system also included woody species and livestock, with sows *harvesting* the corn strips, and strips of trees providing a windbreak for livestock as well as fuel, hardwood, and nuts.

The Frantzens have been influenced by the Holistic Management program of Alan Savory, and Tom even organized a group of farmers who met regularly to apply the HM teachings. Yet Frantzen looks back wistfully to the 1990s, when he had more freedom to be experimental. Even though some consider him to be a marketing pioneer, he expresses the concerns of many small farmers today:

> The on-farm research time was one of having the luxury of time and financial resources to try out radically different ideas even if they had a high risk of failure. It is too hard to make a living today for us to have this free wheeling mode of operation. I could not drop a field off of our existing organic rotation without major problems unlike 10 years ago, when that sudden rotation-of-concept change caused little interference.

Taking the Pulse of a CSA

Farming itself demands many decisions, but consumer-supported agriculture (CSA) farmers also get personally involved with their consumers and with the logistics of nurturing and servicing that "market." Nestled in the hills of a glacial moraine in north central Iowa, *One Step at a Time Gardens* manifests all the complexity and strains of a typical CSA. The proprietors, Jan Libbey and Tim Landgraf, face an additional challenge in that their farm is far from the kind of population center that provides many CSAs with most of their members.

Jan and Tim undertook a quest to answer some basic questions about the functioning of their farm. For two years they tracked their labor for individual crops, and they examined their CSA members' preferences and willingness to pay for a CSA-quality carrot, raspberry, or cauliflower. In this self-examination, Jan and Tim were aided by a set of ratios developed by another

CSA producer, David Washburn, of *Red Cardinal Farm,* in Minnesota. They were also able to compare their farm to two larger and more established CSA farms that have undergone the same analysis.

They found that their farm compared well to the other CSAs in scale-neutral measures such as the ratio of gross income to acres. However, they documented that the small size of the farm meant that fixed costs were eating up a good deal of that gross income. They were also able to document improvement in certain parameters, and they became more proactive in developing alternatives that addressed weak points. Perhaps most significantly, Jan Libbey and Tim Landgraf have now extended the geographic reach of the CSA by enrolling consumers in a population center some distance from the farm.

Without such an analytical approach to their farming situation, Jan and Tim might have made similar changes, but it is likely that it would have taken longer for them to reach those decisions. Moreover, other CSA farmers would not have the record of that self-evaluation to guide them in their own development.

Developing On-Farm Resources

Paul and Karen Mugge farm near Sutherland, in northwest Iowa, where Paul was born. In 1975, when Paul left an engineering job in Seattle to return to farming, he started with a diversified but fairly input-intensive system. But Paul also brought to farming an environmental consciousness that did not square with the farming options available at the time. When Mugge began doing research with PFI in 1989, he was already using ridge tillage, a system that leaves the soil protected by crop residue over the winter. In other respects, Paul's cropping practices were typical of most Iowa farms, with regular expenditures for synthetic herbicides and fertilizers. The first experiment was a side-by-side comparison of chemical and mechanical weed control for soybeans, and trials have progressed from there to include small grains, cover crops, cropping systems, and manure management. By the time Paul decided to certify the crops as organic, he already had the skills and the crop rotation necessary for the transition. He modestly observes, "I mainly have taken small steps and have demonstrated to myself that it would work and that I could do it, which are not necessarily the same thing."

Mugge reports that his research questions "have come from a desire to lighten my environmental footprint." He is now a part-time high school biology teacher, and he brings to class some of the same principles and issues that he deals with in farming. He also hosts classes on the farm, and field days bring many others. Paul Mugge presently serves as an ex officio member of the board of directors of the Leopold Center for Sustainable Agriculture.

For most of Mugge's on-farm research, the simple paired comparison has been the appropriate design, but he has not hesitated to use factorial and split-plot designs where necessary. Though most of his research has been on his own ideas, Paul has also collaborated with local Extension and ISU scientists on many trials. In his experience,

> On-farm research is the most appropriate and cost-effective form of ag research for some questions. It is only as good as the data, however, so attention to detail and scientific method are critical. If you're going to (take) the time to do it, it's counterproductive if you do a sloppy job.

Documenting Discoveries, Testing Alternatives

Observation and experimentation have been part of farming at least since agriculturalists learned that characteristics of plants and animals could be reproduced in their progeny. This goes back to the first plant breeders, women who selected the best plants to propagate when agriculture began 10,000 years ago. However, farmer experimentation took on a new meaning for Iowa farmers in the 1980s. A new paradigm was emerging, thanks to the economic strain of the farm crisis and to concerns about environmental effects of agriculture. There were few off the shelf answers to the new questions, and the organization Practical Farmers of Iowa was formed to allow producers to share information with each other. There was no shortage of information; as noted, the challenge for producers became one of sifting through quantities of options.

A critically important tool appeared at this time thanks to Richard Thompson, who farms near Boone, Iowa, with his wife Sharon, son Rex, and daughter-in-law Lisa. In consultation with university agronomists, Thompson and Exner developed simple and practical methods that farmers can use to carry out their own replicated experiments and analyze results of these trials. The basic design is a paired-comparison, or simple randomized complete block, with the pairs usually repeated six times. The design is farmer friendly, using long narrow strips all the way across the field. A simple *t*-test allows producers to generate an LSD (least significant difference) test of statistical significance using a pocket calculator. This innovation has given new power to farmer-to-farmer information sharing, and is accepted in the university research community (Rzewnicki et al., 1989).

Richard Thompson developed the methodology over the first part of the 1980s, and took the concept statewide with the PFI farmer research network in 1987. Since then, PFI research cooperators have accomplished more than 650 research trials, and Thompson has carried out dozens of experiments on his own farm. Many of these have fine-tuned the use of ridge tillage. It was Thompson who first demonstrated the potential of ridge-till to control weeds without herbicides. He continues to evaluate modifications to the ridge-till planter and cultivator to make them more effective against weeds. The coulter was removed and depth bands moved to the sides of the ridge in 1991; row cleaners were added in 1995 and 2004; heavy-duty residue guards were added in 1999 and modified in 2004; and the sweep was removed in 2003. Rotary hoe trials have also been part of the weed management techniques tested on the Thompson farm.

Richard Thompson has carried out many experiments with fertilizers and the mix of manure and municipal biosolids that he applies to crops. The overall direction has been toward more efficient utilization of the organic sources of fertility—rates, timing, and placement. The manure/biosolids holding structure Thompson built in the 1980s retains potassium and nitrogen that previously were lost from the system. The greater efficiency has improved both soil test levels of nutrients and the results of fertility trials.

In Thompson's estimation, manure has increased next-year corn yields by 15 bushels per acre when applied on hay fields in the fall and immediately plowed under. Seven manure trials during 1994-2003 showed no yield increase in corn or soybeans in the application year when the manure was spring-applied on top of ridges. New studies are in progress to determine if a spring application on ridges before soybean planting will result in a yield increase in corn the following year. This change also has the advantage of moving the spring manuring two years away from the oats, reducing the likelihood of lodging of the oat crop.

Not all of Richard Thompson's experimentation has come in the form of replicated trials. Thompson carries a pocket notebook in which he keeps day-to-day observations. Sometimes these observations contribute to significant changes in the farming system. For example, the moldboard plow is now used once in the five-year rotation because Thompson observed declining yields and leaf nutrients in the second corn crop of the rotation. The decision to make that change was not the result of experimentation; rather, it was based on observation over time, yield records, and leaf tissue and soil tests. Thompson would be the first to agree that replicated trials are not the only route to knowledge.

Richard Thompson is also a scrupulous recorder of the farm as an economic system. Every input and operation of each crop goes into a spreadsheet that allows him to compile budgets for each crop and each year. For comparison, he also constructs a hypothetical system based on the typical Boone County yields, practices, and corn-soybean crop rotation. These records go back to 1988 and make a strong case for the profitability of diversified, sustainable farming. Since 1988, the average advantage for the Thompson corn-soybean-corn-oat-hay rotation over Boone County corn-soybeans is $155 per acre per year. During the first years the advantage was under $100 per acre, but in the last few years the advantage has increased to $200 per acre. He attributes this improvement to the following: moldboard plowing increased first-year corn yields, use of the flex harrow increased hay yields, a seed roller increased oat yields, and use of Bt corn increased corn yields and residue yields.

Thompson is aware that most of agriculture is going in other directions; if he did not have solid numbers, he might well be reluctant to speak out about diversified farming. But Richard and Sharon Thompson have described their approach to thousands of visitors to their farm and to audiences around the United States and beyond.

REPLICATING THE EXPERIENCE

On-farm research has been for PFI an "organizing principle," something that has brought people together under a common set of interests. Other groups rely on different organizing principles. The on-farm research network of PFI coalesced at a moment when farming was facing a paradigm challenge. As mentioned earlier, agriculture was experiencing both environmental and economic challenges, and there were few solutions available from traditional sources. Today information on sustainable agriculture is available from the universities, from Extension, from nonprofit organizations like PFI, and from agribusiness. The environmental and economic crises, if not diminished, are at least things to which we are inured. So is farmer research no longer a useful organizing principle? Here are some questions to test that proposition.

- Do farmers have questions about production, economics, or the functioning of their farm in the environment? In Iowa, as one set of questions was resolved, others arose. As mentioned, the motivator went from production efficiency to markets.
- Are there alternative sources to obtain the information? In the case of sustainable agriculture, the answer is now often yes. However, in most cases the information needs to be customized to the farm. Mere awareness that there is a soil test for nitrate nitrogen isn't sufficient; producers need to know how it fits with their soils, management style, and systems. In addition, on-farm tests are a credible way to get a

technology or practice out to a farming public that likes to "kick the tires."

- Finally, is there a community of interest in which on-farm research can take its place alongside the other kinds of information sharing and communication? In our experience, sustained interest in farmer research is rare without a social context. In 1987, farmer research might have been sufficient to create a social network all by itself. Currently it works best as a contributor to a larger network of information and support.

Assuredly on-farm research is not the only way to get information. Neither is it the only form of observation, and observation is not the only way to learn. In fact, Practical Farmers of Iowa wraps together multiple information sources in most of its farm field days and workshops, combining trial results with the experiences of practitioners and scientists. For PFI, farmer research is an activity that brings people together, allowing for different levels of communication. There are other types of farmer organization that could also give rise to farmer research. For example, in the Midwest there are clubs for no-till farming, there are groups dedicated to Holistic Management, and there are statewide forage/grazing groups. If they meet the above criteria, they are candidates for on-farm research.

For PFI, farmer research has placed the producer "in the center" and also implied a democratic sense that information shared allows everyone in this community to move forward together. It is congruent with the organization's identity and has contributed to the shared sensibility. On-farm research is a methodology that others can employ to build capacity and community around problem solving.

REFERENCES

Exner, R. and R. Thompson. 1988. *The paired-comparison, a good design for farmer-managed trials.* Available at: http://www.pfi.iastate.edu/ofr/ OFR_worksheet. htm [Accessed 4 April 2006].

Rzewnicki, P.E., R. Thompson, G.W. Lesoing, R.W. Elmore, C.A. Francis, A.M. Parkhurst, and R.S. Moomaw. 1989. On-farm experiment designs and implications for locating research sites. *Am. J. Altern. Agric.* 3(4):168-173.

Stender D. and E. Stevermer. 1994. *ISU Swine Research Report, ASL-R1186.* Iowa State University, Ames. pp. 105-112.

Chapter 13

Impacts of Private Foundations on Sustainable Agriculture and Food Systems

Oran B. Hesterman

INTRODUCTION

Private foundations have been important contributors to the growth in sustainable agriculture policies and practices over the past 15 years. Sustainable agriculture as a major foundation funding area emerged out of the farm crisis of the 1980s. Since that time, foundations have invested significant resources to advance a diverse sustainable agriculture agenda. This chapter presents a summary of investments of private foundations to sustainable agriculture efforts since 1988, provides a brief analysis of impacts from those investments, and concludes with two case examples from projects supported by the W. K. Kellogg Foundation and others.

The summary of investments in this chapter is based on data collected by Headwaters Group Philanthropic Services from 25 private foundations. Data sources included interviews of past and current foundation staff and foundation annual reports. A summary of investments is followed by charting the changes in interest by these foundations over 15 years, an accounting of what types of projects have been funded, and a measure of the estimated impact this funding has made on agriculture and food systems. Results of the cluster evaluations financed by W.K. Kellogg Foundation were used to document many of these impacts (Berkenkamp and Mavrolas, 2001; Scheie, 2000). Finally, there is a section on lessons learned in the process, from the point of view of the foundations. This information is useful for the historic record, and

This chapter contains information gathered from interviews with Gary Huber, Jerry Dewitt, Pat Gray, and Robert Karp, including direct quotes and concepts from the discussions and the survey carried out.

also provides some guidelines for both the funders and those seeking support for their research and education programs.

SUSTAINABLE AGRICULTURE FUNDING, 1988-2002

The 25 private foundations in the United States that have been most active in funding sustainable agriculture projects awarded a total of $220 million through approximately 1,850 grants between 1988 and 2002. Total annual funding peaked in 1994 at just over $20 million (Figure 13.1). Over the 15-year period, five foundations (Kellogg, Pew, Joyce, Wallace Genetics, and Northwest Area Foundations) accounted for 69 percent of all funding, and the Kellogg Foundation alone accounted for 35 percent of the total amount (Table 13.1). The foundations awarding the greatest number of grants over this time period were J.S. Noyes Foundation (378), Kellogg Foundation (178), Organic Farming Research Foundation (170), Joyce Foundation (157), and Northwest Area Foundation (107).

Total funding for sustainable agriculture from private foundations was greatest in 1994-1998. An upward trend can be seen throughout the entire 15-year period (Figure 13.1), although there has been a general leveling of support after the initial six years (1988-1994).

TABLE 13.1. Funding amounts and percent of total funding from the 25 private foundations most active in funding sustainable agriculture projects from 1988 to 2002.

Foundaion	Total (in dollars)	Percent of total
W.K. Kellogg Foundation	76,294,677	34.73
Pew Charitable Trust	21,883,100	9.96
Joyce Foundation	20,262,000	9.22
Wallace Genetics Foundation	18,085,800	8.23
Northwest Area Foundation	14,570,515	6.63
Jessie Smith Noyes Foundation	9,405,873	4.28
Farm Aid	9,165,025	4.17
Charles Stewart Mott Foundation	6,500,000	3.53
Foundation for Deep Ecology	6,500,000	2.96
CS Fund	6,034,333	2.75
Ford Foundation	5,682,500	2.59

Foundaion	Total (in dollars)	Percent of total
W. Alton Jones Foundation	3,621,400	1.65
Great Lakes Protection Fund	2,663,500	1.21
Education Foundation of America	2,211,750	1.01
McKnight Foundation	2,188,450	1.00
Columbia Foundation	2,112,000	0.96
Bullitt Foundation	1,945,000	0.89
Turner Foundation	1,925,000	0.88
Nathan Cummings Foundation	1,569,700	0.71
Beldon Fund	1,420,000	0.65
Veatch	1,395,000	0.64
Organic Farming Research Foundation	1,119,010	0.51
Clarence E. Heller Memorial Trust	1,065,000	0.48
George Fund Foundation	645,690	0.29
Mary Babcock Reynolds Foundation	160,000	0.07
Total	219,684,510	100.00

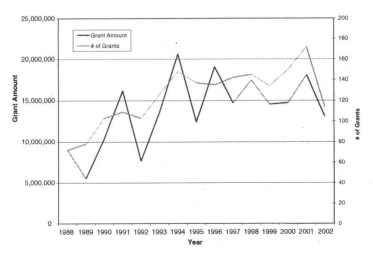

FIGURE 13.1. Sustainable agriculture, total funding and number of grants, 1988-2002.

LOSS OF FUNDERS: IMPACT
ON THE NONPROFIT COMMUNITY

These changes over the past decade can be explained by a range of decisions in different foundations. In the late 1990s, Northwest Area Foundation stopped funding sustainable agriculture, and W. Alton Jones ceased to exist. Pew, Mott, McKnight, Great Lakes Protection Fund, and Turner drastically decreased their funding. The cumulative loss in resources was more than $26 million. Impacts on both grantees and funders include the following:

- Some nonprofits have ceased to exist.
- Other nonprofits have shifted successfully to other funding streams, such as rural state and federal USDA programs, for economic development-related work and for on-the-ground activities that Sustainable Agriculture Research and Education (SARE) grants provide.
- Traditional and new funders have helped fill the void. Some interviewees observed that most of the funders who stopped or reduced support had traditionally funded policy work. With their departure, other funders—notably Kellogg Foundation—began funding policy work. Furthermore, the public sector (SARE and other rural economic development programs available locally) began providing support for on-the-ground work.
- Overall, to successfully apply for rural development monies, organizations diversified their focus and reframed their work to include economic development—a narrow focus solely on sustainable agriculture would limit their potential for funding.
- From this shift toward economic development grew today's focus on local or regional food systems that benefit multiple groups (farmers, workers, consumers, and communities) in multiple areas (health, environment, and economy). This food systems framework holds promise. As one interviewee noted, sustainable agriculture in itself is limiting and can often be a polarizing issue. By reframing the work under *local and healthy food systems,* it is possible to include, rather than exclude, more players and new, more diverse funding resources.

FUNDING TYPES

Funding by Topic and Strategy

All sustainable agriculture grants were categorized by topic—the subject being addressed, and by strategy—how the grant project planned to achieve change. More than half the grant awards (58 percent or $127.4 million) pertained to sustainable agriculture in general, with no specific topics mentioned (Figure 13.2). Of the remaining grants, topics related to pesticides and rural development constituted the next largest group. Analysis of funding trends over time (data not shown) indicated that support of work related to both pesticides and water quality peaked in 1995-1999. One reason given for this peak is the emergence of strong efforts on key policy initiatives (Food Quality Protection Act; Safe Drinking Water Act) during this period.

The two dominant funding strategies in sustainable agriculture grants during the 1988-2002 period were public policy and public education/advocacy (Figure 13.3). Many of the policy-oriented grants focused on federal farm bills. Trends over time (data not shown) indicate that foundation funds for research have declined as a percentage of total foundation funding since 1988 (from 29 to 7 percent) while funds for market development have increased as a percentage of total foundation funding (from less than 9 to 17 percent).

Funding by Beneficiary Region

From 1988 to 2002, nearly half of all foundation funding was considered for national benefit (Figure 13.4). Of all grant funding targeted regionally, the Midwest received the greatest portion (24 percent) and New England received the smallest portion (2 percent). The only region with relatively stable funding since 1990 is the Pacific region, with $5-6 million annually.

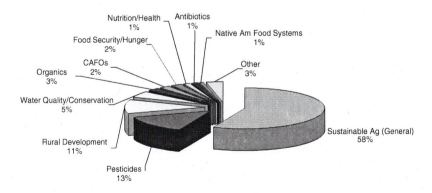

FIGURE 13.2. Sustainable agriculture, funding by topics, 1988-2002.

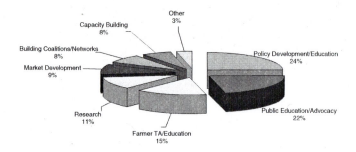

FIGURE 13.3. Sustainable agriculture funding by strategy, 1988-2002.

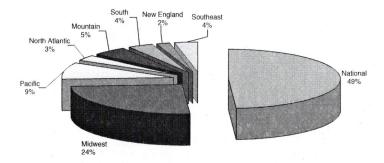

FIGURE 13.4. Amount of funding by beneficiary region, 1988-2002.

Funding by Grantee Type

The majority of all sustainable agriculture funding from foundations from 1988-2002 (85 percent) was awarded to nonprofit organizations [those designated as 501(c)(3) under the Internal Revenue Code]. Institutions of higher education (primarily land grant universities) received 14 percent of all foundation funding during this period.

IMPACTS OF PRIVATE FOUNDATION FUNDING

Impacts of sustainable agriculture projects funded by private foundations over the 15-year time period can be seen in six areas: Public Policy,

Market Development, Land Grant Universities, Farm Practices, Organizational Capacity, and Attention to Disadvantaged Communities.

Public Policy

From the 285 grants totaling $44.3 million invested in public policy development or education, the main impacts were seen in following areas:

- Federal Farm Policy, with major efforts in establishing and expanding the Conservation Title; establishing the Conservation Security Program; expanding support for local marketing options, including farmers markets and community supported agriculture; establishing national organic food standards; and establishing/supporting research and education on sustainable agriculture through the LISA/SARE programs.
- Pesticide Related Policy, with major efforts to enact the Food Quality Protection Act and to challenge state-level efforts to preempt local pesticide laws that were more stringent than state or federal laws.
- Water Quality and Conservation Policy, with major efforts to create protection of groundwater during reauthorization of the Federal Safe Drinking Water Act; state level efforts to protect groundwater from agricultural pollution; and developing a collaborative plan to reduce the amount of runoff fertilizer nutrients contributing to the hypoxia in the Gulf of Mexico.

Perhaps the most compelling progress supported by foundations related to public policy has been in the earlier stages of the change process: policy development, educating policy makers about options for change, engaging a greater variety of individuals in change efforts, and building relationships that are necessary precursors to meaningful policy change. To those ends, many policy-related grants were aimed at capacity building to help grantees become better fundraisers, managers, and communicators, and to improve networking and collaboration among organizations. The Regional Sustainable Agriculture Working Groups, the National Campaign for Sustainable Agriculture, Great Lakes Grazing Network, and the National Coalition for Pesticide Policy Reform are examples of networks that have made important contributions to sustainable agriculture policy and practice at state and federal levels.

Market Development

From the 138 grants totaling $17 million invested to help organizations develop markets for sustainably raised agricultural products, the main impacts were seen in:

- Marketing "locally grown" products, with support for buy local campaigns, farmers markets, and community supported agriculture. Notable efforts include:
  a. CISA (Community Involved in Sustainable Agriculture) has run a successful "be a local hero–buy locally grown" campaign in Western Massachusetts since 1999. Sixty-nine percent of participating farmers report sales increases due to this communications campaign and over 50 percent of residents say they are buying more locally grown food as a result (CISA, 2003; Lattanzi and Berkenkamp, 2002).
  b. Food Routes Network is supporting "Buy Local" food campaigns in ten states across the United States (Maine, Michigan, Northern California, Iowa, North Carolina, Louisiana, Pennsylvania, Montana, and Minnesota) to promote local and sustainable agriculture in their regions.
- Expanded use of United States Department of Agriculture's (USDA) value-added marketing program to support marketing incentives for organic and sustainably raised specialty products. Two examples of products developed from the market-based work follow:
  a. With a line of eco-label products, the Food Alliance based in Portland, Oregon, serves as a nonprofit organization broker to retail and wholesale food to partners such as grocery store chains, restaurants, and institutions. The 200 different eco-labeled agricultural products include fruit, vegetables, wheat, wine, livestock, and dairy products that come from farms and ranches certified as using sustainable practices. To obtain certification, producers must meet a rigorous set of criteria for pesticide reduction and elimination; soil and water conservation; wildlife habitat preservation; safe and fair working conditions; and healthy and humane care for livestock.
  b. Local and ecologically grown products under the *Red Tomato* brand. Started in 1998, Red Tomato, a nonprofit organization in Boston, has partnered with more than 35 families in the Northeast and 15 African-American farmers in the southeast delivering their products three times a week to 40 stores in the greater Boston and Philadelphia areas.
- Partnerships among farmers and universities, schools and restaurants to expand marketing possibilities for farmers while providing high

quality local products to students and diners. These partnerships include both individual privately owned restaurants and large publicly-held distribution companies. The farm-to-school movement is catching on in many locations and successful models can be found across the country. The Community Food Security Coalition has emerged as a leading nonprofit organization in this work.

- Establishment of eco-labels in the food marketplace. The Food Alliance, supported by several foundation grants, launched a new eco- label that certified producers for their production and labor practices. Annual farm gate sales of Food Alliance certified products reached $7 million in 2002, with participating farmers in the Pacific Northwest and Upper Midwest. Food Alliance certified products are also being carried by grocery stores, food service companies, and university cafeterias.

- Two examples of particularly effective partnerships in the market development arena follow:

  a. Center for Agricultural Development and Entrepreneurship (CADE) received a grant to help reverse the decline of agriculture-related businesses in rural upstate New York by assisting farmers and entrepreneurs with new product development, business planning, and market development activities. CADE believes that decentralizing farms to reach regional community markets will enhance farm sustainability and provide residents with fresher, healthier foods purchased using environmental friendly methods. To make the shift, CADE helps farmer-entrepreneurs succeed through its Agriventure business model that emphasizes skill development in production, business development, and marketing. CADE believes that without these combined skills, agricultural businesses will continue to fail. The Agriventure model also provides a checklist for entrepreneurs as they develop and hone their businesses. The CADE approach appears to be working. Success includes helping 150 small farms significantly increase sales.

  b. Shorebank and Oregon State University (OSU), partially funded through foundation grants, have created unusual technical assistance services for entrepreneurs in the struggling West Coast fishing industry. These collaborators consider the project to be a "research and development" arm for small family businesses. New business ventures are supported, financially and intellectually, to help develop, test, produce, market, and distribute their value-added products.

  c. Shorebank and OSU work with entrepreneurs on value-added shrimp, crab, albacore, and oyster products. For example, the project has created an albacore business plan involving 60-80 boats and 8-10 processors, created new products for shellfish and albacore, and are researching new high-pressure processing and logbook technology for implementation by 20 fishermen in 2006. Again, the approach is full-service, comprehensive, and long-term.

d. Shorebank and OSU understand the critical importance of having adequate financial resources to undertake economic development projects. They work closely with Community Development Financial Institutions and other financial institutions in the private sector to provide capital for "sustainable agriculture" entrepreneurs, in this case sustainable fishermen. Their comprehensive approach to finance and business development offers important lessons for nonprofit colleagues with less market-based experience. Ultimately, if food systems change is to occur and be sustained in the market place, nonprofit organizations must think and act like savvy businesses to promote their socioeconomic missions.

Land Grant Universities

From the 195 grants totaling $27.2 million invested to establish and support sustainable agriculture programs at land grant universities, the main impacts included:

- *Increased legitimacy* of sustainable agriculture research and development at universities. This was especially true when partnerships between universities and nonprofit organizations were funded by private foundations. Two notable examples of these partnerships are Iowa State University/Practical Farmers of Iowa and Oregon State University/Shorebank Enterprise Pacific.
- *Underwriting and endowing professorships* in sustainable agriculture. These have provided leadership in moving research into sustainable agriculture applications, fostered interdisciplinary research and outreach on sustainable agriculture topics, and elevated the field of sustainable agriculture in colleges of agriculture and at land grant universities. Sites receiving foundation grants to endow professorships in sustainable agriculture include Michigan State University, The Ohio State University, and Iowa State University. Recently, Kellogg Foundation has endowed the Kellogg chair in Community and Food Systems ethics at Michigan State University, as well as a "sustainable Michigan fund" to help network together a number of endowed chairs across the university to focus on sustainable agriculture and development. This experiment in investing foundation resources with the in-

tent of leveraging interdisciplinary effort in sustainability will be an interesting experiment to follow.

• Between 1993 and 1997, we identified 19 new positions that were created at ten land grant universities that were involved in Kellogg Foundation-funded Integrated Farming Systems projects. These included positions in California, Georgia, Iowa, Kansas, Kentucky, Michigan, Minnesota, Montana, North Carolina, and Ohio. New programs ranging from research on grazing and organic vegetable production, to new academic programs in sustainable agriculture, were also begun at eight of the participating universities.

Farm Practices

From the 220 grants totaling $27.5 million invested to encourage adoption of integrated and resource efficient farming methods, impacts included:

• Adoption of new agronomic practices, such as management intensive grazing, cover crops and reduced tillage. The Great Lakes Grazing Network, supported by several foundations and universities, estimated that in a two-year period (1999-2001) managed grazing systems were adopted on over 100,000 acres in the Midwest.
• Adoption of alternative crops (e.g., peas and lentils in Montana; fruit and vegetable crops by grain and tobacco producers).
• Adoption of whole-farm planning, including the use of longer rotations, integrating crops and livestock, and transition to organic production.
• Two cluster evaluation reports, both available from Kellogg Foundation (www.wkkf.org), provide details of farm practices changes related to Integrated Farming Systems and 34 projects funded by Kellogg Foundation between 1992-2000.

Organizational Capacity

According to our interviews and analysis of foundation annual reports since 1988, foundations have invested $27.8 million in building the capacity of individual organizations and networks of organizations. Though little formal evaluation work has been done on either the individual or collective level, some impacts can be observed by the programmatic outcomes achieved. These include:

- The aforementioned efforts of several coalitions to successfully address public policy concerns, such as the national organic standards, conservation funding in successive farm bills, and funding for the USDA sustainable agriculture research and education (SARE) program.
- Increased capacity of sustainable agriculture organizations and leaders to build partnerships and work across organizations. In Montana, Alternative Energy Resource Organization (AERO) worked with farmers, rural residents, and rural nutrition advocates to encourage the State Department of Health and Human Services to permanently fund the Women, Infants, and Children's (WIC) nutrition program at farmers markets across the state.
- In Pennsylvania, the Pennsylvania Association for Sustainable Agriculture successfully engaged regional planners, economic development officials, and county governments in economic development planning for Southwestern Pennsylvania. These county governments have since taken responsibility for incorporating agriculture into their economic development plans.
- Enhanced capacity and proficiency at using the media and other communication channels to educate the public about issues important to their organization.
- As one of the few funders to invest extensively in evaluation, the Kellogg Foundation has found that project- and cluster-level evaluation has contributed to building organizational capacity for reflection, learning, and changing course leading to more effective program implementation.

Attention to Disadvantaged Communities

The amount of foundation funding targeted to organizations working on sustainable agriculture in disadvantaged and minority communities has been small. A limited number of grants (about 1 percent of all grants awarded) have been awarded to organizations comprised of or working on behalf of American Indian, African-American, Hispanic, or new immigrant farmers. Capturing impacts beyond individual project reports is difficult; however, highlights to date include:

- Greater attention being paid by USDA to the continuing decline of African-American owned farms and farmland. The most recent farm bill authorized a 150 percent increase in funding for the Outreach and Assistance for Socially Disadvantaged Farmers Program (from $10 million to \$15 million) and created a USDA Assistant Secretary for Civil Rights.

- Greater attention and initial funding from USDA and the Federal Department of Health and Human Services to the needs of immigrant and refugee farmers, many of whom are from Southeast Asia, Africa, and Latin America.

- A focus in Native American communities on reclaiming their traditional (and sustainable) food systems for economic, health, and cultural reasons. For example, First Nations Development Institute and Kellogg Foundation have been supporting Native Food Systems projects in the following communities:

Chickaloon Village Traditional Council	Chikaloon, AK
Kenaitze Indian Tribe	Kanai, AK
Louden Tribal Council	Galena, AK
Native Village of Kotezebue	Kotzebue, AK
Aloha 'Aina Health Center	Kane'ohe, HI
Oneida Tribe of Indians of Wisconsin	Oneida, WI
Pinewoods Community Farming, Inc.	Gowanda, NY
Wampanoag Tribe of Gay Head Aquinah	Acquinnah, MA
Cheyenne River Youth Project, Inc.	Eagle Butte, SD
Indigenous diabetes Education Alliance	Rapid City, SD
Standing Rock Sioux Tribe	Fort Yates, ND
Columbia River Inter-Tribal Fish Commissions	Portland, OR
Makah Tribal Council	Neah Bay, WA
Native American Community Board	Lake Andes, SD
A:shiwi A:wan Museum and Heritage Center	Zuni, NM
Corporation of New Sogobia	Austin, NV
Developing Innovations of Navaho Education, Inc.	Flagstaff, AZ
Pyramid Lake Paiute Tribe	Nixon, NV
Santa Fe Indian School	Santa Fe, NM
Tohono O'odham Community Action	Sells, AZ
Vechiji Himdag Mashchamaku	Sacaton, AZ
Yauapai–Apache Nation	Camp Verde, AZ
Native Resources Developer, Inc.	Paga Paga, AZ

- Inclusion of issues faced by farm workers as part of the focus on sustainable agriculture. Without foundation funding, this issue (and many of the others listed in this section) would likely receive even less attention as part of the work in sustainable agriculture.

WHAT HAVE WE LEARNED?

With an investment of over $200 million during the past 15 years, we would hope that some significant learning for the philanthropic community would result. Based on the interviews and analyses conducted to gather information about funding patterns and impacts, some important lessons can be shared. These are lessons that future funders might take into account as they establish programs and consider future grant proposals in sustainable agriculture. Lessons are listed by topical area.

Public Policy

- Invest in grantees' long-term organizational capacity and sustainability. Policy and institutional change work demands new skills and necessitates engagement with those who are beyond familiar circles and who may historically have been perceived as opponents.
- Simultaneously support activists to push for policy and institutional change from the *outside* and allies to influence change from the *inside* of key institutions.
- Support collaborations with organizations and constituencies outside of sustainable agriculture to enhance success in the policy arena.
- Fund grantees to conduct thorough policy analysis and research. Anticipate that such analyses may identify the need for modification of already-funded projects to enable greatest impact down the road.
- Recognize that policy and institutional change is grounded in the slow and subtle building of relationships and the involvement of new voices.

Market Development

- Recognize that solid markets for sustainably raised products are the key to changing conventional farm practices. Farmers need assurances that they can make a living using sustainable practices; they understandably need and expect added economic benefits. Technical assistance and education on sustainable agriculture alone are not adequate to create long-term change in farmer practices.
- Assist grantees to obtain the professional assistance needed to conduct high quality feasibility studies, develop business plans, sponsor consumer research and other analyses that are critical to the success of market-oriented efforts.

- Support development of organizational infrastructure that nonprofit organizations need to venture into more business-oriented efforts.
- Invest with the understanding that market-based change takes a long time to achieve and that sustained investments (5-8 years or longer) may be needed.
- Encourage organizations to view marketing projects not only as economic development, but also as a vehicle to deepen and diversify leadership and participation in their communities.

Land Grant Universities

- Recognize that different cultures and goals between land grant universities and nongovernmental organizations can create tension. Extra time and effort is necessary to ensure good communication, well-defined roles and strategies, and common goals among the different players.
- Leveling the playing field by placing significant resources in the hands of the nonprofit partners is a necessary and worthwhile strategy for grant making. This approach helps to bolster the nonprofit partners' legitimacy in the eyes of the land grant universities partners and for enabling the nonprofit to function more as an equal partner.
- Partnerships between land grant universities and nonprofit organizations are essential for promoting positive change, but great care must be taken in the identification and cultivation of these relationships. Partnerships that form organically among local players and are grounded in a genuine commitment to work together have, on the whole, proven less conflicting and more fruitful than partnerships that were motivated largely by the availability of funds or the desires of the funder.
- The development of working relationships among these partners often takes much more time, energy, and funding than is anticipated.
- The pay-off from these investments may take years to materialize and may be quite incremental in nature. A realistic sense of what constitutes meaningful change and the likely pace of that change is essential for funder and participants alike.
- From our experience at the Kellogg Foundation, we believe that three key factors must be present for significant change to occur within land grant university environments. These are:
  a. Support by senior level administrators for interested faculty who choose to engage in research and education in the sustainable agriculture arena.

   b. Faculty within the land grant universities who are interested in working at what their colleagues would consider "the margin" in research and education and who will serve as champions for these issues.

   c. Pressure from outside constituency groups that can constructively encourage and promote change within the land grant universities.

Kellogg Foundation believes that each of these factors is necessary, but none in itself is sufficient. One lesson from Kellogg Foundation's experience is that investments made in environments that lack any of the three elements are unlikely to be successful and that even where all three are present, positive changes are far from guaranteed. Expanding support for sustainable agriculture within land grant universities has, in Kellogg Foundation's experience, been a very slow and hard-won process. Change has come as a series of incremental and often isolated changes.

Farm Practices

- Support funding for farmer-to-farmer engagement and education about alternative farming systems. This approach was identified as one of the most powerful strategies for encouraging producers in more integrated systems. Participation by outside "experts" was helpful where they complemented, rather than served as an alternative, to farmer-centered efforts.

- Recognize that investments in formal communication programs strengthen outcomes and greatly leverage the impact of farming systems research. To have an impact on the ground, farming systems data must be locally relevant, conveyed through credible messengers, packaged in a way that enables farmers to make decisions, and placed in the hands of people working on sthe land. Kellogg Foundation has seen very favorable outcomes from investments that help grantees develop their communications capacity and strategically disseminate information about alternative practices.

- Create more effective approaches for reaching beyond the "early adopters." Reaching more conventional and traditional producers is challenging. Funders will need to help the sustainable agriculture community address these barriers if farming systems change is to occur on a broader scale.

- Farming systems change is inextricably linked to the farmer's success in the marketplace. Shifts toward more integrated farming systems are unlikely to occur unless they generate an economically viable product with a viable way of reaching an interested consumer.

Organizational Capacity

- Provide funding that is sustained and flexible, to heighten impact. Trying to pursue long-term change with short-term funding (often cobbled together from multiple funding sources) is not the most effective strategy.
- Building the leadership base and political savvy of the sustainable agriculture community is particularly critical now, given new opportunities to reach out to new allies and constituencies. Leaders who can effectively develop diverse coalitions, speak multiple "languages," and operate in difficult political arenas are particularly needed.

Shared Learning and Evaluation

- Provide opportunities for nonprofit organizations to learn from each other and gain exposure to those outside their customary circles who have skills that these groups need.

Disadvantaged Communities

- Recognize the potential of new farmers and minority farmers as new constituents for sustainable agriculture organizations and encourage new outreach programs to this group.
- Encourage sustainable agriculture funders to work with health funders to understand how food and diet impact disadvantaged communities.
- If the funders want grantees to become more multicultural in their programs and internal operations, they probably need to explicitly fund that work. It doesn't happen on its own.
- Building relationships across diversity is a valuable means of increasing leverage for change. However, it also presents a challenge in that it takes time to truly understand others and manage conflicts which often do not arise with less diversity.
- Projects and organizations strongly rooted in minority communities have much to teach about critiquing and changing the dominant system. Long experienced in surviving in a hostile environment, these communities have had to become skilled at recognizing strengths and weaknesses, threats and opportunities, allies and opponents.

- Information and collaboration often flows first along ethnic, gender, and class lines. This means that organizations and farmers of color often face barriers in accessing information from predominantly white institutional officials and policy-makers; sustainable systems activists are often slow to tap the expertise and analytical skills of organizations of color. Systems change strategists must work intentionally to develop multicultural skills and relationships, to bridge these historic divides.
- Diversify leadership in the overall sustainable agriculture community in ethnicity, gender, age, and geographic representation. Prospects for effecting lasting change could improve if the funding community (and the grant seeking community) made a more strategic, long-term commitment to deepen the pool of leaders, diversify this group, and add more up-and-coming leaders now in their teens, twenties, and thirties.

CASE EXAMPLES OF SUSTAINABLE AGRICULTURE PROJECTS

Community Involved in Sustaining Agriculture: Local Heroes for Locally Grown Food

Building connections between farmers and consumers and building local food systems are key priorities for the Kellogg Foundation. The Kellogg Foundation's support for Community Involved in Sustainable Agriculture (CISA) exemplifies Kellogg Foundation's growing emphasis on harnessing the power of the marketplace to help farmers stay on the land, profitability producing healthy food.

Based in Amherst, Massachusetts, CISA is committed to developing a thriving local food and farming system in Massachusetts' Connecticut River Valley—a food system that is economically, environmentally, and socially sustainable.

First funded by Kellogg Foundation in 1993 when the organization was just forming, CISA was selected in 1997 to participate in the Kellogg Foundation-supported Food Routes Network Initiative (formerly Fires of Hope). The initiative explored the hypothesis that with greater investments in communications, program evaluation, and fund development, sustainable agriculture nonprofits could reach a whole new level of sophistication and scope in engaging consumers to support local food systems.

Once selected for the initiative, CISA received significant grant funding as well as several years of consulting support from communications experts, fundraisers, and an outside program evaluator. Experts from the University of Massachusetts and Hampshire College also made significant contributions to CISA during its first five years. CISA was also able to hire an outside research firm to survey area consumers about their attitudes toward local foods and shopping habits. This research also tested communication messages and identified channels by which consumers were getting information about food.

CISA quickly set about designing a new community-based, multi-media campaign to promote locally grown farm products. With the new consumer research in hand, CISA developed a campaign tagline, "Be a Local Hero, Buy Locally Grown," a logo, and a wide range of campaign materials.

Building from its base among local farmers, CISA began enrolling farmers in the campaign, along with a handful of retail grocers. The campaign was launched in mid-1998 through radio and newspaper advertisements, and bus boards featuring campaign farmers, community events, and point of purchase materials used by farmers and retail partners.

Entering its fifth season in 2004, the Local Hero campaign included 86 farms covering more than 18,000 acres in a three-county area. Forty-six retailers and 15 area restaurants have also joined the campaign. Independent consumer surveys, conducted annually since 1999, show that:

- Nearly 80 percent of surveyed consumers are aware of the Local Hero campaign and can recall the theme and slogan without being prompted.
- Eighty-six percent of those surveyed responded favorably to the campaign and believe it makes a convincing case for buying locally grown food.
- Sixty-five percent of those who are aware of the campaign say they are buying more local food as a result.

CISA farmers also report that the campaign is having a positive impact on their farms through increased sales, new sales accounts, higher prices, and increased production. CISA's campaign model has been shared with organizations across the United States and Europe through a practical how-to manual, "Harvesting Support for Locally Grown Food: Lessons Learned from the 'Be a Local Hero, Buy Locally Grown' Campaign", jointly written by CISA and FoodRoutes Network.

The Local Hero campaign experience offers a host of lessons for both funders and nonprofits working in this arena: [http://www.buylocalfood .com/Local%20Hero.htm] (accessed on 5 April 2006)

- Precampaign consumer research designed to segment the market, understand targeted consumers, and identify compelling messages is critical. Funders need to fund this work and ensure that its results are shared with other communities.

- Post-campaign consumer research to assess impact of consumer awareness and behavior is also a very worthwhile investment. For instance, a "tracking" survey costing less than $10,000 enabled CISA to document the campaign's effectiveness with consumers, show farmers and retailers the campaign's impact, and enlist additional funders. The research also allowed CISA to redirect scarce budget resources to the communication vehicles that had greatest impact.

- CISA has found it essential to bring many stakeholders together to find ways to sustain agriculture in the Pioneer Valley. In addition to farmers and consumers, this has included business people, policymakers, environmentalists, and community planners. Partnering well and being innovative is the key. While the use of more sophisticated communication tools greatly heightened the campaign's reach and impact, CISA has also been committed to keeping the campaign grounded in the needs and aspirations of their community.

- An inclusive approach to engaging farmers has aided CISA's success. The campaign encourages farmers to adopt more earth-friendly practices, but embraces all small- and medium-sized farmers in the effort to sustain the area's local farming base.

- Engaging an external professional evaluator to assess the campaign's effectiveness and identify areas for improvement has also heightened impact. Kellogg Foundation's commitment to sound program evaluation has helped CISA see its campaign objectively, gather and interpret data and lessons learned, and build the organization's capacity to assess and continually strengthen its own work.

Iowa State University and Practical Farmers of Iowa: A Partnership for Innovation

As discussed earlier in this chapter, expanding commitment and support for sustainable farming systems among land grant universities remains a great challenge. However, Iowa State University is one example of a land grant that is complementing its work in more conventional agricultural arenas with new educational programs and services that promote more farmer-driven, conservation-focused research, and community-based initiatives that expand market access for farmers using more sustainable farming methods (see Chapter 7).

This evolution began more than 15 years ago and has its roots in an innovative partnership between Iowa State University and a local nonprofit organization, Practical Farmers of Iowa (PFI). Formally launched in 1987, the partnership between PFI and Iowa State University initially focused on on-farm research. Complementing its efforts with researchers, PFI also worked with the Leopold Center for Sustainable Agriculture to build a farmer constituency for alternative research priorities.

Efforts took a big step forward in 1993 when the Kellogg Foundation awarded a $732,000 grant for a four-year collaborative project between PFI, Iowa State University Extension and the Leopold Center. Called *Shared Visions: Farming for Better Communities,* the program sought to build upon the on-farm research effort by using more comprehensive approaches for addressing the decline in farm income and the rural social fabric experienced by many PFI members.

Through an innovative mini-grant program designed to place funds in the hands of local communities, *Shared Visions* supported 15 community groups across Iowa. Comprised of both farmers and other community members, the groups designed their own programs to support rural economic development and the adoption of farming systems that protect the environment and human health. *Shared Visions* also supported the emergence of more market-oriented approaches that were bubbling up in many Iowa communities and helped build a cadre of leaders across Iowa committed to alternative approaches to agriculture.

Conscious of the need to level the playing field between land grants and much smaller nonprofits, Kellogg Foundation awarded the *Shared Vision* grant to PFI. The grant also required implementation in partnership with Iowa State University. This approach solidified a new way of working together and provided significant resources for the partnership to expand into new territories. The Kellogg Foundation grant also heightened the PFI credibility and influence, and enabled the fledgling nonprofit to build its membership and organizational capacity.

Practical Farmers of Iowa leverage further increased when, in 1998, the organization opened an off-campus office and secured significant new grant funds to expand its work in community based local food system development. In 1998 the Leopold Center and PFI entered into an agreement for the Center to provide stable funding ($50,000/year) for PFI to conduct on-farm research for a five-year period. The hiring of an Executive Director a few years later also gave PFI greater access to Iowa State University administrators and the ability to more readily speak as an independent, outside voice.

In the meantime, a host of innovators within the University was bringing new ideas and approaches to the fore. Among them was Jerry DeWitt, then head of Extension for Agriculture and Natural Resources. As DeWitt recalls,

From the start, we were fortunate to have a number of things going for us. By the late 1980s we had a core group of people within Iowa State University and Extension that supported alternative approaches for Iowa's agriculture. While small in number, we were committed and were able to enlist a second wave of supporters. Also, our Administration didn't put up the roadblocks that innovators at many other land grant universities have experienced. Then the infrastructure began to take shape. The Leopold Center was formed in 1988. Kellogg Foundation, along with Wallace Genetic Foundation, contributed $1.5 million to create an Endowed Chair for Sustainable Agriculture. Over time, more innovative ideas emerged and the funds to advance them fell into place.

Iowa State University has since created a new graduate program in Sustainable Agriculture and funded a new organic specialist position. Today, the partnership is taking on new forms and priorities. Among these are the Agricultural Innovation Network, a project codesigned and co-coordinated by PFI and Iowa State University extension staff. The network consists of extension staff and other agricultural professionals who are trained to assist farmers who want to start alternative agricultural enterprises.

Kellogg Foundation has also awarded a $560,000 grant to the Leopold Center to support "Value Chain Partnerships for Sustainable Agriculture." Implemented in partnership with Iowa State University, PFI, and others, the project focuses on developing markets for small and midsize farmers who use production practices that follow high standards of environmental and community stewardship. As PFI's Executive Director, Robert Karp, observed, "These programs are the outcome of years of work. They are the result of working with rather than working against one another."

Looking back across the 15 years of partnership between Iowa State University and PFI, key stakeholders reflected on the factors that have enabled positive change to occur. Among them are the following:

1. Change within universities is enabled when key administrators create a more accepting environment, when there are champions within the university who are committed and willing to take risks, and when outside constituencies are organized to constructively encourage change from without. Funders need to take all three factors into account in their funding decisions.
2. While champions within the university are key, it also takes champions within those advocacy groups who can mobilize their membership, are committed to making the partnership work for both parties, and who have the organizational wherewithal to stick with it when things get tough.

3. In the case of Iowa State University and PFI, the above factors were supported by foundation funding that was awarded with the requirement that the work be done through partnership and that placed funds in the hands of the less powerful partner. Kellogg Foundation support was also sustained and substantial, enabling both the relationship and joint program activities to deepen and mature.

4. Because change within universities is inherently slow and hard-won, it is critical to focus change efforts where you can get the most impact. Jerry DeWitt and others positioned with Iowa State University put their energy into creating a small core of supporters within the university rather than trying to effect change at the departmental level or in the upper reaches of the administration. As DeWitt expressed it, "The lesson we learned in those early years was that it's essential to focus where you can get the biggest bang for your buck. Often, that's not the Administration. In our case, attitudes among faculty at the department level were often a greater roadblock. Our approach was to mobilize a core set of respected individuals across a number of departments. You will never get everyone on board, but if you work to change key individuals, your college will change."

5. Both the land grant and nonprofit partners must adapt themselves to the needs of the partnership. That may mean learning to listen in new ways and adapting one's style of communication and engagement. This requires a significant time commitment and the willingness to learn from mistakes. For such partnerships to flourish, funders need to invest in this capacity building, even when it can't be tied directly to more concrete "project outcomes."

6. Informal, respectful relationships between key participants have always been at the root of the Iowa State University/PFI partnership. This has meant sustained investment of time and energy and a shared commitment to address differences privately, rather than publicly. When conflicts occur, both parties are committed to working out differences quietly and in ways that build trust.

7. It is essential to expect differences and deal with them constructively. As PFI's Gary Huber expressed, "We are committed to exploring the interests behind the positions. We look for ways that our various interests are compatible, even when the position may not be." He also points out that "You have to accept it when things don't go as you want, but you can't let go of the idea that the partnership remains important."

The Food Project: A Commitment
to Diversity Becomes Reality

Among the beliefs that the Kellogg Foundation tries to advance through its grant-making is the idea that the most innovative solutions to the challenges of our day arise when people with different heritages, belief systems, and cultures work together. Recognizing that bringing diverse voices to the table is much easier said than done, Kellogg staff themselves underwent diversity training in 2001. This training was provided later that year to the Project Directors for all of Kellogg's Integrated Farming Systems projects.

Then, in a joint effort between Kellogg's Food Systems and Youth & Education programs, funds were provided to help make diversity a reality for a grantee that had a multicultural constituency at its core. That organization, The Food Project (TFP), brings together youth and adults from diverse backgrounds to build a sustainable food system in the Boston area. Working with African-American, Cape Verdian, Hispanic, and Caucasian youth, TFP uses leadership and agricultural programs to bridge communities traditionally divided by race and class. A targeted Kellogg grant has helped TFP's youth participants, staff, board, and volunteers to better understand the issues of diversity and make TFP itself a more multicultural organization.

Today, youth, staff, and volunteers grow organic produce on 21 acres of land in suburban Lincoln, MA and two acres in the Dorchester and Roxbury neighborhoods of Boston. They also manage two farmers markets and run a 233-member Community Supported Agriculture program. In 2002, they piloted two food enterprises in their new commercial kitchen and participated in the sale and charitable distribution of more than 200,000 pounds of locally grown produce.

A commitment to diversity has always been central to TFP's mission. However, TFP's path to making their organization truly multicultural has not been an easy one. The vision of four suburban, white individuals, TFP was founded in 1991. By the late 1990s, the organization had stabilized financially and grown in strength. However, from its base in a wealthy, largely white suburb the Project remained geographically distant from the inner-city youth it sought to serve. Even after adopting hiring policies intended to promote diversity of its staff and board, TFP struggled to make those changes a reality. Increasingly concerned that it was not "walking its talk", TFP recognized that racial and ethnic backgrounds of staff and board still did not mirror the diversity found in the Project's youth programs.

The organization also found itself in something of a "Catch 22." As TFP's Executive Director Pat Gray recalls,

Funders said they wouldn't fund us without a diverse staff in place, but they wouldn't fund efforts to try to recruit the right people. So we were delighted when the Kellogg Foundation asked if we would be interested in a new initiative to support the creation of a multi-cultural youth and adult community.

Awarded in 2002, the Kellogg grant supported TFP's on-going effort to "engage youth as full partners in achieving a new level of sophistication and organizational integration around issues of diversity."

This funding enabled TFP to enlist the help of VISIONS, Inc. Headquartered in Arlington, Massachusetts, VISIONS is a consulting group that works with organizations across the country to provide multicultural and diversity training and ongoing technical assistance. VISIONS began its work by conducting a "cultural assessment" that engaged stakeholders within and outside the organization. The assessment explored the current climate at TFP regarding diversity issues, and identified strengths, barriers, and recommendations toward greater multiculturalism. VISIONS staff also led intensive, four-day training sessions to help participants develop a deeper understanding of what diversity is about and explore differences in experiences and beliefs about diversity.

At the same time, TFP:

- formed a board/staff/youth task force on diversity;
- allocated staff time to increase the organization's ability to attract and retain diverse staff members;
- expanded mentoring and diversity training in its youth development programs;
- involved parents of participating youth in efforts to promote diversity; and
- facilitated opportunities for TFP youth to educate others in the community about diversity.

As Pat Gray explains it,

Our work with VISIONS, Inc enabled us to unravel these issues in a way we hadn't been able to do before. It was very complex and very touchy. It took a real investment of time and energy. You can't approach these issues without being willing to take risks and to examine your own beliefs. VISIONS helped our staff and youth develop a common vocabulary. After many years of trying, this enabled us to talk about these challenges in depth. We were also able to articulate a shared set of beliefs about diversity and the ways by which TFP will bridge and build diverse communities throughout our work.

The organization also took concrete steps to rectify some of the institutional and internal policy barriers to more multicultural operations. These included:

- Opening an office in Dorchester neighborhood of Boston that makes working for TFP more feasible and attractive for city residents, and provides a physical presence in the communities that TFP seeks to engage.
- Instituting salary and benefit packages that better reflect the aspirations of a multicultural applicant pool.
- Grounding their recruitment efforts in deeper relationships in the communities they seek to recruit from and engaging people throughout the organization to identify recruits.
- Adopting interviewing practices that engage many staff and youth in recruitment and give applicants the opportunity to demonstrate practical skills.
- Supporting youth participants to design and lead diversity training programs for stakeholders within and outside TFPs.

Since 2000, TFP has evolved from a 12-person staff with no people of color to a staff of two dozen, of whom 15 percent are people of color. This includes several people of color in senior positions of leadership. Pat Gray observes that,

> While we have much more work to do, this shift is really having an effect. Our interns are saying that they are finding new role models in the organization and seeing real evidence that we are walking the talk. We realize that we aren't going to solve anything right away, but we are getting closer to our original vision. Our pursuit of greater multiculturalism is an on-going process, one that we will work toward together.

CONCLUSIONS

Although there are only a few private foundations with clearly-articulated and long-term programs focused on sustainable agriculture, funding in this area continues to increase. This is due in part to an increased awareness about the impact of agricultural practices on the environment, and the stewardship ethic and potential embodied in sustainable agricultural systems. This is also due to expanding connections both between sustainable agriculture and community economic development, and between sustainable agriculture and solutions to the current challenges of food, diet, and obesity. As these connections deepen, contributions from private foundations to sustainable agriculture and food systems are likely to increase.

The primary purpose of this chapter is to elucidate for both the grant seeking and funding communities the contributions and impacts that private foundations have provided for sustainable agriculture and food systems. It was also written with the intention of attracting additional financial resources to this field so critical to our collective future. To that end, I invite inquiries and conversation, especially from those engaged in philanthropic activity. I can be reached at oran.hesterman@wkkf.org.

REFERENCES

Berkenkamp, J. and P. Mavrolas. 2001. *Integrated farming systems, phase II: Cluster evaluation.* Report #2 on Farming Systems Change, W.K. Kellogg Foundation, Battle Creek, MI.

CISA. 2003. Community involved in sustaining agriculture, Update, Summer 2003. [for current information see http://www.buylocalfood.com/Local%20Hero.htm] (Accessed 5 April 2006).

Lattanzi, M. and J. Berkenkamp. 2002. *Harvesting support for locally grown food: Lessons learned from the "Be a Local Hero, Buy Locally Grown' Campaign."* Final Report, W.K. Kellogg Foundation, Battle Creek, MI.

Scheie, D. 2000. *Changing the food and farming system through community collaboration, phase I.* Final Cluster Evaluation Report #1 on Farming Systems Change, by Rainbow Research. W.K. Kellogg Foundation, Battle Creek, MI.

Chapter 14

Motivation Theory and Research in Sustainable Agriculture

Shirley K. Trout
Charles A. Francis
John E. Barbuto Jr.

INTRODUCTION

Social scientists have a unique position in the field of sustainable agricultural research. Not only must we examine the ways in which sustainable agricultural practices impact humans through their communities and families, we must also take responsibility for moving conventional agriculture toward a more sustainable mind-set—both in word and deed. Once the social benefits of a sustainable food system are documented through community development, family sciences, and child development, for example, the field will be left with a discouraging, "So what?" unless we also examine further the complex issues of adoption—putting ideas into practice. Thus, the social side of sustainability research occupies a unique position compared to its partners, environmental/ecological and economic research.

This chapter offers a look at what has been done in one field intimately involved with moving from research to adoption: the field of motivation. We begin our discussion by reviewing the major motivation theories from psychology, diffusion, educational psychology, and organizational behavior. We then provide an overview of the motivation research that has been conducted in the field of sustainable agriculture. After presenting the theories and critical motivation work related to sustainable agriculture, we identify some of the tools available to social scientists to measure motivation, and offer examples of ways in which farmers, researchers, and educators discuss motivation. The quotations used in the "voices" section are taken

Developing and Extending Sustainable Agriculture
doi:10.1300/5709_14

from an evaluation study funded by North Central Region (NCR) Sustainable Agriculture, Research and Education (SARE) (unpublished Final Project Report by S. Trout and C. Francis, 2005, available from the NCR SARE office in Lincoln, Nebraska). We conclude with some challenges to all researchers interested in realizing behavioral outcomes. In this way, we hope to provide researchers of every discipline and advocates who embrace "sustainability thinking" with knowledge and tools that will help increase the adoption of sustainable practices.

One of our goals is to stimulate study and educational programming that will accelerate conversion and adoption rates by farmers and ranchers, researchers, and educators. This outcome will provide an important step toward impacting policy and public opinion in ways that will ultimately bring the sustainable mind-set into the mainstream. Another goal is to reveal tangible measurement tools that are available to researchers and educators as they work to build knowledge about the social science of sustainable agriculture. Finally, we use interview participants' own voices to demonstrate ways in which persons in the sustainable community discuss their motivations. This allows researchers and practitioners to recognize the obvious linkages between theory and practice.

In this chapter, we refer to professional agriculturalists as farmers, ranchers, and producers. We recognize that many in sustainable agriculture dispute the term *producer,* arguing that it depersonalizes individuals and places them as cogs in an industrialized system that is not congruent with the sustainability paradigm. Farmers and ranchers are often perceived as hard laborers and not given the respect they deserve for managing and operating high-value, high-risk, and complex business organizations. We acknowledge the dilemma and conclude for simplicity that all four terms are appropriate to use interchangeably. Regardless of the terminology, we provide this discussion to strengthen sustainable agriculture literature related to motivation and adoption. We intend to lead readers to a better understanding of motivation as a well-established dimension of social science, an understanding that can help increase adoption of sustainable practices, research, and education.

MOTIVATION THEORIES

What is motivation? According to motivation researchers Higgins and Kruglanski (2000), the root of the term motivation is "to move," yet most people equate motivation with the term "to want." They identified six major themes among the research and theories of social/personality motivational scientists over the past half century:

1. basic wants
2. when wants change
3. bridging the gap between knowing and doing
4. getting what one wants
5. knowing from wanting and
6. wanting from knowing.

While considerable debate continues to capture the attention of scholars from various disciplines, rare is the motivation theory that does not, in some way, refer to Maslow's seminal work in 1943 in which he conceptualized a hierarchical construct of five basic human needs—physiologic, safety, love, esteem, and self-actualization. According to his theory, once the lowest, most basic need (physiologic, e.g., food, shelter) is satisfied or is near to being satisfied, that satisfied condition then makes way for the individual to begin "wanting" to satisfy the next higher need (safety). And as that need reaches near-satisfaction, the individual's desires and attention are then directed to the next need (love), and so forth. A basic premise of Maslow's theory is that the next higher need can only become a want—thus a motivation—once the lower ones are satisfied.

Maslow's theory was challenged by Alderfer (1969), who argued that Maslow's work could be condensed to a core of three needs: *existence, relatedness,* and *growth* (ERG). Existence needs include all the various forms of material and physiological desires. One of the basic characteristics of existence needs is that one person's gain is another person's loss, when resources are limited. Relatedness needs include all the needs that involve relationships with significant other people. Growth needs, according to Alderfer, include all the needs that involve people making creative or productive impacts on themselves and the environment. In essence, the growth need can be related to the need to learn or grow intellectually, or in Maslow's hierarchy, the need for self-actualization.

A major difference between Maslow's hierarchy and Alderfer's ERG theory is the order in which these needs are able to provide impact. Maslow believed the lower need must be largely satisfied before the next need would "kick in" and motivate the person to begin "wanting." Alderfer's theory proposes that each need moves to the forefront based on the strength of the need—either existence-based or growth-based—which is ultimately defined by each person's individual preference. Existence-based individuals tend to move between existence needs and relatedness needs, whereas growth-based individuals tend to move between and growth needs and relatedness needs.

McClelland (1987) also approached motivation from a needs theory, but hypothesized that needs change over a person's lifetime and are impacted by life experiences. He believed that most needs could be classified into one of three categories, *achievement, affiliation,* or *power.* McClelland described people with high need for achievement as those who seek to excel and thus tend to avoid both low-risk and high-risk situations. Those with high affiliation need, according to this researcher, need harmonious relationships with other people and need to feel accepted by others. Need for power can be one of two types, McClelland hypothesized—personal and institutional. Those who need personal power want to direct others, while those who need institutional, or social power, want to organize the efforts of others toward common goals.

Pittman (1998) explains that a person studying motivation from a psychological perspective likes to look within the person for desires. For example, they ask themselves, "What is wanted? What is strived for? What will feel satisfying or unsatisfying to the actor?" Pittman supports the notion that these basic characteristics revolve around the assumption that the person is an active participant. Alderfer's work in organizational behavior seems to be supported by Pittman and others, that humans have a basic need to learn or grow.

Pittman contends that one fundamental issue that influences a person's motivation appears to be the task of making sense out of and acting in a world that is extremely complex and uncontrollable. They inherently ask themselves questions like, "How do individuals construct their understanding of their world?" And "How do individuals deal with unresponsive, confusing, and sometimes frightening and hostile environments?"

A self-proclaimed unorthodox researcher, W.F. Whyte (1991) embraces methods across several disciplines. In doing so, Whyte projects one of the facets of motivation science—the role of the active participant.

> It is impossible to explain how it is that many individuals remain committed to certain objectives over a long period of time during which they receive little or no rewards and encounter many frustrations. To explain such commitments, we have to make two assumptions: that humans are not simply passive subjects, but rather active agents; and that they become especially strongly committed to the objectives they set for themselves—or at least play an active role in setting. (p. 177)

The assumption that the individual is an active player is also embedded in Deci and Ryan's (1985) self-determination theory (SDT). This theory maintains that an understanding of human motivation requires a consideration of three innate psychological needs for *relatedness, competence,* and *autonomy.* In SDT, these psychological needs are considered essential for understanding the *what* (i.e., content) and the *why* (i.e., process) of goal pursuits.

Relatedness refers to the desire to feel connected to others—to love and care, and to be loved and cared for—and is present in virtually every motivation theory from Maslow onward. *Competence* indicates a need to know or understand. This is consistent with earlier references to the growth orientation of other theorists (e.g., Alderfer, 1969; Maslow, 1943; McClelland, 1987; White, 1959; Whyte, 1991). Self-determination theory further assumes that this inherent tendency will be actualized so long as the necessary and appropriate conditions are present.

Finally, *autonomy,* in Deci and Ryan's work, refers to an individual's ability to self-organize experience and behavior and to make decisions consistent with one's integrated sense of self. Autonomy is distinct from independence, the latter of which assumes that an individual could exist in an environment totally separate from others. Autonomy concerns the extent to which people authentically or genuinely concur with the forces that influence their behavior. The issue, according to Deci and Ryan is whether people are pawns to those forces, or whether people perceive the forces as being valuable, helpful, and congruent sources of information that support their initiative.

Self-determination theory has received considerable attention and has accumulated a strong empirical support for its propositions. In their work, Deci and Ryan determine that social contexts and individual differences that support satisfaction of the basic needs facilitate natural growth processes. On the other hand, those contexts and differences that forestall autonomy, competence, or relatedness are associated with poorer motivation, performance, and well-being.

Further, when people are experiencing a reasonable need satisfaction, according to Deci and Ryan, they will not necessarily be acting specifically to satisfy their needs. Rather, they will be doing what they find interesting or important. SDT recognizes this natural direction of activity so long as the conditions surrounding the individual support competence, relatedness and autonomy. Under these conditions, the individual will innately work toward the satisfaction of relevant needs.

To help bring some order from the diverse approaches, Leonard et al. (1999) proposed an integrative taxonomy of motivation built on past research efforts in the field. In their model, several motivation theorists' perspectives are integrated, identifying five sources of motivation: intrinsic process, instrumental, self-concept external, self-concept internal, and goal internalization.

Intrinsic process motivation refers to being motivated by the sheer fun of doing the act or work itself, and not the end result. Other researchers have referred to a similar motivation in terms of sensory intrinsic motivation and physiological intrinsic motivation (Bandura, 1986), task pleasure (Deci, 1975), and principles embedded in the impulsive stage (Kegan, 1982; Loevinger, 1976). Piaget (1972) discussed concepts relevant to intrinsic process motivation in his preoperational stage of life cycle.

Instrumental motivation is when a person perceives that his or her behavior will lead to certain tangible outcomes, such as pay, promotion, or time off. This motivation source integrates Barnard's (1968) exchange theory and Katz and Kahn's (1978) external rewards. Piaget's (1972) concrete operational stage, McClelland's (1987) need for power, and Barnard's (1968) material inducements were also instrumental in the theoretical development of this motivation source.

Self-concept external motivation refers to that which is other-directed. Persons with this motivation seek affirmation of their traits, competencies, and values. Affiliation with reference groups is very important. This motivation source embraces Deci's (1975) extrinsic interpersonal motivation and Barnard's (1968) social inducements, conformity to group attitudes, and communion constructs. Similar ideas were discussed in Kegan's (1982) interpersonal stage and Piaget's (1972) early formal operational stage, as well as in McClelland's (1987) need for affiliation, and Maslow's (1943) and Baumeister and Leary's (1995) needs for love, affection, and belonging. Katz and Kahn (1978) also integrated self-concept-external-type concepts when they described the ways in which employees seek membership in organizations and approval from leaders and groups.

Self-concept internal motivation implies that a person is inner-directed and is doing the act because it fits with one's personal definition of the ideal self. This source integrates McClelland's (1987) high need for achievement, Deci's (1975) internal motivation, and Katz and Kahn's (1978) idea of internalized motivation. Similar constructs emerged in the descriptions of Piaget's (1972) full formal operational stage, McClelland's (1987) need for achievement, and Maslow's (1943) need for esteem. Bandura (1986) used similar terms in his description of self-evaluative mechanisms, self-regulation, and personal standards.

Whereas self-concept internal motivation focuses attention on one's ideal self, *goal internalization* refers to the motivation that causes a person to work for the good of the cause, which is outside oneself. That is, the attitude or behavior is adopted because it is congruent with one's personal value system, and not because the individual will directly benefit from the action. This source is similar to Katz and Kahn's (1978) internalized values, Deci's (1975) internal valence for outcome, and Etzioni's (1961) pure moral development. Similar concepts were discussed in Piaget's (1972) post-formal operational stage, by Kegan (1982) in his inter-individual stage of development, and by Maslow (1943) when he described self-actualization. The taxonomy articulated by Leonard, Beauvais, and Scholl (1999) provided the framework for a measurement tool that we will discuss later in this chapter.

To summarize, motivation in the social sciences has been examined from many perspectives, including need-based, intrinsic, social identity, sense-making, value-based, self-determination, self-concept, and to some extent, developmental perspectives. Next, we look at the ways motivation has been studied in sustainable agriculture so we can begin to get a picture of where those studies fit within today's leading motivation theories.

MOTIVATION AND SUSTAINABLE AGRICULTURE

According to Pittman (1998) motivation involves the activation of internal desires, needs, and concerns. He proposes that motivation energizes behavior and sends the organism in a particular direction aimed at satisfaction of the motivational issues that gave rise to increased energy. Along with this definition, Pittman includes Schachter and Singer's (1962) caveat that the connections between the energizing and direction functions are not always simple and straightforward. He seems to have had agricultural producers in mind when he included this caveat.

Making sense of or organizing one's complex world seems to be highly relevant in the study of sustainable agriculture. Producers work with considerable uncertainty, caused by the very nature of their work. An entire year's income is laid out at the mercy of Mother Nature to either nurture over a growing season or obliterate in a single storm. Financial payback for farmers efforts comes in chunks scattered sporadically over the year, rather than in predictable monthly or bi-monthly salary checks. The economic value of their year's efforts is vulnerable to the markets, which are often affected by the whims of persons far-removed from the rural landscape, or by international currency markets distant from the farm gate. Trying to make sense of such a complex world is an important part of the mix when deciding to change production systems to yet another complex way of meeting one's existence needs.

Alderfer's (1969) work offers some intriguing possibilities in studying motivation and sustainable agriculture. Certainly there is an existence need. Bills must be paid. People and animals must be fed. Taxes must be paid. These basic needs are essential to survival and must be worked into the equation as farmers and ranchers make production decisions. As we have already suggested, there is a strong social context in producers' lives that motivates actions. Farmers have a role in the community, which they are expected to maintain; their life's work is in clear view for others to see and judge; their decisions impact—or are impacted by—their families; and their ability to operate successful farms and ranches adds strength and quality of life to rural communities, schools, and faith communities. Growth is also an apparent characteristic of adopters, as can be seen in studies such as the adoption of innovation studies pioneered by White (1959), and by Whyte's (1991) interdisciplinary look at the role of social action in the change process.

While each of Higgins and Kruglanski's (2000) themes is relevant to the discussion of motivation at the individual level, most of the research in sustainable agriculture has focused on Theme 3, bridging the gap between knowing and doing. Some research has been conducted at the level of community impacts. Another cluster of research examines the diffusion of innovation, which is led by Rogers' (1995) 40 years of theoretical and empirical development. His work identifies characteristics of innovators and early adopters, the adoption process itself, how local knowledge is used to influence change, and how formation of social groups leads to the expansion of ideas that lead to exploring new alternatives.

Some attention has been given to understanding the personal motivation of individual producers. Salamon (1992) has conducted studies using sociological methods to address the understanding of motivation. In these, she gathered social characteristics, demographics, and participant comments, then formed conclusions from the aggregated data. While this work is helpful, the qualitative nature of her work prevents the analysis to be generalized to a larger audience. Dollisso and Martin (1999) attempted to get at farmers' motivation to participate in educational programming by using an adult education framework that incorporated McClellan's needs theory. These researchers assessed learners' motivation based on preferences farmers reported regarding learning, such as experiential learning, lecture, and other instructional/learning formats. Their work reflected a reasonable attempt to incorporate pedagogy relevant to this population, but they failed to build from the motivation literature.

In Black's (1995) study of policy issues that foster sustainable agriculture, he mentioned motivation in his concluding statement, that educational programs are hampered without understanding the farmers' motivation. He stopped short of suggesting how motivation should be measured. Allen (1993, 2004) describes the motivation for personal decisions in the food system, based primarily on empirical studies.

Fairweather's (1999) study of decision making among adopters and nonadopters of organic agricultural practices in New Zealand demonstrated that having a motivation is no guarantee that farmers will actually grow organic products. There is a need to understand what motivates some producers to embrace the paradigm of sustainable agriculture and then to act on that motivation and implement production practices consistent with the sustainable agriculture paradigm.

Fairweather conducted a comprehensive New Zealand study in which he used Gladwin's (1989) decision tree model to examine how farmers choose between organic and conventional production. Fairweather's study resulted in some interesting quantitative findings that seemed to work well to quantify the decision-making process among farmers. However, the rational-choice decision model has been shown to have serious limitations when compared to actual behaviors (Yukl, 1998).

A quantitative investigation of 29 Missouri cash grain farmers exploring motivational benefits received from farming (Kliebenstein et al., 1981) revealed that doing something worthwhile and being one's own boss were rated the top two motivations from among a predetermined list of eleven "motivation" options. No explanation was provided as to how these options were linked to motivation theory.

In their evaluation of the SARE Producer Grant program, editors den Biggelaar and Suvedi (SARE, 1998) used various qualitative and quantitative approaches to study all three components of sustainability when they compared farmers who had received SARE funding during the first five years (1992-1996) and those who had applied for—but had not received—such funding. They approached the examination of motivation by asking questions about farmers' goals, problems, and bottlenecks, quantifying those responses into general response-type categories, and cross-tabulating data with a self-reported "sustainability" rating.

While these and other studies demonstrate an attempt to examine the motivation of sustainable agriculture adopters, they have done so largely without articulating a strong theoretical framework. In most cases, their work could have been strengthened even further by the use of stronger measurement tools shown to be reliable in the motivation field.

In our testing of a motivation theoretical framework, we found that including goal internalization motivation, openness to new ideas, and low resistance to those behaviors was the best model for predicting sustainable agricultural behaviors. This study was developed from the theories presented here, and it used one of the measurement tools available to measure participants' motivation. In the next section, we describe several of the instruments and methods used in contemporary studies to measure this illusive concept.

TOOLS OF THE MOTIVATION TRADE

One of the greatest challenges in any study is to determine which measurement tools to use in order to accomplish the research goals. While theories can be built based on careful study of data and intuitive conclusions, the best measurement instruments are designed to withstand rigorous psychometric scrutiny following their use in a variety of studies. Following are several tools that have been developed to measure motivation and comments about their respective structures, strengths, and limitations.

D.C. McClelland, one of the most frequently cited motivation theorists in the sustainable agriculture literature, developed his need theory from the Thematic Apperception Test (TAT). This test of imagination shows subjects a series of ambiguous pictures and asks them to develop spontaneous stories for each picture. The assumption is that the subject will project his or her own needs into the story. From the responses, subjects are scored and determined to be motivated by achievement, power, or affiliation. The TAT has been used extensively, and psychologists have developed fairly reliable scoring techniques (Zurbriggen and Sturman, 2002).

The Miller Motivation Scale (Miller, 1988) is a Web-based scale designed to measure positive and negative aspects of an individual's motivation. It measures responses to 160 statements that describe the subject's behavior, attitudes, and feelings about a situation, or reasons for a given behavior. The instrument was developed from a review of the literature, which led the designer to identify eight domains; four of these domains are usually considered positive (creative, innovative, productive, cooperative), and four are generally viewed as negative emotions (attention, power, revenge, give-up). The development of this scale is relatively thorough, and based on the review by Plake et al. (2003), its greatest weakness is that it has not yet been tested in a general population.

The Work Potential Profile (Rowe, 1997) was developed as a tool for the initial descriptive assessment of individuals seeking employment. The instrument consists of 171 items organized in six areas comprising 19 scales. Most of the items are statements about disposition, preferences, and personal characteristics. One of the six areas measured is motivation, which is identified within work motivation, intrinsic motivation, extrinsic motivation, need for status, and a total motivation score. Two of the greatest limitations of this tool, according to reviewers (Plake et al., 2003) are the wording that makes it specific to Australia and the weak psychometric data available to assess its reliability and validity.

The Motivational Patterns Inventory (Byrd and Neher, 1990) was designed to help members of an organization explore dominant motivations that can affect the way they contribute to the success of that organization

(Conoley and Impara, 1995). The instrument's 20 items are developed to measure the subject's work value system against three competing values systems labeled as *farmers, hunters,* and *shepherds.* Farmers value quality, technical competence, and constant attention to details. Hunters value competition, objective goal achievement, power, and recognition. Shepherds value cooperation, teamwork, and involvement with people. Reviewers (Conoley and Impara, 1995) caution that the instrument is designed primarily as a teaching tool and lacks data to support its reliability and v alidity.

The Motive to Attain Social Power (Good and Good, undated) was developed to evaluate one's probable enjoyment or dislike (motivational preference) for a variety of socialized power activities. The instrument includes 28 true or false statements that aim to discern distinctions articulated by McClelland (1987). While the instrument received relatively high praise for its ease of use and scoring, reviewers were highly skeptical of an instrument that was developed in 1970, yet did not reflect any development or testing in 30 years.

The Motivation Sources Inventory (MSI) (Barbuto and Scholl, 1998) is a 30-item questionnaire that measures five motivation subscales that represent the Leonard et al. (1999) integrative taxonomy, discussed earlier in this chapter. The strength of the MSI (Barbuto and Scholl, 1998) in an agricultural population maintained validity comparable to other studies, which had not included farmers and ranchers (e.g., Barbuto et al., 2000).

Having identified several of the tools that are available to quantitative researchers, we now turn to the most important question: How do producers discuss and describe their personal motivation to engage in sustainable practices? To hear from these experts, we turn to a few of the interviews that were conducted as part of our evaluation of the NCR SARE program in which we interviewed more than 20 SARE grant recipients. We include a sample of quotations here so that readers can begin to see where responses support or depart from the motivation theories discussed earlier.

MOTIVATION IN PRACTICE

In his book *Social Theory for Action: How Individuals and Organizations Learn to Change,* Whyte (1991) summarizes the ever-present paradox that exists in agriculture: Why do farmers and ranchers choose this way of making a living? Several of our study participants seemed to concur with Deci and Ryan's (1985) emphasis on autonomy as one of the primary reasons. "I can be my own boss," and "You don't have to deal with people every day," was a common comment. And while the desire for autonomy rang clearly, the relatedness need also came through in comments such as this woman's as she describes her adjustment from city to rural life: "What I didn't realize is that,[in] farming, you're very isolated." Or the frequent reference to the neighbors: "I don't know what the neighbors think"; "I think the neighbors think I'm pretty crazy for trying some of the things I do"; or "I don't put [my experimental plots] out by the road where the neighbors can see them."

Other needs described in the theories above were easy to hear in virtually every interview, as these examples indicate:

> My wife was pregnant . . . and we had to test our water for nitrates. And we found out that our water wasn't safe for our baby that was going to be born. And we'd thought we'd been doing everything pretty good, here, and [laughs] it really was kind of a catharsis for us and got us really looking at what we were doing. (Safety; Growth)
>
> I just wanted my kids to know what their mother did. I wanted to have them with me. At first it was lonely, but I got some help from a neighbor who was just curious about a woman from the city trying to farm. (Relatedness)
>
> The labor—the finance and the labor was the thing. But I guess I'm like that Taurus we talked about. I kinda like something different or challenging once in a while. And I think that was a big part of it too, I really do. (Safety; Growth)

These and other examples provide ample evidence that the theories described here offer appropriate frameworks for studying motivation in ways that will allow more empirical work that can add to our understanding of adoption of sustainable agricultural practices.

By shifting from conventional, "what I know," agriculture to sustainable, "new territory" practices, producers must be willing to learn how to raise, harvest, store, and market new products. These new areas of learning probably mean using different equipment and facilities, finding new markets for their noncommodity crops or exotic animals, and accepting the new label cast by one's neighbors. A farmer, once proud to be known as "the guy with the straightest rows in the county," must be able to remain self-confident as he may hear himself come to be called "the farmer's market guy." Indeed, our experience interviewing SARE producers revealed that this new territory often means adapting to an entirely new identity. Several interviews uncovered a common opinion that adopters often have to accept a relabeling by their neighbors that, "[those guys] aren't really farmers anymore."

The sources of motivation articulated by Leonard et al. (1999) and measured by Barbuto and Scholl (1998) seem appropriate when studying motivation in agriculture, as seen in representative statements made by participants in our qualitative study. When asked, "What motivated you to get started in sustainable agricultural research?" the following comments were among those we heard: "Having fun! Just have fun! Whatever you're gonna do, just do it and have a jolly time about it!" *(intrinsic process);* "I look pretty heavily at costs;" "It's obviously gotta pay the bills." *(instrumental motivation); "I like it when the neighbors drive by, and then they stop and get a little curious." (external self-concept);* "But with sustainable agriculture, they try a few new things . . . whatever it is. And . . . they start getting excited, because they're actually trying something new and they're learning something!" *(internal self-concept);* "So when I returned to my grandparents' farm, I saw that the renters had just mined the hills. There were no wildflowers anywhere. And it nearly broke my heart. . . . Today, the whole hill is blooming with more than 15 species of wildflowers" *(goal internalization).*

By using the solid theoretical foundations of the motivation field, our understanding of motivation will lead to a better understanding of how advocates can be more effective at enticing their neighbors to explore the opportunities that exist in this new way of operating their farms and ranches.

Research into the social side of sustainable agriculture is quite different from that which is done in the other two legs of the sustainability stool—economic and environmental. The social science domain contains an action component that is critical to expansion of the sustainability paradigm into mainstream production. Contemporary motivation literature has a lengthy history that dates back to Maslow (1943), who articulated his hierarchy of needs. Many of these theories have considerable relevance to the dilemma faced in agriculture to understand "what makes producers tick" and then to use that knowledge to increase their curiosity and awareness of sustainable agricultural concepts and practices. By drawing more strongly from the motivation literature, we will make a greater impact on the adoption of sustainable agricultural practices.

CONCLUSIONS

Implications for Researchers

The next frontier in sustainable agriculture research resides in the application of social sciences related to changes on the farm. It is encouraging to see that motivation science is providing increasingly meaningful theories and tools that can assist researchers in this particular area of the social science exploration. We encourage more social scientists to share their theoretical foundations with their physical science and economic science colleagues. Although these contributions must be pared down in order to make them palatable to a "marginally interested" audience, the discussions are important. As the agronomists, animal scientists, agricultural economists, plant scientists, and others see evidence of the theoretical foundation of this work, we may begin to see more influential integrations of research and outreach efforts. Ultimately, we would encourage collaborative research that can provide a more holistic understanding of the adoption of sustainable practices and the economic, environmental, and social outcomes that result.

We encourage social science researchers to use the most up-to-date, empirically sound research instruments and designs available. In this way, we will be able to present replicable studies that will help us understand social issues and outcomes of sustainability interventions in ways that improve our ability to understand this complex phenomenon. In doing so, we will have the greatest ability to see a meaningful, measurable increase in adoption of sustainable practices throughout mainstream agriculture.

Implications for Educators

Educators interested in opening minds to the possibilities that exist in sustainable agriculture should incorporate motivation science into their pedagogical underpinnings. They need to plan programs that will not only add to producers' knowledge, but also will help farmers and ranchers get a mental picture of how they can organize the new production practices in their already complex world. Programs should be planned to include sessions that address the social realities that will face producers, especially during the experimental stage. Time also should be scheduled for safe exchanges of ideas as interested producers seek ways to resolve their personal uncertainties regarding knowledge, organization, and social issues of greatest interest to them at that moment (Francis and Carter, 2001).

Careful attention also needs to be paid to the language used in promoting these programs. Session titles, key attention-getting words, and session descriptions need to capture the attention and stir the emotions of prospective participants. These words should appeal to the needs identified in the motivation theories presented here.

Finally, when delivering programs, the language and presentation of material should tap into the motivations—the needs—of producers. It should resonate in the emotional center of producers' minds, meeting their current high-priority needs, but also point out the stimulating opportunity for this new paradigm to lead to growth of the individual, and the ways in which that growth will allow a strengthening of relationships important to farmers and ranchers. As more is learned through empirical testing for prevailing motivations within rural populations, the probability of reaching people through the right messages will be increasingly more predictable.

As long-time researcher and advocate of sustainability Cornelia Flora wrote, "The strength of a positive vision exists in its ability to fire people to actively seek these ends" (Flora, 2000). It seems obvious that we're going to stimulate people with the right triggers only when we fully understand their motivation and the context—economic, environmental and social—in which they operate. Social science provides many of our tools to pursue this worthwhile scholarly engagement.

We must also acknowledge the dilemma that exists within this discussion: As researchers and educators, are we impartial and unbiased "scientists," or are we advocates who readily admit our personal bias? If the latter, then does that diminish the credibility of our research or programming? As authors we represent the spectrum of responses to these questions. By working collaboratively, we have not only informed each other of new ways of approaching the subject of motivation, but we have integrated an internal check and balance that may provide one solution to the very real dilemma that exists in almost any research situation. Perhaps there are other solutions as well. We encourage a dialogue that will help us address this important issue.

We cannot afford to lose credibility with experts and decision makers, but we also cannot afford to sit back and be so neutral that our communities, families, and the environment that sustains us are sacrificed. By drawing on the strongest, most contemporary theories, using the strongest tools, and drawing research questions and educational programming most relevant to our target populations, we are confident that our understanding of motivation can be used to increase the number of professional agriculturalists who "discover" and embrace the world of sustainable agriculture.

REFERENCES

Alderfer, C.P. 1969. An empirical test of a new theory of human needs. *Organ. Behav. Human Perform.* 4:43-175.

Allen, P. 1993. *Food for the future.* John Wiley & Sons, New York.

Allen, P. 2004. *Together at the table: sustainability and sustenance in the American agrifood system.* Penn State Univ. Press, Universtiy Park, PA.

Bandura, A. 1986. *Social foundations of thought and action: A social cognitive theory.* Prentice Hall, Englewood Cliffs, NJ.

Barbuto, Jr., J.E., S.M. Fritz, and D. Marx. 2000. A field study of two measures of work motivation for predicting leader's transformational behaviors. *Psychol. Rep.* 86:295-300.

Barbuto, Jr., J.E. and R.W. Scholl. 1998. Motivation sources inventory: Development and validation of new scales to measure an integrative taxonomy of motivation. *Psychol. Rep.* 82:1011-1022.

Barnard, C.I. 1968. *The functions of the executive.* Harvard University Press, Cambridge, MA.

Baumeister, R.F. and M.R. Leary. 1995. The need to belong: Desire for interpersonal attachments as a fundamental human motivation. *Psychol. Bull.* 117:497-529.

Black, A.W. 1995. Policy instruments designed to foster sustainable agriculture: An appraisal. *Res. Rural Sociol. Develop.* 6:123-147.

Byrd, R.E. and W.R. Neher. 1990. Motivational patterns inventory. Reviewed in 1995. *The twelfth mental measurements yearbook,* J.C. Conoley and J.C. Impara (eds.). Buros Institute of Mental Measurements, Lincoln, NE. pp. 645-647.

Conoley, J.C. and J.C. Impara (eds.). 1995. *The twelfth mental measurements yearbook.* Buros Institute of Mental Measurements, Lincoln, NE.

Deci, E.L. 1975. *Intrinsic motivation.* Plenum Publ., New York.

Deci, E.L. and R.M. Ryan. 1985. *Intrinsic motivation and self-determination in human behavior.* Plenum Publ., New York.

Dollisso, A.D. and R.A. Martin. 1999. Perceptions regarding adult learners' motivation to participate in educational programs. *J. Agric. Educ.* 40(4):38-46.

Etzioni, A. 1961. *A comparative analysis of complex organizations.* Free Press, Glencoe, IL.

Fairweather, J.R. 1999. Understanding how farmers choose between organic and conventional production: Results from New Zealand and policy implications. *Agric. Human Values,* 16(1):51-63.

Flora, C.B. (2000). A vision for the northern Great Plains. *Great Plains Sociol.* 12:79-91.

Francis, C.A. and H.C. Carter. 2001. Participatory education for sustainable agriculture: Everyone a teacher, everyone a learner. *J. Sustain. Agric.* 18(1):71-83.

Gladwin, C. 1989. *Ethnographic decision tree modeling.* Sage Publ., Thousand Oaks, CA.

Good, L.R. and K.C. Good. (no date). Motive to attain social power. Reviewed in 2003 *The fifteenth mental measurements yearbook,* B.S. Plake, J.C. Impara, and R.A. Spies (eds.). Buros Institute of Mental Measurements, Lincoln, NE. pp. 598-599.

Higgins, E.T. and A.W. Kruglanski. 2000. Motivational science: The nature and functions of wanting. In: *Motivational science: Social and personality perspectives,* E.T. Higgins and A.W. Kruglanski (eds.). Psychology Press, Philadelphia, PA.

Katz, D. and R.L. Kahn. 1978. *The social psychology of organizations.* John Wiley & Sons, New York.

Kegan, R. 1982. *The evolving self.* Harvard University Press, Cambridge, MA.

Kliebenstein, J.B., W.D. Heffernan, D.A. Barrett, and C.L. Kirtley. 1981. Economic and sociologic motivational factors in farming. *J. Am. Soc. Farm Managers Rural Appraisers.* 45(1):10-14.

Leonard, N.H., L.L. Beauvais, and R.W. Scholl. 1999. Work motivation: The incorporation of self-concept-based processes. *Human Relat.* 52:969-998.

Loevinger, J. 1976. *Ego development.* Jossey-Bass Publ., San Francisco, CA.

Maslow, A.H. 1943. A theory of human motivation. *Psychol. Rev.* 50:370-396.

McClelland, D.C. 1987. *Human motivation.* Cambridge University Press, Cambridge, UK.

Miller, H. J. 1988. Miller motivation scale, publ 1986. Reviewed in *The fifteenth mental measurements yearbook,* Plake, B.S., J.C. Impara, and R.A. Spies (eds.). Buros Institute of Mental Measurements, Lincoln, NE. pp. 573-575.

Piaget, J. 1972. Intellectual evolution from adolescence to adulthood. *Human Develop.* 15:1012.

Pittman, T.S. 1998. Motivation. In: *The handbook of social psychology,* D.T. Gilbert, S. T. Fiske, and G. Lindzey (eds.). McGraw-Hill, New York. pp. 549-590.

Plake, B.S., J.C. Impara, and R.A. Spies (eds.). 2003. *The fifteenth mental measurements yearbook.* Buros Institute of Mental Measurements, Lincoln, NE.

Rogers, E.M. 1995. *Diffusion of innovations* (4th ed.). Free Press, New York.

Rowe, H.A.H. 1997. Work potential profile. Reviewed in *The fifteenth mental measurements yearbook,* B.S. Plake, J.C. Impara, and R.A. Spies (eds.). Buros Institute of Mental Measurements, Lincoln, NE. pp. 1028-1031.

Salamon, S. 1992. Social and cultural factors affecting sustainable farming systems and the barriers to adoption. (SARE Grant LNC92-050.) North Central Regional SARE, Lincoln, NE.

SARE. 1998. *Ten years of SARE: A decade of programs, partnerships and progress in Sustainable Agriculture, Research and Education.* C. den Biggelaar and M. Suvedi, (eds.). USDA Cooperative State Research, Extension and Education Service, Washington, DC.

Schachter, S. and J.E. Singer. 1962. Cognitive, social, and physiological determinants of emotional state. *Psychol. Rev.* 69:79-399.

White, R.W. (1959). Motivation reconsidered: The concept of competence. *Psychol. Rev.* 66:297-333.

Whyte, W.F. 1991. *Social theory for action: How individuals and organizations learn to change.* Sage Publ., Thousand Oaks, CA.

Yukl, G.A. 1998. *Leadership in organizations.* Prentice Hall Publ., Englewood Cliffs, NJ.

Zurbriggen, E.L. and T.S. Sturman. 2002. Linking motives and emotions: A test of McClelland's hypotheses. *Pers. Soc. Psychol. Bull.* 28(4):521-535.

Chapter 15

Future Potential for Organic Farming:
A Question of Ethics and Productivity

Frederick Kirschenmann
George W. Bird

Organic farming is . . . about positive acceptance of the natural order
and the intention to work with its laws.

Philip Conford

INTRODUCTION

During the final decades of the twentieth century, organic agriculture
went mainstream—in the supermarkets and in farmers' fields. This rapid
adoption threw a series of questions into the public domain. What is organic
agriculture? How is it defined? What practices are allowed? How can customers be sure that food labeled organic is indeed produced according to
organic standards?

In the not-for-profit sector, the International Federation of Organic Agriculture Movements (IFOAM) emerged to address these concerns. A set of
international standards was developed and eventually an association was
formed to accredit certifiers of organic foods. This system assured the public that foods carrying an organic label, backed by an IFOAM-accredited
certifier, were produced in accordance with an internationally accepted set
of organic standards. Meanwhile government agencies in various parts of
the world developed their own standards to govern the use of organic labels.
The U.S. Congress passed the Organic Foods Production Act of 1990 to
control the use of organic labels on food sold in the United States.

Developing and Extending Sustainable Agriculture
© 2006 by The Haworth Press, Inc. All rights reserved.
doi:10.1300/5709_15

While such control structures were a necessary development in the organic food system, they had the unfortunate effect of focusing attention on what *had* to be done to qualify for a market, instead of focusing on what *could* be done to improve agriculture.

Given organic agriculture's increasing popularity, it is useful to contemplate its future potential within the larger agricultural arena. A brief look at its history may be helpful in foreshadowing the future. There are large challenges facing agriculture, and there is a need to examine an alternative paradigm that involves an ecological conscience in designing future sustainable systems .

EMERGENCE OF ORGANIC AGRICULTURE

Philip Conford (2001) reminds us that as a set of farming practices, organic agriculture has been around for centuries. What has come to be called organic agriculture was simply one of the approaches used by early agriculturalists to achieve and maintain productivity. Nutrient cycling and soil humus enrichment were its key features. Such organic practices were, however, far from the only methods used by early agriculturalists. In fact, as William Rees (1999) points out, the predominant form of early agriculture probably involved hunting out or farming out a particular area and then moving on to new virgin territories, allowing the depleted area to recover. Conford (2001) goes on to point out that as a *movement,* organic agriculture emerged in the middle of the nineteenth century. That movement developed largely as a reaction to the evolution of industrial agriculture.

The *magnum opus* of industrial agriculture was Justus von Liebig's historic *Chemistry in the Application to Agriculture and Physiology* which was first published in 1840 (Liebig, 1872). Von Liebig argued that the productivity of agriculture could be maintained without resorting to the laborious task of manuring soils. He reasoned that chemical fertilizers could be substituted for nutrient cycling practices, thereby simplifying farming. While von Liebig was never able to fully demonstrate his hypothesis in actual practice, John Bennet Lawes and J. H. Gilbert were able to establish the effectiveness of von Liebig's thesis through field trials, and then manufactured and patented superphosphates and built the first fertilizer factory in 1843.

The ability to substitute chemical fertilizers for nutrient cycling practices encouraged farmers to simplify their production systems, specialize in raising one or two crops, and abandon the mixed farming practices which incorporated green manure and livestock into farming operations. This transition to monocropping was further intensified with the repeal of the British Corn

Laws in 1846, a move which embraced free trade and encouraged farmers in virgin territories to produce cereal grains for export to Great Britain.

Farmers who had been committed to the idea that the soil's ecological capital must be restored through proper manuring practices in order for agriculture to remain productive viewed the chemical revolution in agriculture as a disaster. It prompted farmers in exporting countries to exploit and diminish their virgin soils to take advantage of markets overseas, while simultaneously forcing farmers in Great Britain to abandon ecologically sound practices and industrialize their operations to compete. This philosophical conflict proved to be the breeding ground for the evolution of the organic agriculture movement.

One government administrator recognized that "the great movement of cargoes of feeding stuffs and mineral fertilizers" to Europe and America could not "be continued indefinitely," and was aware of the fact that such production methods had "never been possible as a means of maintaining soil fertility in China, Korea, or Japan." F.H. King, former chief of the Soil Division of the U.S. Department of Agriculture, traveled to Asia in the early 1900s to learn how farmers in those countries could have farmed their fields for 4,000 years without depleting soil fertility, despite the absence of synthetic fertilizers. The result was *Farmers of Forty Centuries,* his classic text for organic agriculturalists, published posthumously in 1911 (King, 1911).

Among the many other agriculturalists who had an issue with the new, industrialized agriculture was Lord Northburne who was, so far as we know, the first author to use the word "organic." The word described a method of farming that celebrated the complex interrelationships of parts into a functioning whole. It viewed a farm as an "organism," arguing that a farm cannot be managed like an input-output factory. Northburne presented his ideas in a book entitled *Look to the Land* published in 1940. Lady Eve Balfour followed in 1943 with her *The Living Soil* (republished as Balfour, 1975). Edward Faulkner's *Plowman's Folly* and Sir Albert Howard's *An Agriculture Testament* both appeared the same year (Faulkner, 1943; Howard, 1943). All emphasized that farming could not remain productive without caring for the humus of the soil, and that maintaining the humus quality of the soil could not be accomplished simply by adding chemical fertilizers. Like an organism, a farm consisted of many parts operating synergistically to optimize production while restoring the natural resources essential for continued productivity.

Almost two decades earlier, Rudolf Steiner also had condemned industrial agriculture and harshly criticized the science and economics that supported it. In his view, science and economics had become far too mechanistic to reflect actual world realities. Agriculture, he claimed, had "been hit especially hard" by these "modern cultural and intellectual trends." The biodynamic agriculture movement, a twin sister of the organic agriculture movement, emerged from Steiner's work (Steiner, 1958).

As organic foods became popular and the food industry began to include organically labeled foods in their trade offerings, organic agriculture also became an *industry*. As an industry, organic agriculture was less focused on the farmers who believed in restorative farming practices. It now became a network of organized, commercial activities using a variety of marketing strategies to increase the sale of organic products. Focused on maximizing sales, the organic industry naturally pressed for market efficiency, which in turn demanded organic standards that were uniform and easily administered. But uniform standards discourage innovation and give the comparative advantage to larger firms with more capital and marketing power. Some now argue that this has led to the "industrialization" of organic agriculture (Pollan, 2001).

Given that organic agriculture was philosophically driven as a *movement* and sales driven as an *industry,* tension between the two was perhaps inevitable. That tension is evident in the organic arena today. Farmers who continue to adhere to the philosophical tenets of caring for the soil and restoring the ecological health of the land feel that the industry is only concerned about enhancing sales. Many of those farmers are now voicing the need to move "beyond organic."

International Federation of Organic Agriculture Movements, hoping at its 2002 International Congress to provide an arena for negotiating the differences between philosophical affirmations and commercial ambitions, recognized the potential for a third trend in organic agriculture—the evolution of an organic *community*. The theme selected for the congress was "Cultivating Communities from the Ground Up." Adding a "community" component to organic agriculture is consistent with the growing awareness in the larger sustainable agriculture movement that in order for agriculture to remain productive it must not only be economically viable and attend to biophysical health on a landscape scale, it also must be socially resilient. As Niels Roling and Janice Jiggins put it, "Ecologically sound agriculture requires change, not only at the farm level, but also at higher agro-ecosystem levels, such as watersheds, biotopes and landscapes . . . not only at the level of the farm household, but also at the level of the institutions in which it is embedded" (Rölling and Jiggins, 1998).

If organic agriculture adopts these community and landscape dimensions as an integral part of its vision, it could position itself to provide the much needed leadership to meet the challenges facing agriculture in the twenty-first century. But organic agriculture adherents would have to embrace a deeper sense of community than is common today.

As agriculturalists become more grounded in the principles of ecology and evolutionary biology, there is a growing awareness that the boundaries of "community" must be enlarged to include more than the human community. As Aldo Leopold suggested, such boundaries need to be expanded to include "soils, waters, plants, and animals, or collectively: the land" (Leopold, 1949). Leopold's holistic view, however, is not yet widely accepted in the agricultural sciences. This is perhaps understandable since the formal science of ecology is basically a post-World War II discipline. The term "ecosystem" made an initial appearance in the literature in the mid-1930s and was catalyzed by the first edition of Odum's *Fundamentals of Ecology* (Odum, 1971). But the majority of the scientific community associated with agriculture is still embedded in the mechanistic world view developed in the seventeenth century, and this serves as an impediment to the development of the kind of effective postmodern agriculture needed to face the challenges ahead.

During the sixteenth and seventeenth centuries, the medieval world view was gradually replaced with a mechanistic paradigm that spearheaded the scientific revolution and advocated that system behavior was based solely on the properties of the parts. This view holds that linear relationships—and the notion that the whole simply represents that sum of its parts—determine the outcome of all human activities. Consequently it assumes that resources are either infinite, or that replacement technologies will continually be available, and that nature therefore can be successfully manipulated by means of control management. This mechanistic view, accordingly, pays little attention to feedback loops or to the dynamic relationships of complex, living, evolving systems. Consequently, mainstream agriculturists pay little attention to quantum physics, evolutionary biology, the science of networks, or the science of ecology. But as Heinz Pagels reminds us, with quantum theory "the world changed from having the determinism of a clock to having the contingency of a pinball machine" (Pagels, 1982). In other words, the mechanistic world view can no longer be justified by modern science.

Systems science, not yet widely adopted in agricultural circles, is based on five principles:

1. all systems have integrated, emergent properties;
2. systems require environmental thinking—for example, systems properties can only be understood in the context of whole systems;
3. systems components are actually patterns within a web of interactions;
4. relationships are primary and pattern boundaries are secondary; and
5. systems consist of nested networks.

According to these principles, the role of science is to discover the approximate knowledge with the concept of strong interference playing a key role. This shifts the burden of human intervention from control management to adaptive management. Feedback loops are fundamental to both cybernetics and living systems. In all natural systems, consequently, any species has the biotic potential for exponential growth. However, through the long evolutionary journey of self-correcting relationships such growth is generally checked through environmental resistance. Exponential runaways generally occur only when a system with these inherent self-correcting properties is severely disturbed. It is therefore under severely disturbed circumstances, emblematic of industrial agriculture, that plants become weeds, insects become pests, and fungi/bacteria/viruses/prions become pathogens. This puts farmers on a treadmill of constantly creating the problem they are trying to solve. As Peter Senge (1990) puts it in *The Fifth Discipline,* "The long-term, most insidious consequences of applying nonsystemic solutions is increased need for more and more of the solution."

In spite of the slow recognition of the importance of systems thinking and the continued determinism of conventional agriculture, there is emerging awareness of this "deeper" ecology of agriculture. Together with the new challenges and uncertainties that agriculture will face in the decades ahead, it is likely that a new agricultural revolution, as significant for the twenty-first century as industrial agriculture was for the twentieth century, will be in the making. Since the fundamental principles of organic agriculture anticipated such deeper ecological dimensions, the organic agriculture movement may be uniquely positioned to help us chart our course for this new future.

CHALLENGES FACING AGRICULTURE
IN THE TWENTY-FIRST CENTURY

As we enter the twenty-first century, agriculture faces at least nine major challenges that are likely to force it to move beyond the industrial paradigm of the twentieth century. They include the loss of biodiversity, a devastated farm economy, unchecked and geographically concentrated population growth, persistent poverty, energy transformation, food security, environmental degradation, climate change, and an unprecedented explosion of infectious diseases.

Herman Daly has reminded us that industrial economies—including agriculture—have been successful during the past century because of the availability of two key resources—the natural resources needed to fuel industrial systems, and the natural sinks required to absorb the wastes of such systems (Daly, 1996). Both of those resources are now in serious decline. Industrial agriculture which enabled us to double and triple the yields of cereal grains, is largely fossil-fuel driven. Crop inputs, the manufacture of farm equipment, traction fuel, and the breeding of crop varieties that are responsive to chemical inputs and irrigation are all highly dependent on fossil fuel energy. As fossil fuel resources decrease, and the energy ratio (the amount of energy produced for each unit of energy expended to obtain it) decreases, this mode of production will become increasingly difficult to sustain. Other natural resources, such as fresh water and virgin soils, also are in serious decline, largely due to industrial agriculture's intense demand on these resources.

At the same time the natural resources that fueled industrial agriculture are declining, the natural sinks that absorbed the accompanying wastes are filling up. According to recent reports there now appear to be at least 150 hypoxic zones on the planet, all of them related to watersheds that support industrial agriculture. Hypoxic zones are not isolated aberrations but visible indicators of the more pervasive environmental degradation which seems to be intrinsic to industrial agriculture.

Masae Shiyomi and Hiroshi Koizumi (2001) have argued that the combination of the decline of fossil fuels and the increased environmental degradation, caused in part by fossil fuel-based agriculture, will in itself *force* agriculture to change in the decades ahead. And they suggest that a shift toward an ecologically based agriculture may well pose the most viable alternative for agriculture's future.

> The present system of agriculture, which depends on consumption of tremendous quantities of fossil fuel energy, is now being forced to change to a system where the interactions between organisms and the environment are properly used. There are two reasons for this transformation. The first is the depletion of readily available fossil fuel resources. The second is that consumption of fossil fuels has induced deterioration of the environment. . . . Is it possible to replace current technologies based on fossil energy with proper interactions operating between crops/livestock and other organisms to enhance agricultural production? If the answer is yes, then modern agriculture, which uses only the simplest biotic responses, can be transformed into an alternative system of agriculture, in which the use of complex biotic interactions becomes the key technology. (Shiyomi and Koizumi, 2001)

This decline in resources and saturation of sinks is taking place just as the United Nations estimates that the world's current population of more than 6.1 billion people will increase to 9.3 billion by the year 2050. Furthermore, most of the additional 3.2 billion people will be added in the developing world, and much of that growth will take place in poor rural areas. Seventy-two percent of the world's poorest people live in rural communities (Brown, 2001).

Currently, 1 billion of the world's population survives on less than $1 a day and an additional 1.6 billion live on less than $2 a day. The number of people living in poverty has increased by 100 million during the past decade and the United Nations predicts that another 100 million people will live in poverty by the year 2015 (Lee, 2002).

A fifth challenge facing agriculture is the fact that food security increasingly is being viewed as a basic human right. Not only has the world evolved into a global economy, it also is becoming a global civic society. That transformation carries with it a heightened awareness that global stability cannot be achieved unless everyone is adequately nourished. Securing food as a basic right for all of the planet's citizens therefore presents an additional challenge that global agriculture must face in the decades ahead.

The precise role that climate change will play in agriculture's future is not yet certain, but indicators suggest some formidable challenges. A recent report from the Soil and Water Conservation Society focused on just one climactic variable—precipitation—and assessed the potential effects which increased precipitation may have on soil erosion and runoff on cropland. The study indicates that anticipated increases in precipitation due to climate change, together with the likelihood of more violent storms, "heightens the risk of soil erosion, runoff, and related environmental and ecological damages" (Soil and Water Conservation Society, 2003).

A 2003 Iowa State University study highlighted similar concerns. Using computer modeling techniques, the study suggested that the Upper Mississippi River Basin (UMRB) is likely to see significant precipitation increases by the decade of 2040-2050. The study reported that the "model system produced an increase in future scenario climate precipitation of 21 percent with a resulting 18 percent increase in snowfall, 51 percent increase in surface runoff, 43 percent increase in recharge and 50 percent increase in total water yield in the UMRB" (Jha et al., 2004). It is unlikely that Corn Belt farmers will be able to continue growing massive quantities of annual, monoculture crops such as corn and soybeans under these conditions.

Additionally, most climatologists seem to agree that globally, climate change will bring much greater climate instability with wide swings in temperature and more violent storms. Since specialized systems depend on relative climate stability to maintain consistent productivity, more resilient production systems based on greater diversity may be required in the future to stabilize productivity.

Industrial agriculture with its emphasis on specialization and simplification has contributed to a dramatic loss of both biodiversity and genetic diversity. This loss of diversity has contributed to an ecologically brittle system that has made both crop and animal agriculture highly vulnerable to numerous pests and diseases. Such brittle ecologies, combined with the species density now demanded by an industrial agriculture's infrastructure, raise serious questions about the sustainability of the system. As David Tilman has suggested, the growing inherent difficulties of these industrial systems may soon make diversified farms the farms of choice in the future.

> When those running massive livestock operations realize that chronic disease and catastrophic epidemics are the expected result of high densities and low diversity, and when society restricts the release of pollutants from such operations, it may again be profitable for individual farms, or neighborhood consortia, to have mixed cropping and livestock operations tied together in a system that gives efficient, sustainable, locally closed nitrogen cycle. (Tilman, 1998)

Industrial agriculture also has contributed to a devastated farm economy. The industrial agriculture infrastructure demands that raw materials be acquired as cheaply as possible. The combination of low commodity prices and high input costs has marginalized net farm income. Net farm income in constant dollars in the United States is now at a lower level than it was in 1929 and Canadian net farm income, on a per-farm basis adjusted for inflation, is now lower than at any time since the 1930s (Duffy, 2003; National Farmers Union, Canada, 2003). This leaves farmers with few options for economic survival, let alone the flexibility to meet new challenges. According to 2002 data from USDA's National Agricultural Statistics Service, in the five-year-period from 1997 to 2002 alone, the United States lost more than 14 percent of its farms with annual sales ranging from $50,000 to $500,000.

Finally, agriculture is presented with yet another challenge. An unprecedented explosion of more than 35 new infectious diseases has appeared in the past 30 years. The Institute of Medicine, a research arm of the federal government, convened a panel of scientists in the summer of 2002 to determine the causes of these outbreaks. They attributed the phenomena to 13 changes that have taken place, and Dr. Anthony Fauci, director of the National Institute of Allergy and Infectious Diseases, suggested that a substantial proportion of those changes "relate to man's manipulation of ecology" (Borenstein, 2003). Agriculture, of course, has been a major contributor to such ecological manipulations.

A NEW PARADIGM FOR AGRICULTURE

So here are the requirements and the major questions we face as society plans for the future, and as professional agriculturists we chart a research and development strategy to meet future needs. What kind of future agriculture will

- meet the requirements of an exploding human population,
- in the face of entrenched poverty in a post-fossil fuel era,
- that must restore the ecological health of the natural resources on which agriculture depends,
- while the climate is changing,
- when global society insists that food is a human right,
- when increased infectious diseases require that we attend to the ecological ramifications of human activities, and
- when farmers must retain a sufficient share of the value of their productivity to be economically viable?

As we approach this daunting task, several operating principles come to mind. First, the one-size-fits-all, single-tactic, one-dimensional technologically driven practices, characteristic of industrial agriculture, probably will not meet the challenge—especially if that system is heavily dependent on fossil fuels and irrigation.

In his insightful book *Foods Frontier: The Next Green Revolution,* Richard Manning (2000) reminds us that if we try to solve production problems simply by introducing new, single-tactic technologies into complex social, ecological and political systems, we will probably fail. Developing one-dimensional production solutions in one place and imposing them in a linear, top-down fashion in another place is not likely to meet the challenges ahead. Manning argues that "if there was a key mistake of the Green Revolution, it was in simplifying a system that is by its very nature complex" (Manning, 2000).

Both culture and ecology vary from place to place, both are complex, interdependent, and dynamic. Consequently, long-term sustainable production solutions must adapt to local conditions—both cultural and ecological—to be successful. All of this suggests a need for closer collaboration among people who live in the communities where agriculture is practiced, and for paying much more attention to nature and the complex ecologies of those communities. Only then can we determine what kind of agriculture can be adapted best to each place to meet the challenges of the future.

Accordingly, the new paradigm would seem to suggest a shift *away* from therapeutic intervention technologies, or intervening in a system with an

outside counter force to solve an isolated problem
These have been the preferred technologies of the
would seem to suggest a shift *toward* a natural syste
attempt to understand complex systems, the *cause*
within the system, and access free ecosystem servic
implement self-regulating solutions (Lewis et al.,
require that we switch from control management t
(Gunderson, 1995).

Such a natural systems approach is also likely to encourage a shift away from *single* gene technologies to what Robert Goodman calls a "metagenomic" approach. In other words, a natural systems approach not only would attend to the interdependent, evolving species relationships at the ecosystem level, it also would try to understand the total genome of all organisms in a system rather than transferring single genes for single effects (Manning, 2000). Ideally this approach would then seek to adapt agriculture to the complex, evolving biotic community in its ecosystem in ways that would enhance the entire community's capacity for self-renewal.

This strategy suggests an ecologically-based agriculture which is consistent with the original core principles of organic agriculture. As we have noted, the original concept of organic agriculture proposed integrating the parts of the system into a functioning whole—in other words, developing synergistic relationships among a diverse arrangement of species—crops, livestock, microorganisms, insects (and, now, the genome of all organisms) to achieve, as much as possible, a self-sufficient, self-regulating, self-renewing production system.

While numerous organic farmers have begun to develop such synergistic multispecies systems, few have reached a significant level of sophistication. Takao Furuno (2001) a farmer in Southern Japan who has developed an integrated duck/rice/fish farming system, demonstrates some of the possibilities that such systems, adequately researched, promise for our challenging future.

Instead of producing rice in a monoculture, dependent on fertilizers and pesticides to achieve acceptable yields, Mr. Furuno developed an elegant, complex, species-interdependent system that has increased his rice yields while producing a full range of other food products, without relying on any exogenous crop inputs.

Here is how it works. Right after Mr. Furuno sets out his rice seedlings into his flooded rice paddies, he puts a gaggle of young ducklings into the paddies. The ducklings (a cross-breed of wild and domestic species) immediately start to feed on insects that normally attack young rice plants. They also feed on weeds and weed seeds. Mr. Furuno also introduces loaches, a variety of fish that is easily cultivated and produces a delicious meat product. He also plants azolla, normally considered a "paddy weed." The azolla, a nitrogen-producing aquatic fern, serves as food for the fish and the ducks. In addition to the insects, the ducks feed on golden snails which also attack rice plants. Since the ducks and fish both feed on the azolla, its growth is kept sufficiently under control so it does not compete with the growing rice, but does serve as a source of nitrogen. The nitrogen from the azolla, plus the droppings from the ducks and fish, provide all of the nutrients needed for the rice. Furuno also grows figs on the periphery of his rice paddies, supplying him with fruit.

In alternate years Mr. Furuno rotates his integrated rice/duck/fish paddies, with a crop of vegetables and wheat. He also harvests duck eggs which he markets along with the rice, fish and duck meat, vegetables, wheat, and figs. Mr. Furuno's productivity is enhanced further by the fact that his rice yields in this system exceed the rice yields of industrial rice systems by 20 to 50 percent. This natural systems design makes Mr. Furuno's six-acre farm in Japan one of the most productive in the world.

The idea behind Mr. Furuno's design is simple, yet profound. The concept, he writes, "is to produce a variety of products within a limited space to achieve maximum overall productivity. But this does not consist of merely assembling all of the components; it consists of allowing all components to influence each other positively in a relationship of symbiotic production" (Furuno, 2001). This design allows Mr. Furuno to produce more food that is more nutritionally balanced on less acreage at a considerably lower cost. In recent decades and largely without the help of experts, farmers have begun to develop such elegant synergistic systems in many locations (Pretty and Hine, 2001) (for paper: http://www2.essex.ac.uk/ces/Research Programmes/CESOccasionalPapers/SAFErepSUBHEADS.htm) (accessed April 10, 2006).

Such systems are highly productive, nutritionally diverse, largely self-regulating, and devoid of exogenous inputs, and they suggest interesting new models to meet the multiple challenges that agriculture will face in the decades ahead. These models can begin to address the combined challenges of increased populations in poverty stricken, nutritionally deficient, ecologically damaged communities. Building on the original principles of organic agriculture and employing the lessons learned from ecology and evolutionary biology combined with insights from quantum physics and metagenomics, we may be able to develop farming systems that restore genetic and biological diversity, enable people in resource-poor areas to feed themselves inexpensively, restore ecological resilience that can cope with some of the vagaries of climate change, dramatically reduce energy inputs, and make farming more profitable for farmers.

A shift from the current dominant mechanistic world view to an ecological paradigm is probably necessary to the adoption of this new postmodern agriculture. An ecological world view is cyclical in nature and is based on the assumptions that resources are finite and that the whole is always greater than the sum of the parts. It consists of local ecological interdependence and partnerships, cyclic patterns of organization, system components with over-lapping functions and multiple feedback loops, and existence within a vibrant community. In this view *development* (realizing potential or bringing to an enhanced state), rather than the presumption of infinite *growth*, is a fundamental attribute of the ecological world view. Development recognizes that the goal is to enhance the capacity of the biotic community to renew itself, which Leopold (1949) reminded us is the true goal of conservation. The goal of development, then, is the enhancement of the self-organizing, interdependent and interconnected networks of living organisms interacting to bring the overall system to a fuller, more self-regenerating, state—which is simultaneously the ultimate goal of conservation. Such a goal requires making value judgments grounded in an ecological conscience.

THE EVOLUTION OF AN ECOLOGICAL CONSCIENCE

Dramatic paradigm shifts seldom occur without an accompanying evolution of a new conscience. While organic farming still harbors some of the values implicit in an ecological conscience, we need to transition to a fully ecologically-based agriculture that can meet future challenges. A more fundamental transformation is needed.

Under the subtitle "The Ecological Conscience," the prescient Aldo Leopold (1949) outlined the parameters for the kind of new conscience we need in his "The Land Ethic" published as part of *A Sand County Almanac.*

Leopold lamented the fact that most of the content of our conservation education consisted of "obey the law, vote right, join some organizations, and practice what conservation is profitable on your own land; the government will do the rest." Leopold goes on to ask, "Is not this formula too easy to accomplish anything worth-while? It defines no right or wrong, assigns no obligation, calls for no sacrifice, implies no change in the current philosophy of values." He then suggests that "Obligations have no meaning without conscience, and the problem we face is the extension of the social conscience from people to land." And finally he contends that "No important change in ethics was ever accomplished without an internal change in our intellectual emphasis, loyalties, affections and convictions" (Leopold, 1949).

Leopold does not discount the importance of public policies and regulation. They were critical in his view. He was simply asserting that, by themselves, regulations are not sufficient. We also need to foster the kind of internal change represented by an "ecological conscience."

Leopold contended that once we realize that we are not separate from the rest of the biotic community—that we are, in fact, "plain members and citizens of it"—then we can begin to appreciate the fact that our own productivity is absolutely dependent on the health of the rest of the biotic community.

In Leopold's view, we seem to be distracted from this important ecological insight because we appraise the worth of individual species based on their economic value. If a member of the biotic community makes no contribution to *our* economic well-being, then we tend to discount it. But as Leopold points out, "most members of the land community have no economic value." Yet the resilience of the entire community absolutely depends on the "continuance" of all of its members (Leopold, 1949).

The industrialization of agriculture is partly responsible for this "economic" devaluation of our biotic community. Once we decided to "specialize" and "simplify" agricultural production, we began to lose any sense of the complexity and diversity of natural ecosystems. We lost the awareness that agroecosystems ultimately operate by the same principles as natural ecosystems. So we ignore the information that ecology, evolutionary biology, and epidemiology can contribute to understanding and managing effective farming systems. Once we decided that we no longer needed to manure our soils because we could manufacture substitutes from fossil fuels—which we seemed to believe were inexhaustible—we devalued all of the soil microfauna that are essential to soil health. Once we decided that we could manufacture all of the inputs we needed to control crop pests and animal diseases, we no longer valued the complex array of flora and fauna and predator/prey relationships that function as a symphony of interdependent organisms. We lost sight of the fact that such colonies of the biotic community, working in concert in a complex system of competition and cooperation, can be organized into enormously efficient and productive farming systems. And so we lost sight of the fact that agriculture, to remain productive, is absolutely dependent on the health of the land.

In the absence of an "ecological conscience," farmers began to measure success entirely on the basis of yield per acre—producing as much as possible—without regard for the self-renewing capacity of the ecosystem that is so essential for sustained, efficient productivity.

If an ecological conscience is essential for transforming agriculture to meet the challenges of the future, how do we go about developing it? Leopold made a powerful, practical, articulate, compelling case for the need for such an ethic more than 50 years ago. Why have we not responded?

Why do we not already have in place an ecological conscience of the kind he envisioned? Why is it not the guiding principle of the USDA and major farm organizations—let alone environmental organizations—across the country?

There are many theories to suggest how cultural change takes place, and therefore why it may or may not come about. Thomas Jefferson believed that if people were free and had access to the best information they would act in the best interests of themselves and their neighbors. Adam Smith believed that people would not be likely to act in their own best interest unless there was some incentive for them to do so. Both views reflect certain aspects of human behavior but neither by themselves seem to bring about major cultural changes.

We do know that natural selection is a powerful force for change even though it is not always well understood. Darwin used the phrase to explain that certain individuals of a species survived better than competing individuals of the same species under the natural conditions in which they found themselves. The human species is apparently not exempt from this powerful force for change.

From this perspective, it should not surprise us that industrial agriculture was successful during the past century nor that its success spawned an industrial conscience. Given the availability of abundant fossil fuels, fresh water, regenerative soils, and absorbent natural sinks—all ecological capital that we inherited from 3.7 billion years of life on the planet—we were easily seduced into believing that every production challenge could be solved simply by using single tactic, one-dimensional technologies to extract ecological wealth and applying it to production agriculture. Conversely it is easy to see why organic agriculture was, at best, a struggling minority position. Natural selection, under the optimal conditions of the time, gave the comparative advantage to the industrialists.

But natural conditions are changing. As we noted earlier, David Tilman believes that given the intense challenges being faced by industrial operations, the advantage may be shifting to diversified farmers. Even from an economic perspective industrial farming is in a very tenuous position. Since the early 1980s, a significant portion of net farm income in the United States comes from government subsidies. In 1993, for example, government subsidies accounted for 143 percent of net farm income (Duffy and Holste, 2005). If it were not for government subsidies, most industrial agriculture operations would fail. This is not a sustainable economic position.

As the multitude of future challenges descends on us, comparative advantage increasingly will shift away from the industrial and toward a more ecologically-based agriculture. In the process it will become increasingly clear to farmers and non-farmers alike that our industrial conscience is misguided. We will gradually recognize that our industrial conscience—which assumes that humans are in control and that we can indefinitely manipulate natural processes to suit our needs with technological cleverness—is dysfunctional. As this awareness sets in, we will begin to take the current sciences of ecology and cosmology, evolutionary biology, quantum physics, and the science of networks more seriously in our research, on our farms, and in our agricultural policies. Gradually an ecological conscience will begin to seep into our culture.

In the meantime, it will be important to have models of the new, ecologically based agriculture in place as practical harbingers of the future. This is where organic agriculture, transformed by the new sciences and the increased sophistication that the research guided by those sciences make possible, could play a significant role in shaping the new agriculture of the future.

All of this suggests that it may be useful to propose a set of ethical principles, consistent with an ecological conscience. The following principles may serve as a starting point:

- food and farming systems must be regenerative in nature and based on cooperating partnerships and ecological interdependence;
- food and farming systems must be based on family enterprises cooperating to maintain vibrant communities that foster intergenerational equity;
- food and farming systems must generate appropriate wealth through work;
- food and farming systems must foster commerce inoculated with morality and politics with principles [adapted in part from Gandhi's "7 Deadly Social Sins"].

A fundamental unanswered question for our culture is whether we will correctly assess our true situation, anticipate the coming changes, and get a head start in adapting to the new natural conditions that will be upon us. In *Guns, Germs and Steel,* Jared Diamond (1997) suggests that those civilizations that were able to assess their current situations, anticipate the coming changes, and get a head start in preparing for those changes tended to flourish, while those that did not failed. There is no reason to believe that our culture is exempt from similar dynamics. If we do heed this call, "natural selection" promises us a bright future. If we insist on continuing to farm as we have in the past century, natural selection is likely to replace us with those to whom it gives the comparative advantage. In any case, the new farms of the future will very likely resemble Takao Furuno's small organic farm in Japan.

REFERENCES

Page 322
4 Bullets

Balfour, E.B. 1975. *The living soil and the Houghley experiment.* London.

Borenstein, S. 2003. Experts: World faces new age of infections *Register,* Monday, May 5, p. 7A.

Brown, L.R. 2001. *Eco-economy: Building an economy for the ea* Co., New York.

Conford, P. 2001. *The origins of the organic movement.* Floris Books, Edinburgh, U.K.

Daly, H.E. 1996. *Beyond growth.* Beacon Press, Boston.

Diamond, J. 1997. *Guns, germs and steel: The fates of human societies.* W.W. Norton & Co., New York.

Duffy, M., and A. Holste. 2005 . Estimated returns to Iowa farmland . *J. Amer. Soc. Farm Managers and Rural Appraisers* Vol. 68(1):102-109.

Faulkner, E.H. 1943. *Plowman's folly.* University of Oklahoma Press, Norman, OK.

Furuno, T. 2001. *The power of duck.* Tagaari Publ., Tasmania, Australia.

Gunderson, L., C.S. Holling, and S.S. Light. 1995. *Barriers and bridges to the renewal of ecosystems and institutions.* Columbia University Press, New York.

Howard, A. 1943. *An agricultural testament.* Oxford University Press, London, U.K.

Jha, M., Z. Pan, E.S. Takle, and R. Gu. 2004. Impacts of climate change on stream flow in the Upper Mississippi River Basin: A regional climate model perspective. *J. Geophys. Res.* 109: D09105, doi:10.1029/2003JD003686.

King, F.H. 1911. *Farmers of forty centuries or permanent agriculture in China, Korea, and Japan.* Rodale Press, Emmaus, PA.

Lee, M. 2002. State of the planet. *Ecologist.* September (cited in F. Kirschenmann. 2003. Technologies for a sustainable future: therapeutic intervention versus restructuring the system. http://nabc.cals.cornell.edu/pubs/nabc_15/chapters/Kirschenman.pdf, accessed April 10, 2006).

Leopold, A. 1949. *A sand county almanac.* Oxford University Press, New York.

Lewis, W.J., J.C. van Lenteren, S.C. Phatak, and J.H. Tumlinson III. 1997. A total system approach to sustainable pest management. *Proc. Natl. Acad. Sci.* 94:12243-12248.

Liebig, J.v. 1872. *Chemistry in the application to agriculture and physiology.* John Wiley & Sons, New York.

Manning, R. 2000. *Food's frontier: The next Green Revolution.* North Point Press, New York.

National Farmers Union, Canada. 2003. *The farm crisis, bigger farms, and the myths of "competition" and "efficiency."* Saskatoon, Saskatchewan Bibenber 20, p. 12.

Odum, E.P. 1971. *Fundamentals of ecology.* W.B. Saunders, Philadelphia, PA.

Pagels, H.R. 1982. *The cosmic code: Quantum physics as the language of nature.* Simon and Schuster, New York.

Pollan, M. 2001. How organic became a marketing niche and a multibillion-dollar industry—naturally. *The New York Times Magazine,* May 13.

Rees, W. 1999. Scale, complexity and the conundrum of sustainability. In: *Planning sustainability,* M. Kenny and J. Meadowcroft (eds.). Routledge Press, New York.

Rölling, N. and J. Jiggins. 1998. The ecological knowledge system. In: *Facilitating sustainable agriculture: Participatory learning and adaptive management in times of environmental uncertainty,* N. Rölling and M.A.E. Wagemakers (eds.). Cambridge University Press, Cambridge, U.K. pp. 283-311.

Senge, P. 1990. *The fifth discipline.* Currency Doubleday Publ., New York.

Shiyomi, M. and H. Koizumi (eds.). 2001. *Structure and function in agroecosystem design and management.* CRC Press, Boca Raton, FL.

Soil and Water Conservation Society. 2003. Conservation implications of climate change: Soil erosion and runoff from cropland. Report, Soil & Water Conservation Soc., January. (Available at: http://www.swcs.org/documents/news/u_climate_change_final_121404104710.pdf, accessed April 10, 2006.

Steiner, R. 1958. *Agriculture: A course of eight lectures.* Biodynamic Agric. Assoc., London.

Tilman, D. 1998. The greening of the Green Revolution. *Nature* 396(6708):211.

Chapter 16

Future Multifunctional Rural Landscapes and Communities

Charles A. Francis

INTRODUCTION

Cooperative Extension has been an opinion leader and a strong moving force in rural United States for nearly a century. Bringing the industrial revolution to farming and helping the agricultural sector to interface with a rapidly urbanizing culture and economy, Extension has played a key role in the development of a modern production system that is the envy of many in the world. As described by McDowell (2001) and discussed in the opening chapter (see Chapter 1), the role of Extension has changed drastically as many other players entered the agricultural advising stage: seed and fertilizer dealers, representatives of pesticide companies, field specialists from the local cooperative, and private crop consultants. When we add the instant access to products and recommendations on the Internet, there is a dynamic new and complex information environment.

There is a multiplicity of sources of contemporary information about new crop varieties, the latest in herbicide or starter fertilizer formulations, larger and more powerful equipment, and electronic access to markets 24 hours a day, every day of the year through a home computer. Often the farmer is overwhelmed by information and competing choices (conventional versus organic versus sustainable), and finds it increasingly difficult to sort out what is best for a specific crop on his or her farm. New technologies come with a hefty price tag to recoup the costs of research, advertising, and marketing. People doing the advertising and promoting of commercial products work for companies that depend on sales to stay in business. There is no parallel in the agricultural input industry to an objective recommenda-

Developing and Extending Sustainable Agriculture
© 2006 by The Haworth Press, Inc. All rights reserved.
doi:10.1300/5709_16

tion given by Extension, although credible private consultants are independent of soil test laboratories and input suppliers and are hired to give recommendations free of commercial ties. With increasing costs of production and low commodity prices, farmers are faced with a continuing dilemma about what technologies are cost effective and feasible for business continuity.

One of the key lessons from the Sustainable Agriculture Research and Education (SARE) programs in the North Central Region is that farmers and ranchers are key sources of information on sustainable practices and systems. In some production strategies such as organic farming, these professional agriculturists may be the best and at times the major source of information. In several chapters we have emphasized the important role of farmers and ranchers as true equals in the design of the regional program and in the decision process on grant priorities and selection. Although we too often call people on the land "producers," the parallel to workers on an assembly line in an industrial production line is inescapable. And the mentality of society toward farmers and ranchers as merely producers of commodities does not lead to respect or support during elections. It is apparent that the industrial system is not working in many ways (see Chapters 6 and 15).

Even in the midst of high levels of productivity per unit of labor and moderately good yields per acre, we observe that most farmers in the United States are not doing well economically. Many leave agriculture to seek other jobs, and their farms often go to neighbors with expanding operations. 80 percent of Midwest farm families have one person working off farm, and 40 percent have both wife and husband with at least part-time jobs to find secure income, health insurance, retirement, and other benefits not available for farmers. With the median age of a farmer being 58 years, at least half of our farm decision makers in this region will retire within a decade. Given current trends, most of these farms will be consolidated into nearby large operations, strongly buoyed up by federal farm supports, dependent on a competitive local industry demand for labor, and locked into low commodity prices and an unpredictable export market. The lack of young entrepreneurs in agriculture and the loss of a stable, multigeneration family farming system do not bode well for the sustainability and security of our domestic food supply (Strange, 1988).

As described in the first chapter, the SARE Program was established to counter these trends and provide practical and profitable alternatives for family farmers in the United States. Subsequent chapters explored farming practices, systems designs, and marketing strategies that offer innovative options to help farming families survive in this competitive commodity environment. Some solutions include looking at new crops or adding value to farm products. This concluding chapter pulls together all the pieces of this complex puzzle, to put our research and education into context across a wide geographical scale, and to make recommendations for the future of Cooperative Extension. The positive recommendations from each chapter are summarized and discussed. The major challenge is how to integrate these ideas into productive and profitable farming systems that are compatible with a healthy environment and a secure and equitable food system for the future. This chapter can serve as a stand-alone vision for the future that offers additional ideas and recommendations.

BUILDING A FOUNDATION

In the opening chapter we described the growing information base that assists farmers working to design and implement more sustainable systems. Many of the books, journals, and other miscellaneous publications are primarily for communication among scientists. This body of literature is important because it represents the increased attention given by the research community to reducing costs of production inputs, while improving water quality and other environmental indicators. Such research has long been a priority for nonprofit groups such as Rodale Institute, Michael Fields Institute, and The Land Institute, and in recent years we have seen sustainable agriculture emerge as a priority in the land grant university system. The broadening of an agricultural and food systems agenda as shown by this change in research emphasis is much in keeping with the priorities described in a recent National Research Council (2003) study. The future direction of public research funding is likely to support projects that result in reducing environmental impacts and increasing social equity of the food system.

SARE research and education projects have focused on seeking answers to questions in small and moderate-sized family farm systems. The farmer research grants have been especially targeted to solve the challenges faced by family farmers in an industry that increasingly rewards large-scale operations. A global food system forces farmers with limited land resources to seek alternative crops or livestock enterprises, or move into niche markets or high-value opportunities. Organic agriculture (see Chapter 15) is one such opportunity because it is not easily practiced on farms with larger land holdings and equipment of a scale that does not allow flexibility and rapid adjustment to new markets. The foundation of location-specific practices and recommendations has been largely built through SARE research and producer grants, and subsequently demonstrated and moved to farmers in cooperation with Extension. We consider this an important contribution to renewing the contract with rural U.S. farmers and communities.

SARE RESULTS IN COOPERATIVE EXTENSION PROGRAMS

Most of the questions researched by farmers and by scientists using the SARE grant support have employed conventional methods and widely accepted experiment designs, including the long drive-through plots described by Exner and Thompson (see Chapter 12). In fact, most research on alternative practices and enterprises has followed traditional reductionist methods of isolating one or two factors for study and maintaining control as much as possible over other parts of the production system. For this reason, results of most SARE projects resemble those from conventional research. Thus they are readily moved into Extension education channels.

MacRae et al. (1989) described three types of change in agricultural systems: increased *efficiency,* the *substitution* of one input for another, and *redesign* of the system. Many applications of SARE research involve improving efficiency of input use: reducing fertilizer application rates, fine-tuning tillage to better manage weeds and reduce soil erosion, and doing a better job of assessing forage quality and adjusting grazing frequencies. These changes are valuable and can easily be accommodated by mainstream Cooperative Extension because this is "business as usual." With the possible exception of such controversial recommendations as partial-rate herbicide applications, there is generally no objection from industry because these changes scarcely rock the boat of conventional agriculture. Further, there is little challenge from the research establishment because these improvements in efficiency have been researched through a conventional approach that isolates individual components of crop or animal enterprises, either on the experiment station or on the farm.

More frequent SARE research questions address the issue of *substitution:* replacing herbicides with cultivation, reducing chemical fertilizer application rates by introducing compost or green manure crops, and changing from a low-profit commodity crop to a value-added alternative crop. Most of these recommendations are also easily accommodated in Extension programs because they represent another enrichment of normal business. The communication of new ideas fits within the scheduled tours, demonstrations, and workshops of Cooperative Extension.

SARE research that results from experiments in *redesign* of systems is the most difficult outcome to assimilate in Extension programs. Even though many farmers and ranchers are struggling with commodity crops and livestock, they often hesitate to risk the adoption of a new system with uncertain results. Most researchers likewise are hesitant to work on systems-level problems, and especially those at the landscape or watershed levels, because of the greater complexity and longer time frame needed to conduct research and draw relevant conclusions. We do not have good statistical tools to handle whole systems, deal with confounding of many components, and help bring understanding to the complexity that must be a part of farm, landscape, or watershed analysis. One example of a major system redesign that has been successful is intensive rotational grazing (see Chapter 4), an approach that can greatly increase pasture productivity compared to conventional large-pasture systems without rotation. When the research is sound, there is no difficulty incorporating the ideas into Extension programs. But there is less attention given in overall programming to issues related to changing systems, compared to our emphasis on individual components of technology such as varieties or fertilizer rates.

FUTURE FARMING PRACTICES AND FARM EVALUATION

The SARE research programs in the North Central Region have provided substantial direction for designing and improving future farming systems that will be profitable, environmentally sound, and socially viable. Integrated pest management studies (see Chapter 2) have expanded our appreciation of how to manage weeds, insects, and pathogens with reduced chemical application, through integrated pest and habitat management, and using innovations such as organic systems. One of the strong keys to future pest management will be increased diversity on farms and within production fields. In a drastic departure from current trends toward larger fields, greater specialization, and monocultures, the results point toward a need for greater biological diversity at several levels in the spatial hierarchy. In Chapter 2, George Bird and Michael Brewer describe the importance of developing a new eco-literacy among our farmers and those in the research community, and how we need systems that are designed to perform well under the unexpected consequences of changes in weather, pest populations, and economic uncertainty.

At the field level, intercropping and strip cropping along with crop rotation have potential to reduce pest populations and facilitate management. Perennial polyculture research at The Land Institute provides another long-term visionary model for using the concept of the natural prairie's diversity to help manage pests. At the farm level, the field boundaries, field size, and frequency of ecotones between contrasting systems can all influence pest populations and nonchemical management. More research is needed on whole-farm pest dynamics, and outreach designed to communicate the importance of enterprise and farm planning to enhance pest management. Virtually no research on pest management at the landscape and regional levels exists, other than on long-range migration of such organisms as corn borer, sorghum greenbug (aphids), or wheat rust pathogens. This direction should be a focus for future SARE projects.

Soil fertility is key to successful agricultural production, and much emphasis in SARE has been given to alternative approaches (see Chapter 3). Crop rotations and cover crops are long-established methods for enhancing soil fertility while reducing needs for purchased chemical fertilizer. The leadership for reducing chemical fertilizer recommendation rates came from Prof. Robert Olson and colleagues in Nebraska whose early research comparisons of soil testing results from different laboratories showed that commercial recommendations were far in excess of levels needed by crops. Reduced nitrogen (N) applications that could result in a major impact on water quality in Iowa were based on late-spring soil tests developed by Fred Blackmer and colleagues, and demonstrated in numerous sites by Cooperative Extension. These are appropriate roles for the university and Extension, and should receive continuing support from the regional SARE programs. Cover crops, compost management and application, crop/animal integrated systems, and manure management are all critical components of a long-term soil fertility strategy. Such nonconventional approaches must be addressed by SARE and the land grant universities, since there are few financial incentives for industry to become involved in this approach to soil fertility, when the result will be less fertilizer sales.

Intensive rotational grazing is one of the shining success stories of SARE projects, particularly the results of producer grants (see Chapter 4). There is compelling evidence for substantially improved pasture and animal productivity, enhanced animal health, and improved environmental conditions at the field, farm, and watershed levels. The increasing demand for leaner beef due to a national U.S. epidemic of obesity and other factors promoting greater nutritional awareness will create growing markets for grass-fed beef in the future. In the long term, this may be one of the key contributions of the SARE programs in the North Central Region.

The methods for testing grazing management as well as component technologies in cropping systems have been advanced by SARE projects that encourage large plots and drive-through designs (see Chapter 12). Results of coordinated trials across several farms have added credence to research results, when specific practices have been tested by several farmers in Practical Farmers of Iowa (PFI) (Thompson, 2003). Beyond the recommendations derived from multilocation trials, these on-farm experiments have confirmed the importance of location-specific and system-specific practices. Such findings challenge our conventional paradigm that provides similar recommendations and wide deployment of uniform systems. The conventional homogeneity of systems across wide areas is one logical consequence of the industrial approach, and it has led to a homogeneity of thinking by farmers and researchers. In contrast, natural ecosystems provide countless models of local specificity and uniqueness of place. For sustainable farmers the better paradigm is learning to seek and adopt varieties and practices that are unique to their farms and to each field.

Whole-farm evaluation has received attention in a few places such as Wisconsin and Kansas, and in the Ontario Farm Plan program, yet this topic is far from common in the majority of programs in Cooperative Extension. The literature review, testing of various tools, and the River Friendly Farms workshops described by Rhonda Janke (see Chapter 5) provide a valuable example of how whole-farm methods for planning and evaluation can be conducted in a Midwest regional context. This experience in Kansas as well as the rotational grazing model (see Chapter 4) are useful examples of how Cooperative Extension can contribute to the long-term economic and environmental sustainability of farms. This dimension of outreach is much needed by Midwest farm families. It represents a strategy that must be addressed by public sector and especially land grant Extension programs, since it is highly unlikely that industry will ever focus on this as a commercial venture. SARE support has been essential for these programs.

One area of future interest to Cooperative Extension will be better programs to serve clients at the rural/urban interface, as more city people find themselves on the border or living in rural areas for quality of life reasons. Although Extension has moved aggressively in addressing some concerns of acreage owners, and geared an increasing number of programs to meet urban challenges, there has been little explicit focus on the boundary areas between these two groups of citizens who often have conflicting goals and expectations for the place they choose to live. The concept of ecobelts (see Chapter 11) is the use of functional, planned, and highly diverse plantings to buffer city limit subdivisions from adjacent farms, and acreages from individual farms sharing property lines. Ecobelt plantings provide a potential to mitigate pollution and other problems, to perform essential ecosystem functions, and even to create economic activity for both neighbors. Ecobelts have been implemented in a number of Midwest communities, and should have even greater potential for expansion, as stormwater management requirements and desire for green spaces increase.

Cooperative Extension should include ideas on multipurpose tree and native grass plantings within the suite of practices they promote for sustainable development and provide expertise to help integrate these plantings into more holistic landscape designs. Economics are crucial to farm family survival, and recent performance across farms has been highly variable. Many smaller and mid-sized farms have gone out of business over the past century due to low profitability in the commodity marketplace, inability to adapt to changing conditions and narrow margins, or federal farm programs that have driven up land values and favored larger farms (Grant, 2002). A lack of adequate multiple valuation criteria for goods and services has worked against farmers interested in conserving natural resources, maintaining viable farm ecosystems and habitat, and valuing social issues such as expecting higher than minimum wages for farm workers.

John Ikerd (Chapter 6) argues strongly for an examination of our current systems and their relative sustainability in a time of disappearing fossil fuels and other environmental constraints. He quotes the late Robert Rodale as insisting that sustainable systems must be resistant, resilient, renewable, and regenerative; these characteristics of biological ecosystems may be equally applicable to economic systems. Ikerd further articulates this message by describing profitable farms that are viable in the long run due to economic diversity; are less reliant on external inputs and impersonal or distant markets; rely less on facilities and equipment that rust, rot, or wear out; and use less nonrenewable energy. Indicators of profitability are important, but no more so than indicators of economic resistance, resilience, renewability, and regenerative capacity (see Chapter 6 for more details).

Until we learn to fully appreciate the productivity, environmental advantages, and social contributions of small and moderate-sized family farms to society as a whole, they will continue to operate at the margin and continue to disappear. To the extent possible, agricultural policy and education programs should seek to reverse this situation and demonstrate concrete alternatives that favor family farms and their contributions to rural communities and a sustainable landscape. Federal or state subsidies are likely to be needed to encourage indirect social and environmental benefits provided by conservation-minded family farmers.

FUTURE OUTREACH PROGRAMS FOR VIABLE RURAL COMMUNITIES AND LANDSCAPES

Beyond the ideas for successful practices on farms, the potential for evaluating whole-farm systems, and the improvement of a viable interface between urban and rural areas, the SARE programs in this region have developed statewide and regional models of cooperation that can serve agriculture and society well for the future. A crucial component of SARE's success is that its programs involve both providers and recipients as equal partners. In contrast with many programs that encourage competition among players in the rural environment, agencies in government, and states in the region vying for scarce resources, the SARE programs have encouraged cooperative program planning and implementation.

A model state program has been developed by Iowa Cooperative Extension in close collaboration with PFI, the Leopold Center for Sustainable Agriculture, and other agriculture and food groups (see Chapter 7). The remarkable success in Iowa gives us pause to reflect on how this program developed, why it was successful, and whether the model could be implemented in other states. The support of nonprofit foundations was certainly one important factor (see Chapter 13). To be sure, there were several key individuals who played a vital role in designing and implementing a statewide program, and these people with vision and dedication were instrumental in its success. Yet these same people were convinced that they and the organization they represented could not do the job alone. Their spirit of cooperation infused the planning efforts and searches for support in a way that was not characteristic of our typical competitive paradigm. Instead of assuming full leadership for outreach, the leaders in Cooperative Extension willingly embraced the active role of a farmer organization, other state and nonprofit groups, and cooperating researchers and educators on campus to help shape the program and build ownership among client groups. They worked closely with University of Northern Iowa on key projects. This explanation is not meant to minimize the credit that must be given to the Cooperative Extension team in Iowa, but rather to point out how effective shared leadership can help achieve shared goals.

Within the Iowa model was embedded a vital concern beyond the farm boundaries—the development of viable communities on which so much of rural quality of life depends (see Chapter 10). The development of social capital in rural communities to guide future improvements in farming systems and enhance the institutions that serve them is an essential component of a sustainable development strategy. To develop a strategy for an economically successful multifunctional agriculture, a primary focus must be on people, especially promoting the emergence of leadership by sustainable agriculturists who internalize the multiple goals of farm families and communities, and who have the tools and the willingness to act to fulfill those goals. Sustainable agriculture and its interdependence with community development could be considered a social movement to enhance the rural areas of the United States. Beyond sustainable agricultural production, programs such as SARE are meant to build viable economies in rural communities that will lead to more stable businesses, schools and faith communities, and health facilities that will improve quality of life for rural residents. Such a movement will require cooperative efforts of urban and rural people with shared goals and vision, unique new alliances that bring together the human resources needed to effect change, and the financial and institutional resources to make it all happen.

One key source of funding that has been essential to many programs in the North Central Region and across the United States has been the non-profit foundation community. Many of these organizations have embraced the goals of sustainable agriculture and rural development and sustained their support over the past two decades (see Chapter 13). Much of this support has gone to land grant institutions in a quest to help them renew their historical contract with the rural United States, to become once again a major influence on development, and to promote the success of family farms and small businesses that have been the strength of our rural communities. Foundations have also provided essential support to nonprofit farmer groups, and especially to coalitions of organizations working together to improve the economic and social viability of rural areas.

One important lesson learned in the North Central Region was that multistate and regional activities could be more resource-efficient in achieving educational goals than a series of independent state programs. To be sure, there have been Extension programs that crossed state lines in the past, but rarely to the extent as those promoted and funded by the regional SARE program (see Chapter 9). The wise decision by the Administrative Council and regional advisors not to allocate all funds through a conventional formula approach resulted in a highly successful regional sustainable agriculture training program over a period of four years (see Chapter 8). These planning and training workshops brought together educators, specialists, farmers, ranchers, and researchers from public, private, and nonprofit groups to share experiences, resources, and success stories in sustainable agriculture. This sharing process created a network in the region that has proved effective in designing additional multistate programs. We are certain there has been a long-term impact on programming from the friendships and collegial relationships among people in the region that have increased cooperation in other programs in the years since these workshops were implemented in the 1990s.

Many innovations in the SARE program across the United States were initiated in the North Central Region (see Chapter 9). These included programs such as the farmer research grant program and the thesis/dissertation grant program for graduate students. Creative programming was enhanced by establishing a regional producer grant coordinator position, and outreach was promoted by naming a regional communications specialist. The search for complementary goals in research and outreach was enhanced by the historical collaboration in realizing these activities within the land grant university system. Administrative coordination was assured by the selfless dedication of Extension and research directors at University of Nebraska – Lincoln who made an extensive time commitment and assumed some professional risk to strongly support a program that was outside the mainstream of university programs and activities.

One additional area essential to success in regional programs is the study of motivation of farmers, ranchers, educators, and researchers to participate in this innovative program. In order to understand how people in these several groups were attracted to embrace the challenges of designing new activities in sustainable agriculture, the regional SARE program funded a project to assess the motivations of project leaders and farmers in applying for grants and the impacts these activities had on their subsequent professional activities. As described in Chapter 14, motivations among participants in SARE grant activities do not differ greatly from those of the general farmer and researcher populations that were surveyed as a control group. The farmers involved with SARE were less concerned than the general population about what neighbors think, and were motivated by a broader range of factors in the social and environmental arena and beyond short-term economic returns. We learned that the SARE grants moved people in new directions, or helped them research questions of great interest for which they had been unable to get support from other sources. Their motivation and desire to continue seeking innovations to develop a more sustainable agriculture were apparent from personal interviews. The fact that researchers and farmers who received funding do not differ greatly from the general university and farm populations suggests that there is a wide potential to spread the message about long-term viability and multifunctional agricultural activities to a wider audience, and that should be one goal of Cooperative Extension.

Personal ethics and values play a large role in farmers' decisions to move from the mainstream to some alternative strategy such as organic farming (see Chapter 15). Those farmers who take seriously the Leopold ethic about creating and sustaining a viable rural landscape composed of environmentally sound agroecosystems need to consider innovative approaches that reduce chemical pesticide use, minimize excessive chemical fertilizer applications that currently damage aquatic ecosystems adjacent to farmland, and seek environmentally compatible farming practices and systems that promote biodiversity and connected habitat for wildlife (Jackson and Jackson, 2003). An example of effective diversity on the farm of Takao Furuno was described in Chapter 15. Although small scale and organic farming do not assure these results, they provide creative approaches to helping farmers more closely adhere to the Leopold ethic.

FUTURE PRIORITIES FOR COOPERATIVE
EXTENSION

The many projects funded by SARE for research at universities, in non-profit groups, and by farmers on their own fields have served well for education of multiple clients in agriculture. Projects have also helped to increase the information base for sustainable agriculture in an impressive array of production and marketing areas. In the chapters of this book, many of the key leaders in the North Central Region have provided an articulate history of programs and impacts over the past 18 years. In addition to the increased awareness of the importance of sustainable agriculture in the farming community, there is a greater degree of acceptance of the importance of sustainability in our universities' research and education agendas, as described by Oran Hesterman in Chapter 13. Yet there is still much to be done. This section is focused on the challenges that will face rural U.S. agriculture and food systems in the future, and how Cooperative Extension resources can be mobilized to meet those challenges. We recognize that neither SARE nor Cooperative Extension can be all things to all people, but we can enhance programs to better help our clients meet their goals. Foundations can provide seed monies, but programs need to be self-sustaining. We can mobilize our internal resources in public and private organizations to make best possible use of outside funding from the foundation community to work in cooperation toward common goals. Future directions should consider the following priorities.

Production Initiatives

- *Promote diversity in cropping systems,* beyond crop rotations, including in-field diversity, more diverse rather than uniform cultivars, multiline varieties of cereals, and spatial and temporal diversity within fields that can create more biocomplexity on farms and across landscapes.
- *Develop diversity in crop/animal systems* that go beyond merely having animals and crops on each farm to enhance nutrient cycling. Grazing crop residues, integrating multiple livestock species, providing an internal "market" for nonsalable or otherwise damaged crops, and rotations of pastures and cropping lands are some of the feasible alternatives.
- *Research integration efficiencies* based on physical location of crop and animal enterprises on most appropriate lands, for example, grazing on marginal areas and intensive cropping on the best lands, and better planning of land use across the farm.

- *Extend whole-farm planning* initiatives to landscape, community, and watershed levels with cooperation among farmers to best place enterprises in appropriate fields, and to share activities such as residue grazing to allow a degree of specialization which at the same time uses total community resources most efficiently.

Economic Directions

- *Focus on creating new opportunities for adding high-value crops to the farm mix,* such as field and greenhouse-grown medicinal and culinary herbs, greenhouse-grown salad greens in winter, gourmet vegetables, marketing to restaurants and other high-end markets, and pick-your-own operations.
- *Optimize land use and internal/external resource use,* through such analysis tools as linear programming that takes into account all costs and benefits of each enterprise and their combinations.
- *Focus on innovative marketing* and adding value on the farm, in the neighborhood, and in the local community by sharing equipment, marketing together, or otherwise accomplishing some economies of scale while maintaining family farm ownership.
- *Establish local food systems and community supported agriculture (CSA) systems* as a viable alternative to current global markets—studies have shown consumer willingness to buy local food, often at a premium price, to support the local economy and farmers whose values they share.
- *Improve food security* at the local level by increasing consumption of locally grown crop and animal products, and in the aggregate increase state, regional, and national food security and reduce transportation costs with local systems.
- *Move toward a multidimensional valuation system* that takes into account the longer-term and wider costs of production on the environment and the resource base, costs that today are often externalized to society or the future by conventional economic valuation.
- *Think outside the box,* exploring new income opportunities such as bed and breakfast with a farm family, farm vacations for city folks such as the dude ranch approach, internships to reduce labor costs and increase education, and to inform the public about sustainable farms and ranches.
- *Focus on changes in public policy* that currently favor large farms and commodity production. University researchers and Extension specialists have viewed themselves as policy-neutral and stayed out of policy

discussions, yet there is urgent need for serious discussion in agricultural circles about the impacts of current policy and the alternatives that could lead to a more sustainable food system (see Chapter 13).

Environmental Focus

- *Introduce an effective cycling of materials* used in food production, processing, and marketing and incorporate these costs into the current price of food in the marketplace.
- *Encourage an effective cycling of materials* from every step in the food system and its processes (production, processing, marketing, consumption) and put these materials back into the natural environment or into further agricultural production; full life cycle analysis provides tools to assess success in this endeavor.
- *Embrace the farm and its agroecosystems* as an effective ecological habitat to maintain biodiversity, connectivity needed for larger species, and *set aside areas* for native ecosystem preservation—but integral to some part of the farm operation.
- *Develop longer-term perennial agroecosystems* such as agroforestry, permaculture, and diverse prairie analogs and other perennial polycultures that mimic the structure and function of natural ecosystems while providing for harvest of food, feed, fiber, or other agricultural products.
- *Study and demonstrate landscape-level approaches* for restoring environmental services from the land that include multiple farms across the landscape in order to optimize performance of sustainable farm practices on site as well as benefits to society as a whole.

Social Dimensions

- *Promote improved nutrition and health* through integrated programs that educate people and supermarket chain managers about the sources of their food and their value to a balanced diet; increasing support for carefully structured consumer and commodity education programs should be readily available.
- *Focus on youth education* since changes are most likely to start with younger people in kindergarden through twelfth grade, college, and university who are conscious about image, sports, and popularity; we can come at the nutrition issue from where they are, and appeal to age-specific goals.
- *Concentrate on local food systems* so that Cooperative Extension programs will have maximum impact on farm profitability for small and

medium-sized farm families who can produce for local consumers and put a local identity on their products.

- *Make communities the focus* of some future agriculture and food system programs, with emphasis on local value-added activities in growing, processing, and marketing food and thereby increasing the circulation of food dollars at the local level.
- *Support more collaborative, holistic research* between local producers and local communities that can lead to a better understanding of how to strengthen both according to clearly defined "growth" indicators.

CONCLUSIONS

Our conclusions on the various dimensions of research and education in sustainable agriculture are contained in the first fifteen chapters of this volume. The above recommendations for future activities in Cooperative Extension are directed toward *renewing the contract* with rural U.S. farm and ranch families and communities. In this process, we must focus on building partnerships and alliances with other like-minded groups, including state and federal agencies, nonprofit foundations and other organizations, communities, and especially the farmer and rancher groups that have emerged over the past two decades. It is only through cooperation that we can mobilize scarce resources and make a real impact on the agriculture and food system in this country. Here are several specific conclusions that need to be considered as a part of the renewal effort.

- Explore new sources to sustain and increase funding to meet the real demands of farmers and ranchers for research results and education. Currently a small fraction of total support goes to sustainable agriculture, and most funding actually goes to programs that distort markets and push family entrepreneurs in agriculture out of business.
- Embrace the mission and goals and form working alliances with mainstream environmental organizations that share our concern about habitat, wildlife populations, and strategies that can lead to multifunctionality of rural landscapes; these groups have exceptionally good resources that could be used in research and education in agriculture.
- Articulate and implement practices and details of production in a broader context of fields, farms, and landscapes as we determine the impacts of alternative technologies on farm production, economic return, and social equity. Narrow valuation based on short-term economic returns is inadequate to fully measure impacts.

- Recognize that people challenges are the largest and most complex ones, as we study motivation and look for meaningful ways to enhance education in sustainable agriculture, and that programs should always be designed that are specific to local audiences.
- Expand our understanding of and programming for the needs of young people, who are essential as next-generation farmers and ranchers, but who have social and educational needs of their own.
- Recognize that it is essential to expand our analysis of the global marketplace, evaluate the ways that local systems may provide a viable alternative to the present rush to globalization, and include an assessment of food system security.
- Seek unique markets for the highly differentiated products that can be elaborated and sold directly by small and medium-sized farm families, based on local consumer needs, and identified with local farms.
- Establish long-term design and evaluation, plus multifactor valuation schemes that do not discount the future and externalize costs and impacts to other regions, countries, and to our children.
- Take lessons from the commercial market about ways to manipulate public opinion, including image marketing and emotional advertising, then secure resources to promote agriculture in ways that promote the positive qualities and outcomes, rather than discouraging the hearts of those who would love the opportunity to farm and raise their children in a "working rural" environment.
- Involve more female scientists and other underrepresented minorities, educators, and ranchers and farmers to reflect the improving gender and ethnic balance in agriculture and food system decision making—and take advantage of any unique biological potential that women may bring to the study of systems and larger issues.
- Develop programs to reach out multiculturally to the growing number of Hispanics and Asian farmers starting small family farms.
- Conduct serious research on the efficiency of scale in agriculture and the potential benefits versus negative long-term consequences of industrial agriculture. Is our current research agenda promoting an efficient food machine that will continue to pressure family farms from the rural landscape, while providing cheap food to urban residents? Or is this a nonviable model as many of the chapter authors suggest in *A New Social* Contract? Give thoughtful consideration to the current concerns about food security, and whether our only viable strategy must be to provide unilateral military protection from the United States for long supply lines that bring us fossil fuels, food, and other goods over extended transportation distances. Are there local food

system alternatives that can add value to our own natural resources and establish a secure system that is equitable and healthy for all citizens?

- Continue to expand both our spatial and our temporal thinking about the future, to see over the horizon and anticipate changes as we design a more sustainable food system as part of an inter-woven global community.

These are some of the directions that seem indicated for professionals in the public and private sectors to make a meaningful impact on agriculture and food systems in the future. When goals and needs of people are in conflict with those of an industrial system, it is up to public sector and nonprofit organization specialists to carefully analyze alternatives and their long-term impacts on society and sustainability. In science, we need to move beyond the narrow confines of our disciplines, embrace the need for cooperation among organizations, and seek common goals and strategies to achieve them to create a productive and equitable food system. This is a mission that would help us to renew the contract with people in the rural United States, and improve our local food security in a time of great international uncertainty. Surely we owe our children and their children no less than our best efforts to design and implement a sustainable path toward their future.

REFERENCES

Grant, M.J. 2002. *Down and out on the family farm: Rural rehabilitation in the Great Plains, 1929-1945.* University of Nebraska Press, Lincoln.

Jackson, D.L. and L.L. Jackson (eds.). 2003. *The farm as natural habitat: Reconnecting food systems with ecosystems.* Island Press, Covelo, CA.

MacRae, R.J., S.B. Hill, J. Henning, and G.R. Mehuys. 1989. Agricultural science and sustainable agriculture: A review of the existing scientific barriers to sustainable food production and potential solutions. *Biol. Agric. Hortic.* 6:173-219.

McDowell, G.R. 2001. *Land-grant universities and extension into the 21st century: Renegotiating or abandoning a social contract.* Iowa State University Press, Ames.

National Research Council. 2003. *Frontiers in Agricultural Research: Food, Health, Environment, and Communities.* Committee on Opportunities in Agriculture, Washington, DC. 268 p.

Strange, M. 1988. *Family farming: A new economic vision.* University of Nebraska Press, Lincoln.

Thompson, R. 2003. *Thompson on-farm research.* Practical Farmers of Iowa, Boone.

Index

Page numbers followed by the letter "f" indicate figures; those followed by the letter "t" indicate tables.

Balfour, Eve, 58, 309
Bandura, A., 294
Barbuto, John E. Jr., 21, 289,
    300-301
Barnard, C. I., 294
Bass, B. M., 162
Bauer, Lisa, 186
Baumeister, R. F., 294
Beauvais, L. L., 293, 295, 299,
    300-301
Beef cattle
    animal equivalents, 87
    forage quality, 79
    grazing versusversus feedlots, 69
    organic meats, 115
Bell, Michael, 151
Bentrup, Gary, 20, 225, 229
Berry, Wendell, 247
Berton, Valerie, 158, 186
Big Spring Basin water pollution, 143
Bimodal farming model, 4-5
BioEconomy Working Group,
    211, 212
Biological monitoring of IPM, 26-29,
    26f
Biological pest control
    conservation, 30
    described, 30
    manipulative, 30-31
    natural, 30
Biotechnology, genetically modified
    crops, 31
Bird, George, 1, 19, 21, 25, 35, 39-40,
    307, 329
Bison, 76, 80
Black, A. W., 296
Blackmer, Fred, 330
Bloat in animals, 79
Books
    An Agriculture Testament, 309
    agroecology series, 10
    *Chemistry in the Application to
        Agriculture and Physiology*,
        308
    *Concepts in Integrated Pest
        Management*, 27
    Family Farming, 154
    Farmers of Forty Centuries, 309
    *Field Crop Pest Ecology and
        Management*, 27

Books (continued)
    The Fifth Discipline, 312
    *Foods Frontier: The Next Green
        Revolution*, 316
    Fundamentals of Ecology, 311
    green book series, 197
    Guns, Germs and Steel, 154
    Haworth Press, 10, 12, 13
    Home and Consumer Horticulture, 12
    *IPM in Practice: Principles and
        Methods of Integrated Pest
        Management*, 27
    IPM literature, 27
    The Living Soil, 58, 309
    *The Living Soil: Exploring Soil
        Science and Sustainable
        Agriculture with Your Guide,
        the Earthworm*, 16
    Look to the Land, 309
    national SARE publications, 12
    Nebraska Green Books (University
        of Nebraska Extension), 12
    organizations, 12
    Our Sustainable Future (University
        of Nebraska), 10, 11
    Plowman's Folly, 309
    Silent Spring, 29, 141-142
    *Social Theory for Action: How
        Individuals and Organizations
        Learn to Change*, 299
    *Sustainable Food, Fiber and
        Forestry Systems*, 12
    *Sustainable Horticulture: Today and
        Tomorrow*, 17
    A Thousand Acres, 142-143
    university book series,
        10-13
Brewer, Michael, 19, 25, 329
British Corn Laws, 308-309
Brome, 83
Brown, LeRoy, 150
Buffalo, 76, 80
Buffalo no-till planters and cultivators, 15
Bulletins
    Extension, 8-10
    Michigan Field Crop Ecology, 8
    *Michigan Field Crop Pest Ecology
        and Management*, 8
    reporting of test results, 3
Butler, Lorna Michael, 20, 203

Order a copy of this book with this form or online at:
http://www.haworthpress.com/store/product.asp?sku=5709

DEVELOPING AND EXTENDING SUSTAINABLE AGRICULTURE
A New Social Contract

_____in hardbound at $69.95 (ISBN-13: 978-1-56022-331-3; ISBN-10: 1-56022-331-6)

_____in softbound at $49.95 (ISBN-13: 978-1-56022-332-0; ISBN-10: 1-56022-332-4)

342 pages plus index • Includes illustrations

Or order online and use special offer code HEC25 in the shopping cart.

COST OF BOOKS_____

POSTAGE & HANDLING_____
*(US: $4.00 for first book & $1.50
for each additional book)*
*(Outside US: $5.00 for first book
& $2.00 for each additional book)*

SUBTOTAL_____

IN CANADA: ADD 6% GST_____

STATE TAX_____
*(NJ, NY, OH, MN, CA, IL, IN, PA, & SD
residents, add appropriate local sales tax)*

FINAL TOTAL_____
*(If paying in Canadian funds,
convert using the current
exchange rate, UNESCO
coupons welcome)*

☐ **BILL ME LATER:** (Bill-me option is good on
US/Canada/Mexico orders only; not good to
jobbers, wholesalers, or subscription agencies.)
☐ Check here if billing address is different from
shipping address and attach purchase order and
billing address information.

Signature_____

☐ **PAYMENT ENCLOSED: $**_____

☐ **PLEASE CHARGE TO MY CREDIT CARD.**

☐ Visa ☐ MasterCard ☐ AmEx ☐ Discover
☐ Diner's Club ☐ Eurocard ☐ JCB

Account # _____

Exp. Date_____

Signature_____

Prices in US dollars and subject to change without notice.

NAME_____

INSTITUTION_____

ADDRESS_____

CITY_____

STATE/ZIP_____

COUNTRY_____ COUNTY (NY residents only)_____

TEL_____ FAX_____

E-MAIL_____

May we use your e-mail address for confirmations and other types of information? ☐ Yes ☐ No
We appreciate receiving your e-mail address and fax number. Haworth would like to e-mail or fax special
discount offers to you, as a preferred customer. **We will never share, rent, or exchange your e-mail address
or fax number.** We regard such actions as an invasion of your privacy.

Order From Your Local Bookstore or Directly From
The Haworth Press, Inc.
10 Alice Street, Binghamton, New York 13904-1580 • USA
TELEPHONE: 1-800-HAWORTH (1-800-429-6784) / Outside US/Canada: (607) 722-5857
FAX: 1-800-895-0582 / Outside US/Canada: (607) 771-0012
E-mail to: orders@haworthpress.com

For orders outside US and Canada, you may wish to order through your local
sales representative, distributor, or bookseller.
For information, see http://haworthpress.com/distributors

(Discounts are available for individual orders in US and Canada only, not booksellers/distributors.)

PLEASE PHOTOCOPY THIS FORM FOR YOUR PERSONAL USE.
http://www.HaworthPress.com   BOF06